ANNUAL REVIEW OF MATERIALS SCIENCE

EDITORIAL COMMITTEE (1992)

NEIL BARTLETT
JEROME B. COHEN
MERTON C. FLEMINGS
CURTIS W. FRANK
JOSEPH A. GIORDMAINE
ROBERT A. HUGGINS
ELTON N. KAUFMANN
JOHN B. WACHTMAN, JR.

Responsible for the organization of Volume 22
(Editorial Committee, 1990)

NEIL BARTLETT
JOSEPH A. GIORDMAINE
ROBERT A. HUGGINS
JOHN B. WACHTMAN, JR.

Production Editor   SANDRA H. COOPERMAN
Subject Indexer   STEVEN M. SORENSEN

ID
ANNUAL REVIEW OF MATERIALS SCIENCE

VOLUME 22, 1992

ROBERT A. HUGGINS, *Editor*
Stanford University

JOSEPH A. GIORDMAINE, *Associate Editor*
NEC Research Institute

JOHN B. WACHTMAN, JR., *Associate Editor*
Rutgers University

ANNUAL REVIEWS INC 4139 EL CAMINO WAY P.O. BOX 10139 PALO ALTO, CALIFORNIA 94303-0897

ANNUAL REVIEWS INC.
Palo Alto, California, USA

COPYRIGHT © 1992 BY ANNUAL REVIEWS INC., PALO ALTO, CALIFORNIA, USA. ALL RIGHTS RESERVED. The appearance of the code at the bottom of the first page of an article in this serial indicates the copyright owner's consent that copies of the article may be made for personal or internal use, or for the personal or internal use of specific clients. This consent is given on the conditions, however, that the copier pay the stated per-copy fee of $2.00 per article through the Copyright Clearance Center, Inc. (21 Congress Street, Salem, MA 01970) for copying beyond that permitted by Section 107 or 108 of the US Copyright Law. The per-copy fee of $2.00 per article also applies to the copying, under the stated conditions, of articles published in any *Annual Review* serial before January 1, 1978. Individual readers, and nonprofit libraries acting for them, are permitted to make a single copy of an article without charge for use in research or teaching. This consent does not extend to other kinds of copying, such as copying for general distribution, for advertising or promotional purposes, for creating new collective works, or for resale. For such uses, written permission is required. Write to Permissions Dept., Annual Reviews Inc., 4139 El Camino Way, P.O. Box 10139, Palo Alto, CA 94303-0897 USA.

International Standard Serial Number: 0084-6600
International Standard Book Number: 0-8243-1722-X
Library of Congress Catalog Card Number: 75-172108

Annual Review and publication titles are registered trademarks of Annual Reviews Inc.

∞ The paper used in this publication meets the minimum requirements of American National Standard for Information Sciences—Permanence of Paper for Printed Library Materials, ANSI Z39.48-1984.

Annual Reviews Inc. and the Editors of its publications assume no responsibility for the statements expressed by the contributors to this *Review*.

TYPESET BY BPCC-AUP GLASGOW LTD., SCOTLAND
PRINTED AND BOUND IN THE UNITED STATES OF AMERICA

PREFACE

This is the twenty-second volume of the *Annual Review of Materials Science*, one of the 26 annual review series published by Annual Reviews Inc., a nonprofit scientific publisher dedicated to promoting the advancement of the sciences.

From its beginning, this series has sought to publish significant, timely, and critical appraisals of various areas within materials science, with special emphasis upon recent progress and current trends. It is hoped that these publications will help provide definition and coherence in this field, and be of significant assistance to the increasing number of those involved in materials education, as well as to those actively involved in the many rapidly moving areas of materials research and technology.

An important and widely recognized feature of materials science has been the recognition of the fundamental similarities underlying the structures, phenomena, and properties of a wide variety of interesting and technologically important materials. This has led to an unusual degree of cross-fertilization of ideas and approaches among several different traditional scientific and engineering fields.

The Editorial Committee selects topics and authors and organizes the presentation of their contributions in these volumes so as to promote this amalgamation of concepts, tools, and techniques, and thereby to further this multifaceted approach to the understanding and control of physical phenomena and technological properties in a wide range of substances. This is reflected in the arrangement of the Table of Contents: The chapters are collected into the categories of Experimental and Theoretical Methods; Preparation, Processing, and Structural Changes; Properties and Phenomena; Structure; and Special Materials. We are also delighted to honor senior members of the materials science community, as well as to benefit from their wisdom, by including an invited Prefatory Chapter in most volumes.

We wish to express our sincere appreciation to the authors who have contributed to this volume, and to Sandra Cooperman of the Annual Reviews staff for her fine help in bringing it all to fruition.

We hope that these volumes are indeed useful, and welcome suggestions for topics that would be of interest in future years.

ROBERT A. HUGGINS
FOR THE EDITORS AND THE EDITORIAL COMMITTEE

Annual Review of Materials Science
Volume 22, 1992

CONTENTS

PREFATORY

Perfecting the Solid State, *Hans J. Queisser*   1

EXPERIMENTAL AND THEORETICAL METHODS

X-Ray Tomographic Microscopy (XTM) Using Synchrotron Radiation, *John H. Kinney and Monte C. Nichols*   121

Z-Contrast Transmission Electron Microscopy: Direct Imaging of Materials, *S. J. Pennycook*   171

RMC: Modeling Disordered Structures, *R. L. McGreevy and M. A. Howe*   217

PREPARATION, PROCESSING, AND STRUCTURAL CHANGES

Transient Liquid Phase Bonding, *W. D. MacDonald and T. W. Eagar*   23

Microwave Sintering of Ceramics, *Joel D. Katz*   153

Phase Transitions of Gels, *Yong Li and Toyoichi Tanaka*   243

Plasma Ion-Assisted Evaporative Deposition of Surface Layers, *S. Pongratz and A. Zöller*   279

PROPERTIES AND PHENOMENA

Ostwald Ripening, *P. W. Voorhees*   197

SPECIAL MATERIALS

Controlled Permeability Polymer Membranes, *W. J. Koros, M. R. Coleman, and D. R. B. Walker*   47

Design and Properties of Glass-Ceramics, *G. H. Beall*   91

INDEXES

Subject Index   296

Cumulative Index of Contributing Authors, Volumes 18–22   301

Cumulative Index of Chapter Titles, Volumes 18–22   303

SOME RELATED ARTICLES IN OTHER *ANNUAL REVIEWS*

From the *Annual Review of Earth and Physical Chemistry*, Volume 43 (1992):

Miceles and Microemulsions, D. Langevin

Transition State Spectroscopy of Biomolecular Chemical Reactions, D. Neumark

Equilibrium and Dynamic Processes at Interfaces by Second Harmonic and Sum Frequency Generation, K. Eisenthal

Molecular Electronics, M. Ratner and C. A. Mirkin

Phase Decomposition in Polymers, C. Han and A. Z. Akcasu

Polymer Dynamics in Electrophoresis of DNA, J. Noolandi

Transport Properties of Polymeric Liquids, R. B. Bird and H. C. Öttinger

Dynamics of Rigid and Semirigid Rodlike Polymers, R. Pecora and M. A. Tracy

Classical Dynamics Methods for High Energy Vibrational Spectroscopy, J. M. Gomez Llorente and E. Pollak

Stimulated Emission Pumping: Applications to Highly Vibrationally Excited Transient Molecules, T. Sears and F. J. Northrup

Phase Transitions Monolayers, C. Knobler and R. Desai

Chemical Reactions in Energetic Materials, G. Adams and R. Shaw

Electron Density for X-Ray Diffraction, P. Coppens

Mobile Ions in Amorphous Solids, C. A. Angell

ANNUAL REVIEWS INC. is a nonprofit scientific publisher established to promote the advancement of the sciences. Beginning in 1932 with the *Annual Review of Biochemistry*, the Company has pursued as its principal function the publication of high quality, reasonably priced *Annual Review* volumes. The volumes are organized by Editors and Editorial Committees who invite qualified authors to contribute critical articles reviewing significant developments within each major discipline. The Editor-in-Chief invites those interested in serving as future Editorial Committee members to communicate directly with him. Annual Reviews Inc. is administered by a Board of Directors, whose members serve without compensation.

1992 Board of Directors, Annual Reviews Inc.

J. Murray Luck, Founder and Director Emeritus of Annual Reviews Inc.
  Professor Emeritus of Chemistry, Stanford University
Joshua Lederberg, Chairman of Annual Reviews Inc.
  University Professor, The Rockefeller University
James E. Howell, Vice Chairman of Annual Reviews Inc.
  Professor of Economics, Stanford University
Winslow R. Briggs, *Director, Carnegie Institution of Washington, Stanford*
W. Maxwell Cowan, *Howard Hughes Medical Institute, Bethesda*
Sidney D. Drell, *Deputy Director, Stanford Linear Accelerator Center*
Sandra M. Faber, *Professor of Astronomy, University of California, Santa Cruz*
Eugene Garfield, *President, Institute for Scientific Information*
William Kaufmann, *President, William Kaufmann, Inc.*
Daniel E. Koshland, Jr., *Professor of Biochemistry, University of California, Berkeley*
Donald A. B. Lindberg, *Director, National Library of Medicine*
Gardner Lindzey, *Director Emeritus, Center for Advanced Study in the Behavioral Sciences, Stanford*
Charles Yanofsky, *Professor of Biological Sciences, Stanford University*
Richard N. Zare, *Professor of Physical Chemistry, Stanford University*
Harriet A. Zuckerman, *Professor of Sociology, Columbia University*

Management of Annual Reviews Inc.

John S. McNeil, Publisher and Secretary-Treasurer
William Kaufmann, Editor-in-Chief and President
Mickey G. Hamilton, Promotion Manager
Donald S. Svedeman, Business Manager
Richard L. Burke, Electronic Communications Manager

ANNUAL REVIEWS OF
Anthropology
Astronomy and Astrophysics
Biochemistry
Biophysics and Biomolecular Structure
Cell Biology
Computer Science
Earth and Planetary Sciences
Ecology and Systematics
Energy and the Environment
Entomology
Fluid Mechanics
Genetics
Immunology
Materials Science
Medicine
Microbiology
Neuroscience
Nuclear and Particle Science
Nutrition
Pharmacology and Toxicology
Physical Chemistry
Physiology
Phytopathology
Plant Physiology and
  Plant Molecular Biology
Psychology
Public Health
Sociology

SPECIAL PUBLICATIONS

Excitement and Fascination of Science, Vols 1, 2, and 3

Intelligence and Affectivity, by Jean Piaget

For the convenience of readers, a detachable order form/envelope is bound into the back of this volume.

Hans Joachim Queisser

PERFECTING THE SOLID STATE

Hans J. Queisser

Max-Planck-Institut für Festkörperforschung, 7000 Stuttgart 80, Germany

KEY WORDS: single crystals of semiconductors, crystal defects, crystal growth, solid-state materials characterization, semiconductor device principles

INTRODUCTION

Scientific mastery of materials in the twentieth century has overcome the purely empirical endeavors of earlier times. Stone, bronze, and iron ages were based on remarkable craftsmanship, but modern technology, especially in the field of semiconductors, could never have been initiated and completed without a profound scientific basis of materials science (1). Microelectronics technology relies on deliberately and meticulously controlled introduction of imperfections into nearly perfect crystals. Future successful technical applications will continue to be based on materials of crystallographic integrity and chemical purity, which can be characterized to fine details and then altered deliberately on increasingly smaller scales, today down to atomic dimensions.

Witnessing this fascinating development, actively participating in the research, and nurturing manifold applications are both challenge and opportunity, and I am grateful for my own chance to participate in this exciting pursuit and have never regretted my choice of materials science. I also gratefully acknowledge the offer to contribute an article to this review series in which I describe my personal involvement within this dynamic realm of perfecting and culturing solid materials.

APPRENTICE OF THE AMORPHOUS

Goettingen University physics students in the 1950s usually opted for thesis work in nuclear and particle disciplines. Few chose solid-state physics, although Goettingen can muster a glorious local tradition of Robert Pohl's pioneering color center work (2). I elected the study of solids, partly because of early stimulating and encouraging encounters with Ernst Ruska, Max von Laue, and their collaborators, but also because materials science promised more individuality of research work than the anonymity of rapidly growing teams at accelerators and reactors—a distinction that has remained until now. Rudolf Hilsch, one of Pohl's most prolific and imaginative experiment-oriented disciples, continued the tradition of solids during those days. I obtained my diploma and doctoral theses under his umbrella, but learned my lessons with purposely imperfected solids: amorphized thin films.

Buckel & Hilsch (3) made an astonishing discovery in the then still darkly mysterious field of superconductivity: bismuth films, rapidly quenched from the vapor onto a substrate at helium temperature, were superconductors, whereas crystalline bulk Bi was not! Hilsch hypothesized that superconductivity might really be an imperfection-controlled phenomenon. This idea seemed to be supported by observations of increased stabilization of the amorphous, superconducting modification by admixtures such as Cu, Sb, and LiF into the quenched films (4). Hilsch's working hypothesis may have been motivated by the strong impact on everybody involved with color centers, which concluded that such an evident quality as the color of a crystal was not a primary property of the pure alkali halide, but rather dominated by imperfections, namely small concentrations of halogen vacancies, which are occupied by electrons.

Bismuth films and binary systems of alkali halides (5) were my samples, which I investigated with Debye-Scherrer X-ray analysis and saw how the broad bands of the amorphized materials gradually reverted to sharp crystal reflexes upon annealing. The newly found isotope effect of superconductivity and the early, exciting publications from Bardeen's school convinced me—not necessarily my thesis advisors as yet—that not impurities, but phonons played the vital part for superconductivity. The Bi anomalies have just recently been reinvestigated; it has been suggested (6) that surface phenomena might be responsible for the onset of superconductivity in such films.

I looked at the kinetics of the return to equilibrium of my amorphized materials and suggested a "kinetic phase diagram," upside down in temperature, where the liquid-like amorphous state was stable only at lowest temperatures and stepwise achieved equilibrium, as Ostwald in his "step rule" had similarly assumed. Figure 1 shows one result for two alkali halides with different innate structure (5). When co-evaporated, they show an amorphous liquid-like state that can even lead to CsCl, occurring not in its regular lattice, but in the NaCl structure. The Goettingen team had

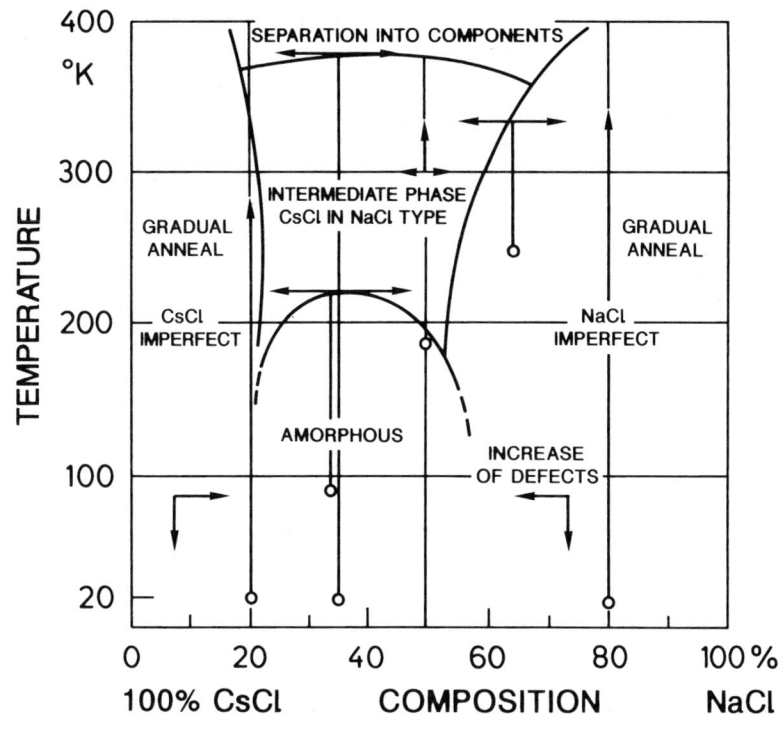

Figure 1 Phase diagram of mixed CsCl and NaCl evaporated onto a substrate at 20 K. The diagram is inverted in temperature from regular phase diagrams, it shows the stepwise annealing of the disordered films (see Reference 5).

an amazingly effective method of freezing defects into solids and then unleashing them to recover equilibrium.

We specialists in quenching condensation were then quite alone, few people were interested in our thin defect-enriched films. Their condition seemed very ill-defined, not easily reproducible, and hardly accessible to quantitative theory. Just like my own specimens, I longed for a more definitive state of approachable, predictable order. Those newly harnessed semiconducting crystals (7) enticed me. I left Goettingen in 1959 to join Shockley Semiconductors, a small company in Mountain View, California. His rented apricot barn was the cradle of what is now called Silicon Valley—an industrial landscape based on crystalline perfection (1).

DISLOCATED DISCOURSE

The dislocation was my first California assignment and an exciting one it was indeed. Shockley's quest for the fastest possible transistor with minimal base width led him to consider an isolated dislocation in an otherwise perfect silicon crystal. The presumed specific dislocation-related electronic acceptor states (8), possibly originating from unsaturated dangling bonds, might be used to generate a cylindrical p-n junction. An extremely small and contactable base region for an ultrafast, ultimate bipolar transistor seemed in reach. One-dimensional electronic conduction along the dislocation was in sight, and all this in 1959, when silicon was not even an established material and lithography or epitaxy were an unknown vocabulary! The concept of introducing and utilizing just one isolated, well-defined defect with non-zero Burgers vector (9) into a highly perfected solid material was elegant, but its reduction to practice demanded a much improved materials control.

Generating just one isolated dislocation was, and still is, really too difficult. Hence we grew bicrystals of Si with two slightly misaligned seed crystals (10). A reasonably well-defined small-angle grain boundary develops in such a crystal. A regular row of edge dislocations is expected to establish such a tilt boundary. The dislocation spacing d is controllable by the tilt angle φ for small angles, since

$$d = |b|/\varphi, \qquad 1.$$

where b is the magnitude of the dislocation's Burgers vector.

Diffusion of dopant impurities emerged at that time as a technique of definitive defect introduction for p-n junctions and was much superior to the coarse alloying previously utilized. The interplay of diffusion and dislocations became our major research topic (10). Dopants diffuse much more rapidly along dislocations, as shown in Figure 2, which we interpreted by point defect equilibria altered at the dislocation core. The strain field of a dislocation attracts impurities and other point defects. This attraction is easily understood because the crystal lowers its energy when, for example, an oversized impurity atom reaches the dilatation region of the dislocation, which is near the edge of the missing lattice plane.

Dislocations were feared to cause short circuits in transistors (11), but also we hoped to use very small junction structures produced with grain boundaries (12). These junctions yielded inferior electrical properties, too many recombination centers spoiled the current/voltage-relation; good devices demand highly perfected silicon. But dislocations may gainfully be introduced into idle portions of a device, on the backside, for example, where they serve as "gettering" sinks for detrimental metallic impurities harmful for junctions (13). The "physics of gettering" remains an exciting challenge for materials science on behalf of semiconductor economics, since device yield decides between black and red figures on the bottom line of balance sheets.

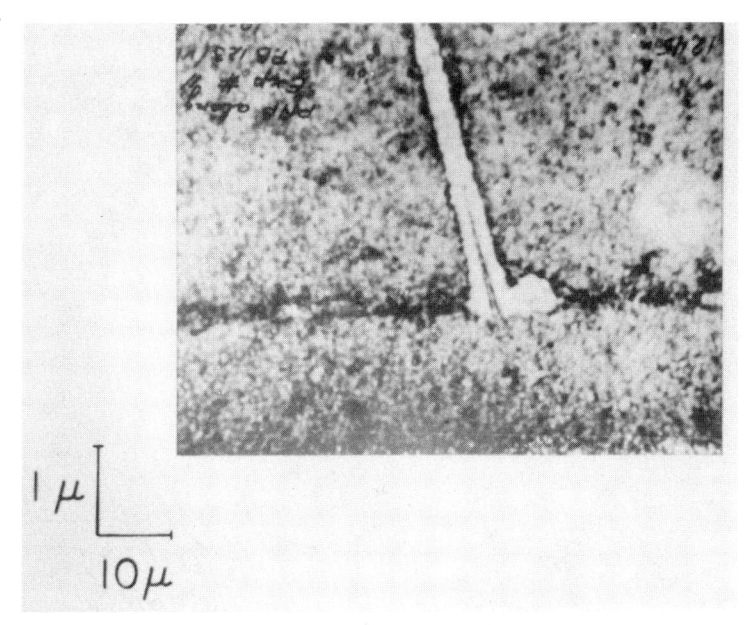

Figure 2 Example of enhanced diffusion of dopant impurities near a small-angle tilt boundary in a silicon bi-crystal. Electrochemical staining reveals p-n junctions; the p-type regions appear dark. The sample is angle-lapped (compare Reference 12).

Dislocations thus enhance impurity diffusion; on the other hand, diffusion of impurities can greatly enhance the dislocation content of a crystal. An undersized acceptor atom, such as boron, replaces silicon atoms and thus strains the lattice so severely that dislocation generation starts as soon as the stored elastic energy gets too large. We found beautiful looking but really fearsome patterns of the introduced slip by boron diffusion into silicon (14). Figure 3 gives an example. This appearance of misfit dislocations in semiconductors was the first one; a long series followed, with the fatal dark-line defects in laser diodes, hetero-epitaxy defects, and strained-layer superlattices being the most prominent cases (15, 16). Shockley and I started patent applications for a method to avoid such dislocations by doping simultaneously with oversized and undersized atoms—a method now used—but Shockley's firm was defunct before we succeeded.

The idea of acceptor states caused by dangling bonds in the dislocation core (8) was much too simple, as we quickly discovered (12, 15). I suggested that impurity clouds and possibly bond reconstruction dominate dislocation properties, a view not generally shared in the 1960s (12), but now agreed upon, at least for silicon. Dislocations are much less harmful than

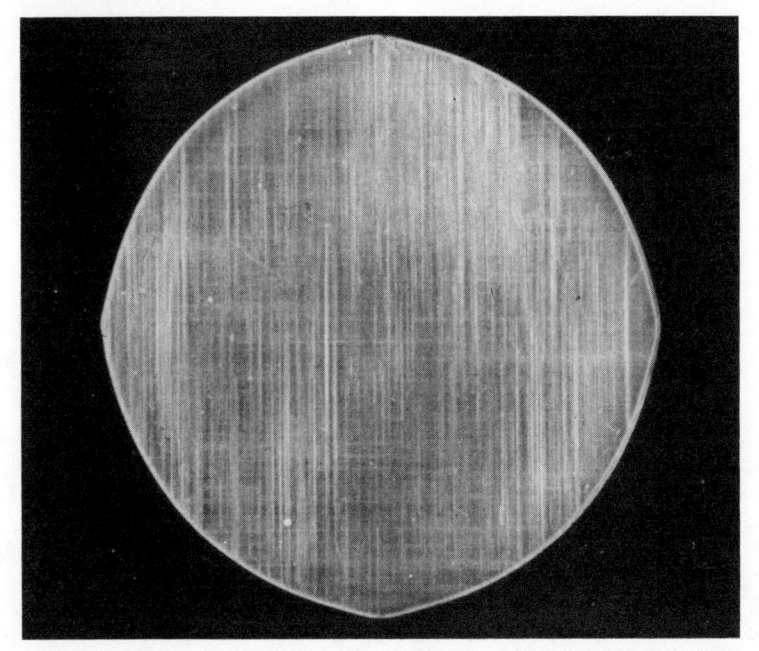

Figure 3 Slip-pattern apparent after etching a silicon wafer, which underwent a diffusion resulting in a high level of boron acceptor concentration. Undersized boron atoms generate slip via generation of misfit dislocations, X-ray topogram (see Reference 26).

first anticipated. This innocuousness is quite fortunate for us and for the yield of integrated circuits; imagine the problems arising from dislocations really being highly conductive miniature wires (11). A narrow base layer or channel would be prone to be short-circuited, and junctions could then not withstand high electric fields when reverse biased.

Dislocations as one-dimensional defects and their interplay with zero-dimensional point defects have continued to fascinate me and many other materials scientists. Luminescence is now a powerful method for their study because dislocations in Ge and Si produce a set of characteristic emission lines (17). Weronek & Weber at Stuttgart (18) recently closely followed the interplay of dislocations and impurities by studying the spectroscopic features of silicon. Still, it remains a tantalizing task to discern between impurity-cloud effects vs true dislocation-core phenomena. Only a few sites, certainly not all core atoms, seem to contribute to the specific radiative emission. Are these kinks or jogs, or other special atomic configurations? Or are impurities to blame, since even extremely dilute three-dimensional concentrations easily saturate the active centers at the one-

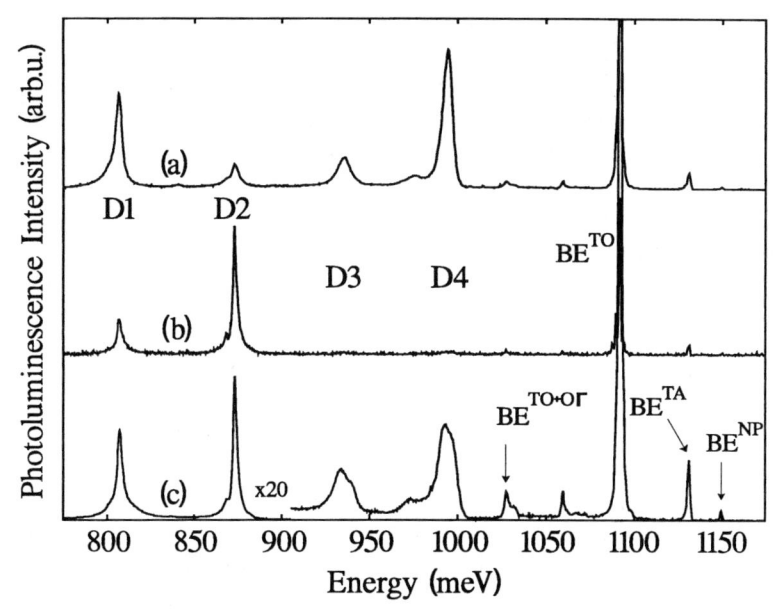

Figure 4 Characteristic photoluminescence spectrum of a silicon single crystal with dislocations. Two pairs of emission lines represent a characteristic feature typical for dislocations.

dimensional dislocations? Figure 4 presents a typical photoluminescence spectrum for a clean, highly-defined dislocation Si sample; the detailed interpretation is still a formidable challenge.

The dislocation is a mediator: flexible adaption to a desired macroscopic shape is reconciled with the rigid regularity of microscopic atomic order in the lattice. The dislocation is the locus of imperfection and elastic energy; it provides the plasticity of metals, which was the decisive property used in the bronze and iron ages. In the age of silicon, dislocations are really undesired, they are destined to be suppressed rather than domesticated.

IMAGED IMPERFECTIONS

Detection and identification of lattice imperfection are basic elements upon which a science of materials could be erected. Here is an exciting field for the tournaments of experimental physicists with crystallographers, device engineers, and others in materials science. Techniques for in-situ observations, non-invasive contactless methods, and tools for atomic resolution are particular challenges, as recognized both by recent Nobel bestowals

(19), and by hard-pressed process engineers responsible for the yields of semiconductor fabrication lines.

Optical emission, excited by incoming photons or electrons, gives plenty of detail about local electronic structure and perfection. Luminescence is, in principle, an extremely sensitive method to detect very small concentrations of impurities. It exceeds the detectability of optical absorption measurements by orders of magnitude. The photon-generated minority carrier rapidly loses energy and becomes captured by an impurity. Luminescence spectra hence reveal mostly the defects, the band-to-band radiative recombination is usually completely absent. This principle can also be expressed thermodynamically by the principle of detailed balance, which links emission rates to absorption rates multiplied by the Planck function of black-body radiation.

The spatial resolution of luminescence analysis, especially for crystals of high quality, is often determined not so much by the imaging method, but by the diffusion length of the minority carriers. Photoluminescence scanning was first demonstrated on GaAs in a thesis by Heinke (20), an industrial crystal grower, who acted in cooperation with our institute (a result is given in Figure 5).

Much more detailed spatial information can be gleaned from electron-optics excitation. The cathodoluminescence method has been especially powerful. It was used by Bimberg, initially at Stuttgart and Aachen, and then developed to high perfection, especially in image-handling, at Berlin, much of it in close collaboration with industry in the United States and Japan (21). The ease of scanning a crystal surface at very high resolution compensates for the somewhat coarse mode of excitation by electrons of energy far above the band gap E_g of the material.

Electron-beam induced current collection is another beautiful way to scan a semiconductor for imperfection. A major step in theory was achieved by Donolato (22) while in Stuttgart. The efforts continue worldwide, including work at our laboratories (23).

Electron microscopy, down to atomic resolution, has become a most impressive and indispensable tool that can be used to scrutinize crystalline perfection. High-quality epitaxy evolves in symbiosis with electron microscopy and tolerates the necessary sacrifice of the sample under the onslaught of the ion milling machines to thin the crystal down to a few layers of atoms.

The very early days of transmission electron microscopy of semiconductors I remember well. We had silicon under good control at the Shockley labs, but we also knew how to introduce dislocations by boron diffusion and stacking faults by oxide-afflicted vapor-phase epitaxy. Elec-

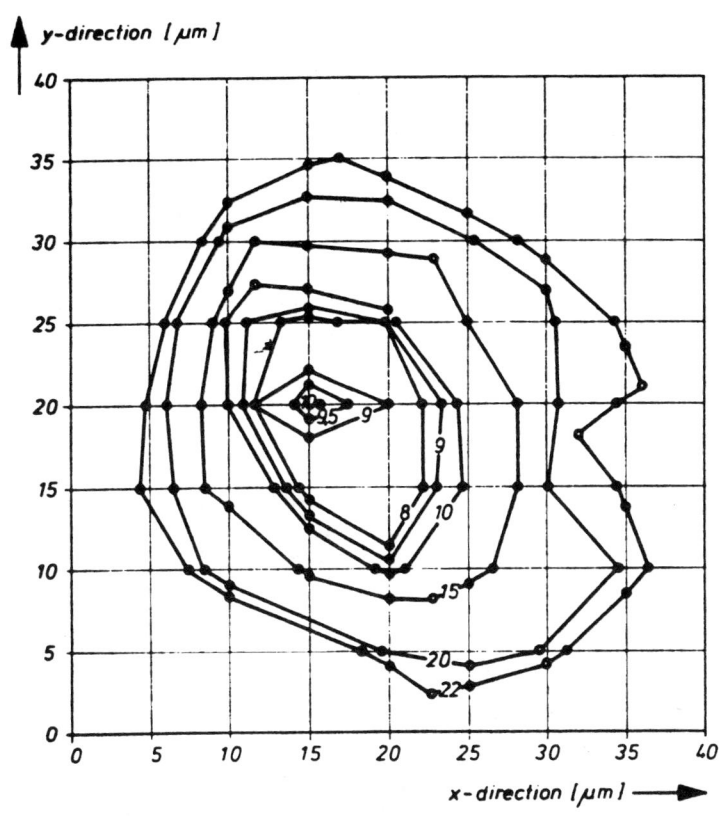

Figure 5 Example of spatially resolved photoluminescence scanning of GaAs. Lines are loci of equal emission intensity, here shown near a dislocation (see Reference 20).

tron microscopy with metals had started to be routine in the early 1960s. I inquired of the microscopists in the United States and Europe and offered samples, but polite disdain prevailed. My silicon seemed dull, uninterestingly perfect. The little Shockley lab, of course, had no electron microscope of its own. So I took samples, nicely thinned with a chemical etching method we had just developed, to the University of California at Berkeley and pretended to Washburn & Thomas that Shockley had sent me (not true, but he sanctioned the subterfuge upon seeing the resulting micrographs), and we started the activity of looking into silicon (24). I shall never forget the first session with Jack Washburn in the basement of the Hearst building. Only the last of my fifteen specimens showed anything:

a beautiful stacking fault (see Figure 6). Washburn's taxed patience revived when I told him that nobody had seen one before in a microscope. Berkeley's instrument was a Siemens Elmiskop 1A, developed by the future Nobel laureate Ernst Ruska. He was my boss when I worked as a lab technician after finishing high school because post-war Berlin had not enough places yet to study physics. Ruska was happily surprised to receive a copy of our first silicon study. I have since continuously tried to enhance this cooperation between electron microscopy and semiconductor crystal growth; I persisted as an irrepressible advocate for my microscopy colleagues in helping to secure rather hefty financing for instruments in Stuttgart, including a megavolt microscope to be delivered in 1991. Our early cooperation at Berkeley is documented on a cover of *Fortune* magazine, although the editors did not quite intend it that way (25).

Transmission electron microscopy devours its specimens, hence we also need nondestructive methods to detect imperfections, especially dislocations and other defects locally affecting the lattice parameters. X-ray methods are of lower spatial resolution, yet beautifully gentle and informative, especially topography, where the crystal is scanned to provide an image in the light of one Laue reflex. Gunther Schwuttke, then at

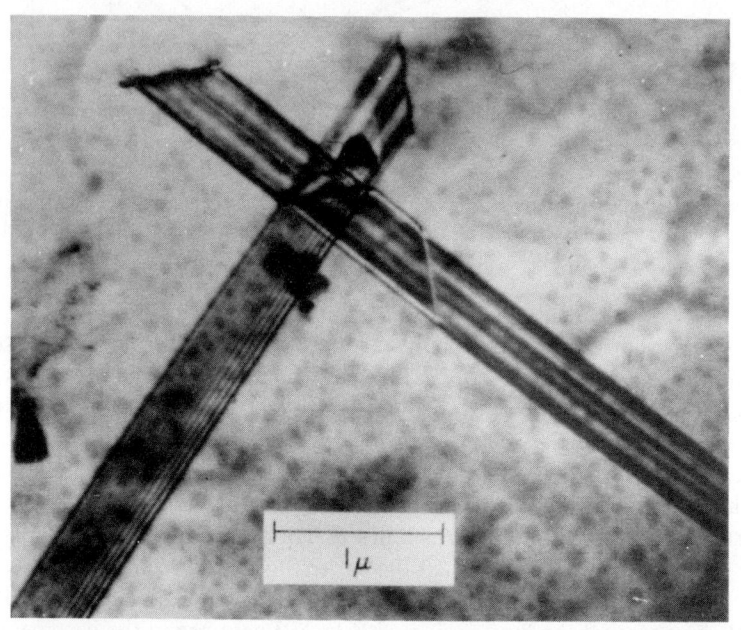

Figure 6 Earliest example of an image of a stacking fault, introduced by vapor-phase epitaxy at an oxide-contaminated Si surface, observed in transmission electron microscopy (see Reference 24).

General Telephone Labs, had a nice Lang topography system, and we cooperated with him across the United States (26). We really proved and characterized the dislocations induced by impurity diffusion. Fifteen years later, in Frankfurt and Stuttgart, we tried in earnest to elevate topography into a real time observational technique (27). Werner Hartmann and I anticipated future powerful synchrotron light sources and devoted attention to the detector end of "live X-ray topography." We succeeded and watched dislocations in silicon move and multiply under a mechanical load (28). We saw in real time how Si surface stress develops during the oxidation process and noted the surprising influences of water vapor (29). Heteroepitaxy of Ge on GaAs was followed in real time; we watched the initial layers grow with tetrahedra distortion then, after a sufficiently thick layer was deposited, saw that misfit dislocations rapidly formed and multiplied (30). These observations were part of the Stuttgart thesis of Werner Hagen, who then teamed up with electron microscopist Horst Strunk, to suggest the now famous Hagen-Strunk mechanism of misfit dislocation multiplication (31), which also explained the pairwise occurrence of these dislocations, which was puzzling to us in the early California work. An example of this arrangement is reproduced in Figure 7.

Atomic resolution can now be attained with the tunneling microscope (19), but its resolution is almost too high to extract useful feed-back for a crystal grower. Surface growth steps can, surprisingly to many, be resolved to the level of atomic step heights by pushing the Nomarski interference contrast microscopy to its limits, as shown in Figure 8. Beautiful photographs can be made over large areas to resolve step patterns, such as spirals emitted from a dislocation (32).

Early in this century, atoms were considered irreal or imaginary by a majority of scientists. Today, the imagery, meaning the art of making images, of defects in crystals, has helped us to realize even the most imaginative ideas of useful atomic arrangements inside a crystal lattice.

INTERFACIAL IDEALITY

Perfecting the electrical performance of p-n junctions in silicon was a major task in the early days of Silicon Valley. In germanium, with its smaller band gap, junctions are easily made to conform to the ideal rectifier equation (7):

$$I = I_o(\exp(qV/kT) - 1), \qquad 2.$$

where current I depends exponentially on voltage V at temperature T; q is the electronic charge. Most of the electronic transport is truly by particles

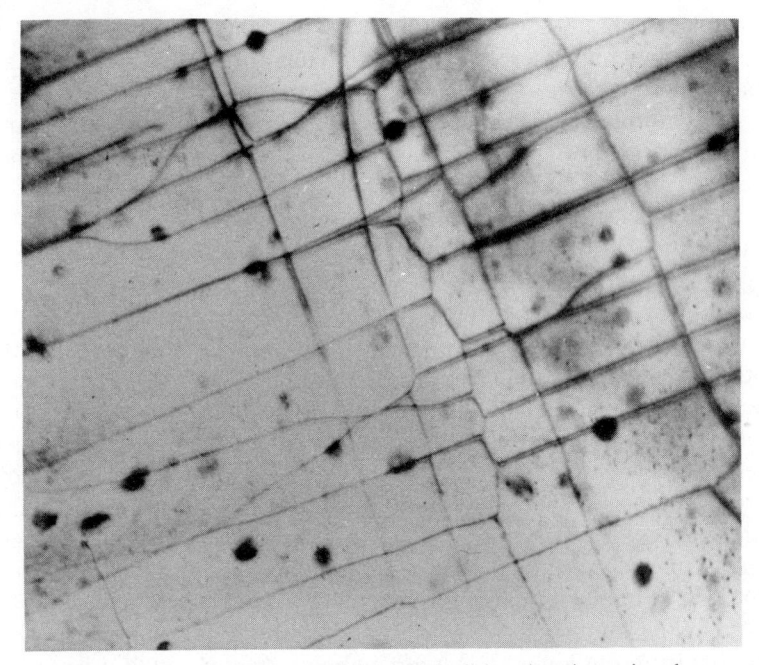

Figure 7 Example of pair-wise occurrence of misfit dislocations in semiconductors, generated by the Hagen-Strunk mechanism of dislocation multiplication (compare Reference 31).

surmounting the potential barrier through thermal activation, as assumed in theory.

Silicon is less obedient than its predecessor germanium. Electrons from the n-type side can encounter holes by recombining at defects within the junction's space charge layer. The seminal paper by Sah, Noyce & Shockley (33) considered this additional current path and derived a modified relation for forward bias:

$$I \sim \exp(qV/nkT),\qquad 3.$$

where the new parameter n could rise to $n = 2$, caused by recombination to give an undesirable, less steep current increase for such imperfect junctions.

A major detriment of such junctions is the reduced efficiency in photovoltaic solar cells. Solar cells were, in the 1960s even more than now, almost neglected, unexcitingly simple mono-junction devices, but the significance of the principles involved interested us. Shockley and I developed a truly thermodynamic theory of conversion efficiency based on the principle of detailed balance and on the generality of the Carnot cycle (34).

PERFECTING THE SOLID STATE 13

Figure 8 Observation of individual steps on a silicon surface. Steps with a height of just one atomic layer are resolvable with visible light in Nomarski interference contrast (photomicrograph by E. Bauser; also see Reference 32).

14 QUEISSER

Maximal energy conversion arises from the ideal, imperfection-free situation. One needs good material to prevent the photon-generated carriers from disappearing before they have reached the contacts. Longest minority-carrier lifetimes are those where only radiative recombination processes occur, as demanded by detailed balance. Figure 9 gives a survey. Therefore, the best solar cell is one where electrons and holes couple strongly with the solar photons, but not at all with the lattice. The crystal lattice thus must be an ideal container, not incurring any losses by recombining carriers before they reach the battery contacts.

The open-circuit voltage V_{op} is the decisive parameter:

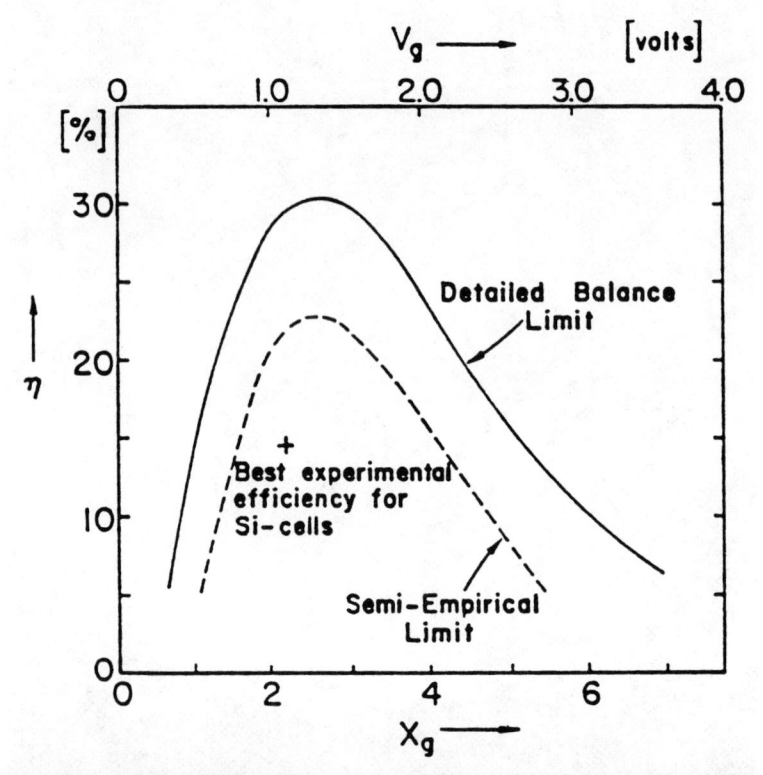

$$V_{op} = (kT/q) \ln(I_{sc}/I_o + 1), \quad\quad 4.$$

Figure 9 Theory of solar cell efficiency, first published in 1961. Top curve shows efficiency vs normalized band gap $x_g = V_g/kT_s$ (T_s is sun temperature) for thermodynamic limit, which arises when lifetimes are only determined by detailed balance. Dashed curve gives a conventionally quoted limit, now already surpassed. The cross indicates the best efficiency recorded in 1960 (see Reference 34).

where I_o, I_{sc} are reverse saturation current and photon-induced short-circuit currents, respectively. The quantity V_{op} indicates essentially the non-equilibrium spread

$$\Delta \zeta = \zeta_n - \zeta_p \qquad 5.$$

of the electron and hole quasi-Fermi levels. The ideal crystal permits the maximal, long-lived carrier non-equilibrium. Present cells of high efficiency, such as the passivated and emitter rear cell (PERC) (35), use this principle by protecting the surfaces and by making contacts as small as possible because contacts force $\Delta\zeta$ to zero. The junction ought to be ideal, too; n = 1 in Equation 3 is desirable (33, 36). Figure 10 shows this connection between forward characteristics and cell efficiency.

Real junctions, whether of the p-n variety or Schottky barriers, almost never attain n = 1. An empirical practice arose to classify interface qualities by this parameter, *n*, an extension beyond its original meaning (33), and to call *n* ideality, and unfortunate misnomer because a large *n* indicates a less ideal case! Werner and co-workers at the Stuttgart Max-Planck-Institute (37) recently demonstrated that $n \neq 1$ originates, in most cases,

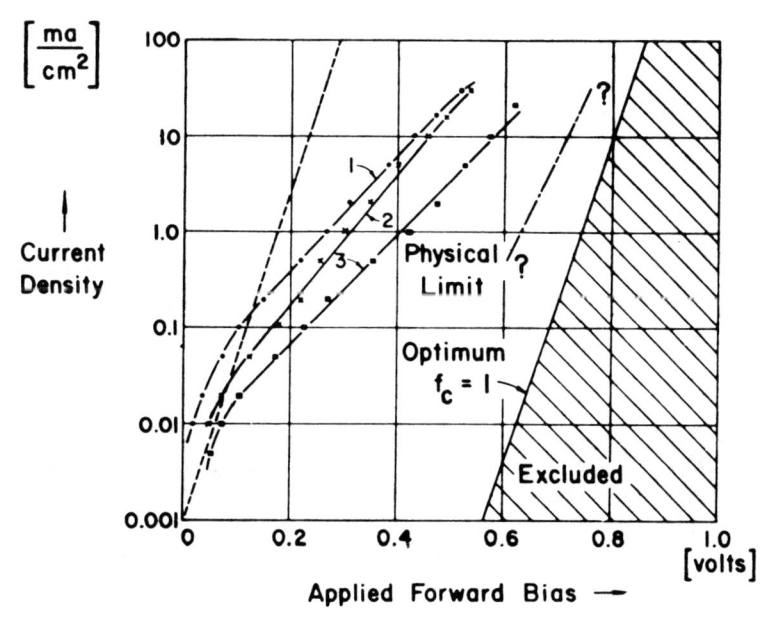

Figure 10 Junction perfection and solar cell efficiency compared (from Reference 34). Shaded area is excluded by the principle of detailed balance. Optimal cell obeys the current vs voltage relation marked $f_c = 1$, with ideality *n* also equal to unity.

from fluctuations in the barrier height V_b of the junction, which in turn changes upon varying the applied bias V_a:

$$(1 - 1/n) = \Delta V_b / \Delta V_a. \qquad 6.$$

Ideal interfaces thus require a very high degree of perfection throughout their entire area: i.e. flatness, no steps with differing crystallographic orientations, homogeneous doping, and perfected interfaces, which today can only be approached through careful heteroepitaxy.

EXPEDIENT EQUILIBRIUM

Perfection of a crystalline specimen is best achieved by growing the crystal slowly, in a pure environment, and as closely as possible to thermal equilibrium. The melting point is usually far from the operating temperature of a solid-state device. Even the otherwise advantageous and well-controlled gas-phase epitaxy techniques are still fairly far from equilibrium. Crystallization from a liquid comes closer, especially in liquid-phase-epitaxy (LPE), where a semiconductor, such as GaAs, is produced by slowly cooling a metallic melt, such as Ga, to precipitate a crystalline layer.

My first and dramatic experience regarding this significance of equilibrium was in working with Morton B. Panish at Bell Laboratories in 1965. We suddenly found thousandfold luminescence yields in LPE GaAs over that observed in regular crucible-grown GaAs (38). Liquid-phase epitaxy is much closer to equilibrium than growth of a crystal from its own melt; a gallium arsenide solution in gallium will furthermore not suffer from gallium deficiency, which generates defects that cause nonradiative recombination. These vague concepts were, of course, later verified by systematic phase-diagram determination.

The huge increase of luminescence yield in LPE GaAs brought plenty of directors and vice presidents as visitors into Panish's and my labs and gave all of us renewed courage that a laser diode operating at room temperature might yet be achievable, contrary to the pessimistic predictions uttered and written in those days (39). Indeed, laser diodes are realities now and, once again, they are achieved by highly controlled structuring of a highly perfected material (40). In spite of the victorious development of molecular beam epitaxy as a supremely controllable growth method, LPE is still the major method today for the mass production of light-emitting semiconductor devices; closeness to equilibrium during crystallization ensures perfection, which is the prerequisite for high conversion efficiency of electron current into a stream of photons.

History does repeat itself, at least whenever thermodynamic principles are involved. Our recent Stuttgart experiments on LPE of silicon from a metallic solution onto Si substrates have produced outstanding solar cells with exceptional efficiencies (41). The relation between these light-absorbing devices and the light-emitting diodes is obviously close: both those junction devices are optimal when crystal perfection suffices to suppress all nonradiative processes to leave only the inevitable radiative and, for high doping, Auger (42) recombination. Solar cells from amorphous Si have in principle low efficiencies and, not surprisingly, degrade under illumination (43), which puts these cells at a severe non-equilibrium disadvantage in photovoltaics.

Nonradiative recombination can be suppressed by localizing electrons and holes, thus preventing them from reaching deadly surfaces or defects. This is the principle of GaAs diodes, heavily doped with Si or other group IV elements, where deep band tails are the localizers, and the emission is shifted markedly toward the red (44). The elements of column IV, such as Si or Ge, are amphoteric in the III-V compounds, especially GaAs. Silicon can thus act as a donor when it replaces a gallium atom or as an acceptor when substituting for an arsenic atom. High solubility facilitates heavy doping with Si or Ge and creates extensive densities of electronic states reaching into the forbidden energy gap. Compensation is strong because donors as well as acceptors are produced, and this displaces the Fermi level from the bands toward the center of the gap. Hence both bands with their tails of localized states are free of carriers and can capture electrons and holes efficiently. Only radiative recombination remains as the route towards equilibrium, and the radiation is appreciably lower in energy than the gap and thus is not reabsorbed.

I found this large efficiency increase (due to the presence of group IV elements) by accident in one sample, when my crystal grower colleague K. L. Lawley happened to position a quartz slide between the sample and the graphite susceptor, and I had mixed up the top and bottom sides of this sample in my luminescence measurements. Such IR diodes are now mass-produced by the millions per month, TV remote controls use them. My AT&T patent application (45) delayed and severely shortened my scientific publication on this topic (44). This patent application, however, seems to have been filed in every industrialized country; I had to appear many times at the United States consul's office in Frankfurt to affix my signature.

This trick of band-tail localization is no contradiction to the rule that semiconductor devices of optimal yield, high sensitivity, low noise, and long life will hardly be found with materials that are far from thermodynamic equilibrium. Semiconductors, after all, now reign after having conquered the worst offender against equilibrium: the vacuum tube with its white-hot cathode, insatiable for energy. Those who had to work with these tubes and knew thermodynamics, like Schottky (46), Shockley, or

Bardeen, became the strongest proponents for solid-state electronics. Electronics nowadays is no longer a composite, non-equilibrium affair as it was in the heyday of the vacuum tube. It is now an integrated technique with the crystal lattice of the perfected material serving as a reference scaffolding, available for all kinds of atomic substitutions to internally generate electronic functions and to execute these functions close to thermal and chemical equilibria (46).

PERSNICKETY PERSISTENCE

A peculiar phenomenon, intimately related to imperfections and inhomogeneities in semiconductors, has recently attracted a remarkable degree of renewed attention. This phenomenon of "persistent photoconductivity" was actually known for a long time and for a surprisingly wide range of solid-state materials. The topic was dormant until the vital interest in alloys of gallium arsenide with aluminum arsenide and their desired applications in very fast electronic devices suddenly aroused it.

Persistent photoconductivity arises at sufficiently low temperatures, usually below the temperature of liquid nitrogen. A conductivity induced by illumination does not decay rapidly, e.g. within milliseconds, as is usually observed. Rather it persists for almost an immeasurably long time after the illumination is switched off. This photoconductivity thus appears to be stored inside the crystal. The lifetimes of the photo-generated carriers are somehow incredibly extended. Some sort of energetic barrier with energy E_{rec} apparently opposes the usual carrier recombination. The barrier is, however, thermally surmountable:

$$\tau = \tau_o \exp(E_{rec}/kT), \qquad 7.$$

which lengthens the regular τ_o at temperatures T insufficient for barrier crossing. In this perspective, we must understand the nature of the barriers that prevent recombination (47).

Early work on this effect was done on many semiconducting samples; most of these numerous observations were critically documented in a review by Sheinkman & Shik (48). Their general interpretation was based on macroscopic inhomogeneities, which separated the electrons from the holes spatially to preclude recombination. Such inhomogeneity might be a doping fluctuation, a precipitate, a fluctuation in alloy composition, or it might be a junction, a contact, or even a surface. The essential contribution of the inhomogeneity must be a local potential gradient. The similarity to a p-n junction solar cell is evident; there the potential is deliberately introduced to suppress carrier recombination. This general explanation was quite universally accepted, for example, for radiation-damaged semiconductors (49), where the defect-clusters form spherical junctions, which trap one type of carrier and leave the other type free to perambulate. Persistent photoconductivity during those days was certainly nothing to be proud of unless the crystal grower had deliberately introduced junctions or barriers. In any other bulk crystal, such persistence primarily indicated inhomogeneity.

At our Stuttgart labs, we wanted to measure screening of ionized impurities in GaAs by light-induced carriers. The mobility of electrons and holes can rise dramatically by screening of this predominant scattering mechanism at low lattice temperatures (50). (Claiming to have perfected a piece of compound semiconductor by citing record high Hall mobilities is only justified if the sample is carefully left in the dark!) Our own experiments became so severely hampered by persisting carriers that we abandoned our original plans and first tackled the problem of persistent conduction. Theodorou & I (51) succeeded with a well-defined GaAs junction sample to quantitatively explain the gradual build-up of charge under illumination. A perfected, well-characterized specimen served as a model to verify the idea of spatial carrier separation. Later, we (52) could even probe the distribution of the localizing deep hole traps with this technique on a GaAs sample with deliberately shaped trap doping; a result is given in Figure 11.

Experiments on epitaxial layers of $Al_xGa_{1-x}As$ mixed crystals, grown on semi-insulating GaAs substrates, became very exciting during that time; important new heterojunction concepts arose (53). But a strong persistent photoconductivity bothered the device engineers because current control by the gate of a transistor was impeded by trapped charge, and everything became electronically slow because of persistence. Work by Lang and colleagues (54) pointed to a different potential barrier against recombination. It is a microscopic barrier, located at a donor-related center. This imperfection has been baptized DX-center, it is bistable and assumes different lattice positions for different charge conditions. Despite many years of tremendous experimental and theoretical efforts to understand this DX-center, there is still enough ignorance and controversy to convene international conferences for this one crystal defect. Deep levels simply are difficult to grasp because of their essential electron-phonon coupling (47, 54).

With the persistent photoconductivity now linked to an atomic property of a specific imperfection, the situation suddenly changed for a wide gamut of semiconducting materials. No longer was persistence an embarrassing deficiency for a sample and for its creator, the crystal grower. Instead, the DX paradigm ennobled all observed similar effects toward basic solid-state physics of lattice relaxation and phonon interaction. Journal referees were overwhelmed by the analogy carried over from (Al,Ga)As to all other kinds of materials of lesser perfection.

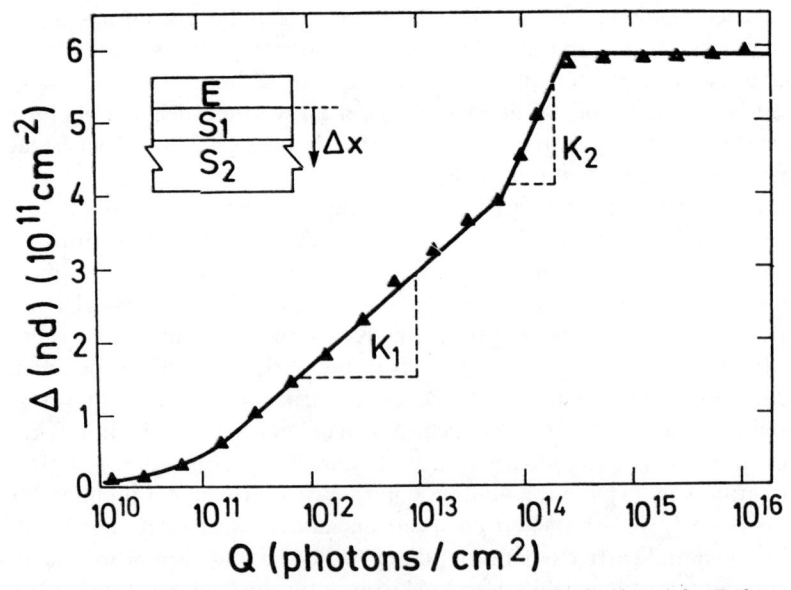

Figure 11 Persistent photoconductivity utilized to probe trap distributions (after Reference 52). Plotted is the persistent areal electron density Δ (nd) stored in epitaxial layer E as a function of cumulative photons by illumination. The two layers, S_1 and S_2, have differing trap densities and cause differing slopes, K_1 and K_2, of the curve.

Semiconductor physics provided a major success in correlating properties of a material to individual atomic impurities. Conductivity, absorption and emission, carrier lifetimes, and spin resonance can all be linked unambiguously and quantitatively to dopant and trap admixtures. This principle is a basic guideline, and it was assured only by meticulous research on highly perfected, homogeneous, single-crystalline specimens, predominantly with Ge, Si, GaAs, and GaP. It is tempting to extend this forceful principle to other materials of inferior purity and impaired lattice perfection, but this venture may become a dangerous transgression. The inhomogeneities and their resulting interfaces may in reality be the controlling elements for electrical and even optical materials properties. Comfort and assuredness in dealing with today's semiconductor science rest on the Herculean labors in perfecting these materials.

CONCLUSIONS

About four decades of intense work on perfecting solid materials resulted in the present semiconductor microelectronics; it gave birth to optoelectronics with sources, detectors, and astoundingly pure transmitting fibers; it improved dielectrics and ionic conductors. The new cuprate superconductors have as yet not nearly attained this degree of perfection, which complicates experimental and theoretical attempts on these materials.

The perfection of the semiconductors opened the way toward dimensional reduction and with it the new quantum devices emerged. The residence of the electron within the lattice is willfully restricted ever more tightly. Quantum wells (53) intentionally confine the carriers spatially and define their energies by this confinement. We have now reached atomic dimensions, for example, in the delta-doping of single atomic layers in epitaxy (55). All these achievements in spatial control, just as in submicron silicon circuits, rely completely on the perfection of the materials employed. Only a few semiconductors and dielectrics are so well-controlled, hence quantum devices are now restricted to a few elements and compounds. This convergence is often derided and deplored by chemists, who feel rejected in their offerings of densely populated families of new materials. But the demands on substances for quantum miniaturization are exceedingly severe: they must be so close to perfection that their own materials parameters begin to lose significance. Their distinct properties start to fade into the background; it is merely the geometrical dimension of a quantum well or a tunnel barrier that governs the electronic phenomena. The vessel for a quantum device can only be a perfected material.

Acknowledgments

I feel indebted to many who guided and accompanied me along my own path in attempting to understand, improve, and utilize solid-state materials: teachers, colleagues, and students, whose friendly spirit of cooperation I gratefully acknowledge.

Literature Cited

1. Queisser, H. J. 1988. *The Conquest of the Microchip*. Cambridge, Mass: Harvard Univ. Press. 200 pp. (Transl. from German *Kristallene Krisen*, 1985. Muenchen, Germany: Piper. 320 pp.)
2. Seitz, F. 1946. *Rev. Mod. Phys.* 18: 384–408
3. Buckel, W., Hilsch, R. 1954. *Z. Physik* 138: 109
4. Barth, N. 1955. *Z. Physik* 142: 58
5. Queisser, H. J. 1958. *Z. Physik* 152: 495–506, 507–20
6. Moodera, J. S., Meservey, R. 1990. *Phys. Rev.* B42: 179–83
7. Shockley, W. 1950. *Electrons and Holes in Semiconductors*. New York: van Nostrand. 558 pp.
8. Shockley, W. 1953. *Phys. Rev.* 91: 228
9. Shockley, W. 1961. *US Patent No. 2979427*; 1960. *US Patent No. 2937114, 2954307*
10. Queisser, H. J., Hubner, K., Shockley, W. 1961. *Phys. Rev.* 123: 1245–54
11. Queisser, H. J. 1963. Dislocations and semiconductor device failure. In *Physics of Failure in Electronics*, ed. M. F. Goldberg, J. Vacarro, pp. 146–55. Baltimore: Spartan. 255 pp.
12. Queisser, H. J. 1963. Versetzungen in Silizium. In *Festkörperprobleme*, ed. F. Sauter, II: 162–87. Braunschweig, Germany: Vieweg. 323 pp.
13. Goetzberger, A., Shockley, W. 1960. *J. Appl. Phys.* 31: 1821–24

14. Queisser, H. J. 1961. *J. Appl. Phys.* 32: 1776–80
15. Queisser, H. J. 1983. Electrical properties of dislocations and boundaries in silicon. In *Defects in Semiconductors II*, ed. S. Mahajan, J. W. Corbett, pp. 323–42. New York: North-Holland. 582 pp.
16. Holt, D. B. 1979. *J. de Phys.* 40(Suppl. 6): C6–189
17. Drozdov, N. S., Patrin, A. A., Ikachev, V. D. 1976. *Pis'ma Zh. Eksp. Teor. Fiz.* 23: 651–53
18. Weronek, K. 1992. PhD thesis. Stuttgart Univ.
19. Nobel Lectures, 1986. *Revs. Mod. Phys.* 59: 615–38
20. Heinke, W., Queisser, H. J. 1974. *Phys. Rev. Lett.* 33: 1082–85
21. Grundmann, M., Christen, J., Bimberg, D., Hashimoto, A., Fukunaga, T., Watanabe, N. 1991. *Appl. Phys. Lett.* 58: 2090–92
22. Donolato, C. 1979. *Appl. Phys. Lett.* 34: 80–84
23. Jakubowicz, A. 1985. *J. Appl. Phys.* 58: 4354
24. Queisser, H. J., Finch, R. H., Washburn, J. 1962. *J. Appl. Phys.* 33: 1536–39
25. *Fortune*. September 1965. Cover photograph shows Sather Gate at Berkeley, CA. (author of this article seen on center left edge, returning from electron microscopy experiment)
26. Schwuttke, G. H., Quessier, H. J. 1962. *J. Appl. Phys.* 33: 1540–43
27. Hartmann, W. 1977. Live topography. In *X-Ray Optics*, ed. H. J. Queisser, pp. 191–218. Heidelberg: Springer. 227 pp.
28. Hartmann, W. 1980. X-ray TV imaging and real-time experiments. In *Characterization of Crystal Growth Defects*, ed. B. K. Tanner, D. K. Bowen, pp. 497–506. New York: Plenum. 602 pp.
29. Hartmann, W., Franz, G. 1980. *Appl. Phys. Lett.* 37: 1004–6
30. Hagen, W., Queisser, H. J. 1978. *Appl. Phys. Lett.* 32: 269–72
31. Hagen, W., Strunk, H. 1978. *Appl. Phys.* 17: 85–87
32. Bauser, E. 1983. Crystal growth from the Melt. In *Festkörperprobleme*, ed. P. Grosse, pp. 141–64. Braunschweig, Germany: Vieweg. 326 pp.
33. Sah, C. T., Noyce, R. N., Shockley, W. 1957. *Proc. IRE* 45: 1228–34
34. Shockley, W., Queisser, H. J. 1961. *J. Appl. Phys.* 32: 510–19
35. Blakers, A. 1990. High efficiency crystalline silicon solar cells. In *Festkörperprobleme*, ed. U. Rössler, pp. 403–24. Braunschweig, Germany: Vieweg. 470 pp.
36. Queisser, H. J. 1962. *Solid-State Electron.* 5: 1–10
37. Werner, J. H., Güttler, H. H. 1991. *J. Appl. Phys.* 69: 1522–33; Queisser, H. J. 1963. *Z. Phys.* 176: 313–22
38. Panish, M. B., Queisser, H. J., Derick, L., Sumski, S. 1966. *Solid-State Electron.* 9: 311–14
39. For reviews of estimates on laser diode possibilities, see Ref. 40, Vol. A
40. Casey, H. C., Panish, M. B. 1978. *Heterostructure Lasers*. New York: Academic. Vol. A, 272 pp.; Vol. B, 330 pp.
41. Blakers, A. W., Werner, J. H., Bauser, E., Queisser, H. J. 1991. *Proc. Tenth Photovoltaic Conf. Eur. Commun.*, Lisbon, pp. 692–94. Dordrecht: Kluwer
42. Auger recombination as unavoidable was first recognized in Ref. 34; it is now generally realized as essential, especially for high doping levels. For solar cells, see Blakers, A. 1990. "High Efficiency Crystalline Silicon Solar Cells." In *Festkörperprobleme*, ed. U. Rössler, p. 465. Braunschweig, Germany: Vieweg
43. Staebler, D. L., Wronski, C. R. 1977. *Appl. Phys. Lett.* 31: 292–95; Krühler, W. 1991. *Appl. Phys.* A53: 54–61
44. Queisser, H. J. 1966. *J. Appl. Phys.* 37: 2909
45. Queisser, H. J. 1966. *US Patent No. 3387163*
46. Queisser, H. J. 1986. *Siemens Forsch. Entwicklungsber.* 1: 272–80 (special issue in commemoration of Walter Schottky)
47. Queisser, H. J. 1984. Persistent photoconductivity in semiconductors. In *Proc. 17th Int. Conf. Phys. Semiconductors*, ed. J. D. Chadi, pp. 1303–8. New York: Springer. 1580 pp.
48. Sheinkman, M. K., Shik, Y. Ya. 1976. *Fiz. Tek. Poluprovodn.* 10: 209–14
49. Gregory, B. L. 1970. *Appl. Phys. Lett.* 16: 67–71
50. Bludau, W., Wagner, E., Queisser, H. J. 1976. *Solid-State Commun.* 18: 861–64
51. Queisser, H. J., Theodorou, D. E. 1979. *Phys. Rev. Lett.* 43: 401–4
52. Theodorou, D. E., Queisser, H. J., Bauser, E. 1982. *Appl. Phys. Lett.* 41: 628–31
53. Dingle, R. 1975. Confined carrier quantum states in ultrathin semiconductor heterostructures. In *Festkörperprobleme*, ed. H. J. Queisser, 15: 21–48. Braunschweig, Germany: Vieweg
54. Lang, D. V., Logan, R. A., Jaros, M. 1979. *Phys. Rev.* B19: 1015–23
55. Schubert, E. F., Ploog, K. 1985. *Jpn. J. Appl. Phys.* (2)24: L608–10

TRANSIENT LIQUID PHASE BONDING

W. D. MacDonald and T. W. Eagar

Department of Materials Science and Engineering, Massachusetts Institute of Technology, Cambridge, Massachusetts 02139

KEY WORDS: diffusion, diffusion welding, activated diffusion bonding, isothermal solidification

INTRODUCTION

Transient liquid phase (TLP) bonding is a joining process that has been applied to many metallic systems throughout the ages, and yet it still holds promise as a technique for joining in aerospace and semiconductor applications. The TLP process produces a strong, interface-free joint with no remnant of the bonding agent. It differs from diffusion bonding in that the formation of a thin liquid interlayer eliminates the need for a high bonding or clamping force. The interlayer can be provided by foils, electroplate, sputter coats, or any other process that deposits a thin film on the faying surfaces.

A schematic illustration of the process, shown in Figure 1, indicates that by placing a thin interlayer of an alloying metal containing a melting point depressant (MPD) between the two pieces of parent metal to be joined and heating the entire assembly, a liquid interlayer is formed. The liquid may form because the melting point of the interlayer has been exceeded, or because reaction with the parent metal results in a low melting liquid alloy. The liquid then fills voids formed by unevenness of the mating surfaces and can sometimes dissolve residual surface contamination. With time the MPD diffuses into the parent metal resulting in isothermal solidification. Upon cooling there remains no trace of the liquid phase, and ideally the joint becomes indistinguishable from other grain boundaries.

It is illustrative to use a phase diagram to explain the process as was done by Tuah-Poku et al (1). In their paper, four stages corresponding to composition regimes on the phase diagram were delineated, although in

Stage 0: Interdiffusion during heatup.

Stage I: Dissolution

Stage II: Widening

Liquid layer widens to maximum value at the end of Stage II.

Stage III: Isothermal Solidification

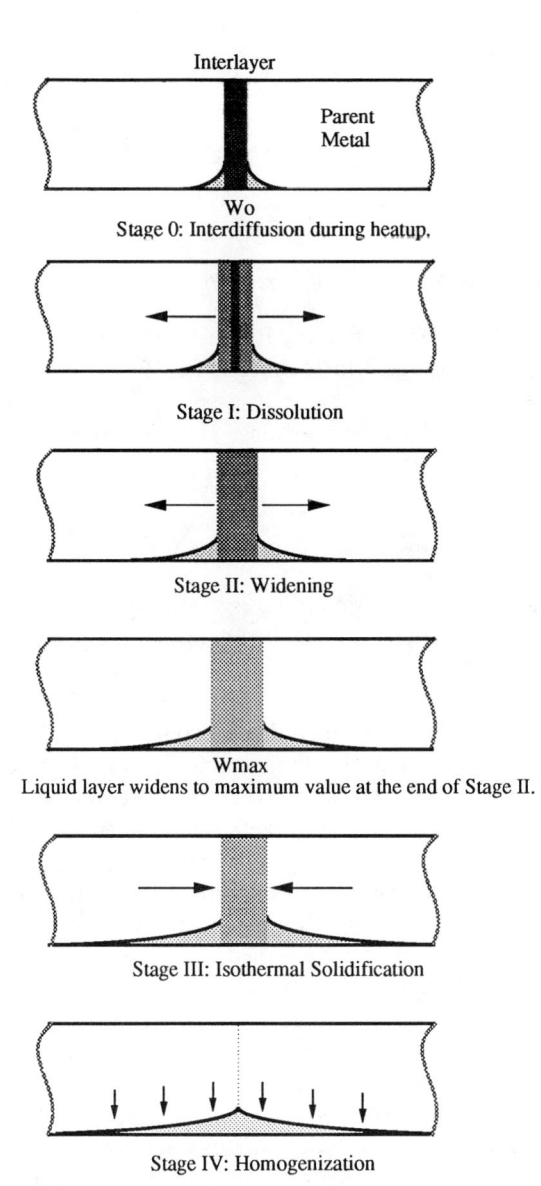

Stage IV: Homogenization

Figure 1 Four stages of bonding as defined by Tuah-Poku et al (1). The grey shading indicates the concentration of the melting point depressant (MPD). The arrows indicate the direction of movement of the solid-liquid interface.

reality there are competing reactions occurring on different time scales. Figure 2 gives a schematic phase diagram of a simple binary eutectic system such as copper-silver; in this case, the silver acts as the MPD. A layer of pure silver is sandwiched between a copper structure. Upon heating to the bonding temperature, the silver interlayer and copper parent metal undergo diffusion to form a liquid phase. As the dissolution progresses, the composition moves from $C_{\beta S}$ to $C_{\beta L}$, read as concentration of the MPD at the beta solidus and liquidus, respectively. In the following, C will be the concentration of the MPD unless otherwise noted. Widening follows as the interlayer is further diluted to $C_{\alpha L}$. At this point, solid state diffusion dominates, and the interlayer begins to narrow as the silver continues to diffuse into the copper. This is stage III or the isothermal solidification step. Once solidification is complete, holding the joint at the bonding temperature will lower the maximum silver concentration from $C_{\alpha S}$, which results in homogenization, termed stage IV.

The TLP process is not limited to binary eutectics, but can be applied to any system where the parent metal or alloy will form a relatively low melting temperature phase and has solubility for the MPD. The concept can also be applied to other systems whereby a chemical or other driving force inherently leads to solid state equilibrium. To be practical, however, the diffusivity of the MPD must be above 10^{-8} cm^2 sec^{-1} in order to

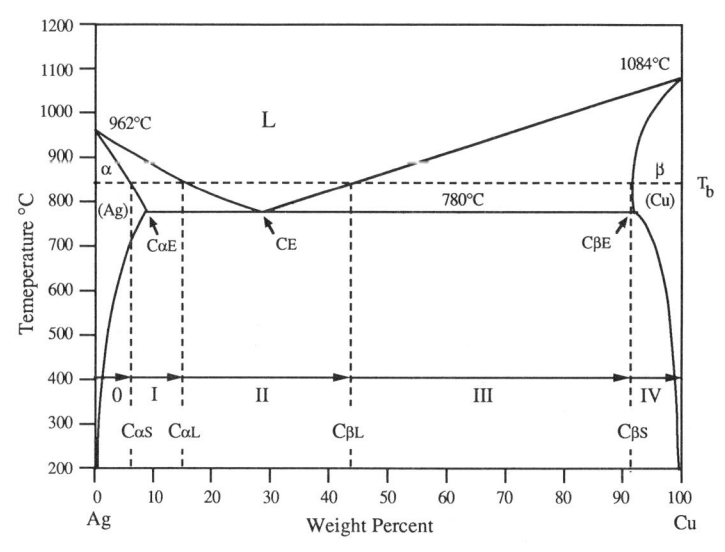

Figure 2 Stages of TLP bonding occurring when silver is bonded with a copper interlayer. This is a model system that reflects all of the important aspects of the process.

achieve complete solidification in reasonable times (2). In essence, any system wherein a liquid phase disappears by diffusion, amalgamation, volatilization, or other processes is a candidate for TLP bonding.

The joining of metals using a transient interlayer dates back to ancient times. Granulation, as described by Cellini (3) in the sixteenth century, used copper oxide paint as the interlayer along with some tallow or glue to hold small gold balls on to a gold article such as a vase and act as a flux. The article was then heated in a reducing flame, which rendered the copper available to form an eutectic with gold. After homogenizing in the flame, an invisible diffusion bonded structure resulted. Recipes for the process are given in the twelfth century work by Theophilus, entitled *De Re Deversis Artibus* (4). Reference is made in the eighth century *Mappae Clavicula* (5), or the *Little Key to Medieval Arts*, to a similar process. It has been suggested by Smith (6) that the ornamentation on King Tutankhamen's gold dagger dating from 2500 B.C. was created using the process of granulation. The Etruscans were able to join decorative gold beads to gold articles using the same process. It is interesting to note that this first use of the TLP process was based on an isomorphous azeotropic system, whereas modern theory has been developed for binary eutectic systems. A very similar process was patented in 1933 by Littledale (7), whereby a mixture of oxide, fish glue, and water was used as the bonding agent. The glue allowed small parts to be cemented and also provided carbon to reduce the oxide at the bonding temperature.

A description of the theoretical aspects of the TLP process, especially the kinetics, is presented below. This is followed by a discussion of experimental measurements of the interface motion, which do not always agree with the theoretical predictions. Modern applications are then described in terms of the alloy system for which the process was developed and, finally, potential new uses are discussed.

THEORY

Several models have been proposed to describe the kinetics of the process. These are analytical descriptions based on Fick's laws using mass balance arguments, generally limited to a description of stages I and III, dissolution and isothermal solidification, respectively. In the following, the analytical models are reviewed with a discussion of numerical results where applicable.

Thin Film Approach

Wells (8) used the thin film diffusion equation as a first approximation to the TLP process for joining titanium with copper where the concentration profile, $C(x, t)$ was given as

$$C(x, t) = \frac{\alpha}{2\sqrt{\pi D_S t}} \exp\left(-\frac{x^2}{4D_S t}\right), \qquad 1.$$

where α is the surface concentration in g/cm^3, D is the diffusivity in cm^2 sec^{-1}, x is the distance from the centerline in cm, and t is the time in seconds. From Equation 1 the maximum concentration at position $x = 0$ is C_{max}, where

$$C_{max} = \frac{\alpha}{2\sqrt{\pi D_S t}}. \qquad 2.$$

Through a series of fortuitous errors, Wells found that Equation 2 accurately predicted the correct bonding time, whereas in reality the actual bonding time using the correct values is almost 15 times longer. The thin film equation fails because the concentration of the MPD rapidly drops to the concentration at the α-rich liquidus by the process of dissolution. For this reason the thick film equation (9) also does not accurately describe the process. Figure 3 is a comparison of the predicted C_{max} values, for the silver-copper system, as a function of time using the thin and thick film equations and the stage model. It is apparent that if an alloy interlayer is used, the thin film equation will predict a decrease in the concentration immediately, whereas in reality the concentration in the interlayer remains constant during the solidification stage.

Discreet Stage Approach

The TLP process was first clearly described as consisting of four stages by Tuah-Poku et al (1), although other workers had already separated the process into individual stages. Here, Stage 0 is introduced to include the effects of heat up.

STAGE 0 Niemann & Garrett (10) were concerned with the loss of an electroplated copper interlayer experienced during heat up. The samples were aluminum-boron composites designed to be bonded in a sandwich structure. Following Darken & Gurry's analysis (11), the total amount of copper, the MPD, lost to the aluminum matrix was determined to be

$$W_{loss} = 8.25 C_{\alpha S}(D_S t)^{1/2}, \qquad 3.$$

where W_{loss} is the coating loss. This analysis is valid for constant temperature conditions, and actual coating losses were determined experimentally by electron microprobe. These experimental results can be compared with those attained by determining an effective diffusion coefficient, D_{eff}, following the method of Shewmon (12), where D_{eff} is given by

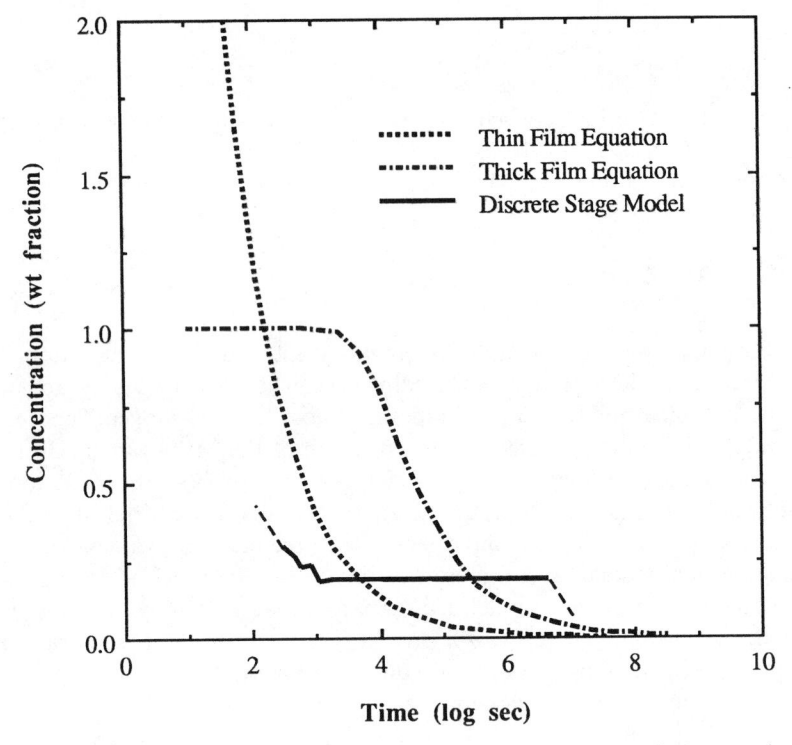

Figure 3 Comparison of bonding times using the thin and thick film diffusion equations and a composite of results (stage I and II) and calculations (stage III) from Tuah-Poku et al (1). The analysis is based on a 79-μm copper foil in silver at 820°C.

$$D_{\text{eff}} = \frac{\int_0^{t_0} D(t)\, dt}{t_0}, \qquad 4.$$

where t_0 is the total time of heat up. Substituting D_{eff} in Equation 4 results in an overestimate of the coating loss.

It is interesting to note that for a constant heating rate the average diffusion constant is independent of the rate, as was discussed by Mac-Donald & Eagar (2). The relationship of temperature as a function of time for a constant rate is

$$T = \tau t + 0.8 T', \qquad 5.$$

where T' is the bonding temperature, and τ is the heating rate.

Shewmon (12) has shown that below 80% of the bonding temperature the contribution to the diffusive flux is negligible. Substituting this equation along with the standard Arrhenius form of the diffusion constant into Equation 4 gives

$$\frac{D_{\text{eff}}T'}{5\tau} = \frac{1}{\tau}\int_{0.8T'}^{T'} D_o \exp\left(-\frac{Q}{RT}\right) dt, \qquad 6.$$

where D_o is the pre-exponential, Q is the activation energy for diffusion, and R is the gas constant. From Equation 6 it is apparent that the heat up rate cancels. The analytical solution of the integral is a slowly converging series function given by

$$D_{\text{eff}} = \frac{D_o}{\dot{T}}\left[\frac{\exp\left(-\frac{Q}{RT}\right)}{T} - \frac{Q}{R}\left[\ln\left(\frac{1}{T}\right) - \left(\frac{Q}{RT}\right) + \frac{1}{2\cdot 2!}\left(\frac{Q}{RT}\right)^2 \right.\right.$$
$$\left.\left. - \frac{1}{3\cdot 3!}\left(\frac{Q}{RT}\right)^3 + \cdots\right]\right]_{0.8T'}^{T'}. \qquad 7.$$

STAGE I: DISSOLUTION The process of dissolution requires the formation of four separate solid-liquid interfaces as sketched in Figure 4. The initial liquid composition is probably indeterminate, but must lie between the two liquidus lines. A compositional gradient develops in the liquid, which

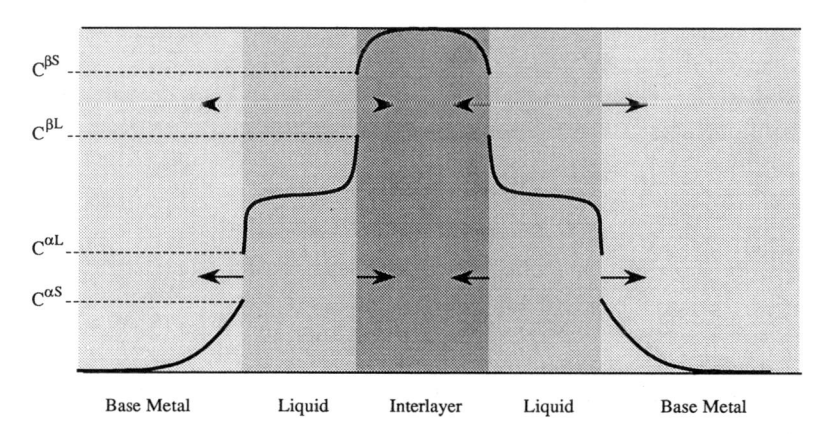

Figure 4 Dissolution on stage I occurs at four solid-liquid interfaces as indicated. At the end of dissolution, the width of the liquid will be greater than the initial interlayer thickness. The concentration profile in the liquid is unknown, but probably consists of two boundary layers and a well-mixed region.

in turn produces a rapid diffusive flux. Therefore early in the process, the liquid composition is controlled by atomic movement at the interface rather than diffusion through the liquid interlayer. Tuah-Poku et al (1) modeled stage I, assuming liquid diffusion control, using a general square root law, such that the movement of the interface was given as

$$Y_1 = K_1\sqrt{4D_L t}, \qquad 8.$$

where Y_1 is the position of the solid liquid interface, K_1 is a constant derived from the phase diagram, and D_L is the diffusivity in the liquid, which follows from the solution to Fick's second law. Accordingly, the width of the liquid zone becomes equal to the width of the interlayer when Y_1 is equal to the half width such that

$$\frac{W_o}{2} = K_1\sqrt{4D_L t}, \qquad 9.$$

where W_o is the initial interlayer width. Upon substituting appropriate values using a 79-μm foil, a dissolution time of about three seconds was determined. This model assumes that the dissolution front grows into the interlayer and that the process is complete once the interlayer is liquid. That this assumption is not valid follows from consideration of the phase diagram, which indicates that to obtain a completely liquid phase, the copper concentration of the interlayer must be lowered to $C_{\beta L}$. Hence, at 820°C, the interlayer must widen by over two and one half times before it is completely liquid. In reality, there are four solid-liquid interfaces as sketched in Figure 4.

Nakao et al (13), using the Nernst-Brunner theory (14), determined an activation energy for dissolution in stage I. The concentration dependence of the interlayer with time was given as

$$C_L = C_{\beta L}\left[1 - \exp\left(-\frac{KAt}{V}\right)\right], \qquad 10.$$

where K is a dissolution rate constant, A is the surface area of solid, and V is the volume of liquid interlayer. Equation 10 predicts that at long times the concentration of the diffusing species in the liquid phase will asymptotically approach the equilibrium value as given by the phase diagram. By differentiating Equation 10 and substituting in the appropriate constants, an expression relating the dissolution rate and the interface displacement is

$$P = Kt = \frac{W_o}{2} \left[\ln \frac{W_{max}\left(W + \frac{pW_o}{2}\right)}{\frac{pW_o}{2}(W_{max} - W)} \right], \qquad 11.$$

where P is the dissolution parameter, W is the instantaneous interlayer width, W_{max} is the maximum interlayer width, p is the ratio of liquid to solid density, and W_o is the initial interlayer width. A linear dependence of the dissolution parameter with time was demonstrated for the superalloy MM 007 with a Ni-Cr-B interlayer. However, Equation 11 predicts that at long times dissolution will continue without bound. The Nernst-Brunner theory was formulated for a semi-infinite solid being dissolved in a finite liquid bath, whereas Equation 10 indicates that at long times the liquid will saturate with respect to the remaining solid. A linear relation is valid only until all the solid has dissolved, but since there are two solids involved, it is surprising that a linear relation is obtained. The thermal dependence of K given as

$$\ln K = -\frac{Q}{RT} + \ln K_o, \qquad 12.$$

where Q is the activation energy in J mol^{-1}, R is the gas constant in J mol^{-1} K^{-1}, and T is bonding temperature in Kelvin. An activation energy of 682 J mol^{-1} was obtained from a plot of $\ln K$ vs $1/T$, which is greater than would be expected for a diffusion controlled process. But since K does not correspond directly to a physical process, the activation energy that is derived may not have much meaning. A straight line fit to the data is unwarranted, since at high and low temperatures K will go to infinity and zero, respectively, as dictated by the phase diagram. Also, in the same paper an activation energy of 226 kJ mol^{-1} is reported for diffusion of boron into the nickel base alloys. Nakagawa et al (15) suggested that the Nernst-Brunner theory does not hold when the width of the required thin boundary layer is wider than the liquid interlayer, and thus there is no bulk liquid. At this time it is unclear what mechanisms control the dissolution process.

Liu et al (16) have outlined an analytic model that accounts for the fact that the dissolution front proceeds into the interlayer as well as the parent metal. A mass balance on the system gives the velocity of each interface as

$$\frac{dY_1^\alpha}{dt} = \frac{D_L}{C_{\alpha o} - C_{\alpha L}} \left.\frac{\partial C_L}{\partial y}\right|_{Y_1^\alpha}, \qquad 13.$$

$$\frac{dY_1^\beta}{dt} = \frac{-D_L}{C_{\beta o} - C_{\beta L}} \left.\frac{\partial C_L}{\partial y}\right|_{Y_1^\beta}, \qquad 14.$$

where Y_1^i refers to the position of the ith interface. Dissolution is controlled by the rate at which element B diffuses into the quiescent liquid as controlled by Fick's second law

$$\frac{\partial C_L}{\partial t} = D_L \frac{\partial^2 C_L}{\partial y^2}, \qquad 15.$$

which has the general solution,

$$C_L = U_1 + U_2 \, erf\left(\frac{y}{2\sqrt{D_L t}}\right), \qquad 16.$$

subject to the appropriate boundary conditions. Equations 15 and 16 imply that the interface movement can be described as

$$Y_1^\alpha = 2G_1^\alpha \sqrt{D_L t} \qquad 17.$$

and

$$Y_1^\beta = 2G_1^\beta \sqrt{D_L t}, \qquad 18.$$

which in turn gives an expression for the interface velocity as

$$\frac{dY_1^\alpha}{dt} = G_1^\alpha \sqrt{\frac{D_L}{t}} \qquad 19.$$

and

$$\frac{dY_1^\beta}{dt} = G_1^\beta \sqrt{\frac{D_L}{t}}. \qquad 20.$$

By rearranging and combining these equations an expression for the concentration gradient at the interface becomes

$$\left.\frac{\partial C_L}{\partial y}\right|_{Y_1^\alpha} = \frac{U_2 \exp(-G_1^\alpha)^2}{\sqrt{\pi D_L t}} \qquad 21.$$

and

$$\left.\frac{\partial C_L}{\partial y}\right|_{Y_1^\beta} = \frac{U_2 \exp(-G_1^\beta)^2}{\sqrt{\pi D_L t}}. \qquad 22.$$

By substituting the expressions for the interface velocities, Equations 19 and 20, and concentration gradients, Equations 21 and 22, into the mass balance, Equations 13 and 14, two explicit functions of the dimensionless growth constants are obtained as

$$\Lambda_o G_1^\beta \exp(G_1^{\beta^2} - G_1^{\alpha^2}) - G_1^\alpha = 0 \qquad 23.$$

and

$$\Lambda_1 G_1^\alpha \sqrt{\pi} [erf(G_1^\beta) - erf(G_1^\alpha)] - \exp(-G_1^{\alpha^2}) = 0, \qquad 24.$$

where

$$\Lambda_o = \frac{C_{\beta o} - C_{\beta L}}{C_{\alpha L} - C_{\alpha o}}, \qquad 25.$$

and

$$\Lambda_1 = \frac{C_{\alpha o} - C_{\alpha L}}{C_{\alpha L} - C_{\beta L}}. \qquad 26.$$

The simultaneous equations are solved numerically to obtain values for the growth constants. Thus the time at which the interlayer is consumed is given by

$$t_1 = \frac{W_o^2}{4G_1^{\beta^2} D_L}, \qquad 27.$$

and the half width of the liquid interlayer at this instant is given as

$$W_L W_o + 2G_1^\alpha \sqrt{D_L t_1}. \qquad 28.$$

Using Liu et al's equations, the time for complete dissolution of the foil is still on the order of seconds, as was calculated by Tuah-Poku et al (1).

Nakagawa et al (15) developed a numerical model of heat up, dissolution, and widening using an explicit forward finite difference approach. The model was devised to distinguish between constant and unrestrained interlayer widths. A constant interlayer width results when some form of an inert spacer is inserted between clamped faying surfaces. The time to completion for unrestrained dissolution was found to be about five times longer than for constant interlayer width. In each case, dissolution time was found to be proportional to the square of the filler metal thickness. The advance of the solid-liquid interface is approximately related to the inverse of the square root of the solute diffusivity in the liquid and not to the diffusivity in the solid. The influence of heating rate above the eutectic temperature was found to be dependent on the thickness of the filler metal. Decreasing the heating rate resulted in less dissolution for a thin interlayer and more dissolution for a thick interlayer.

STAGE II: WIDENING   Analytical models have not been developed to fully describe the widening process and, in fact, there is no discrete boundary in time between stages I and II. Once the interlayer is dissolved, the width of the liquid zone will be greater than the initial interlayer width. Hence the process of widening occurs concurrently with dissolution. Tuah-Poku et al (1) state that the governing equations will have the form

$$Y_2 = K_2 \sqrt{D_L t}, \qquad 29.$$

which can be cast into the more general expression

$$Y_2 = at^n. \qquad 30.$$

Using their results on the silver-copper system, a straight line fit yielded the expression

$$Y_2 = 6.92 \cdot 10^{-4} t^{0.46}. \qquad 31.$$

In order to get a better fit using an exponent of 0.5, an effective weighted diffusion coefficient was fit to the data that reflected the two processes occurring in parallel, namely diffusion in the liquid interlayer and into the surrounding solid. The relationship thus determined was

$$D_e = D_L^{0.7} D_S^{0.3}. \qquad 32.$$

Liu et al (16) showed that the diffusion of boron in the liquid phase could be described by Fick's second law subject to initial conditions that are a function of the previous dissolution stage. It was suggested that a solution to the governing equation would have to be obtained numerically.

STAGE III: ISOTHERMAL SOLIDIFICATION   Isothermal solidification in stage III is controlled by solid state diffusion as has been demonstrated by a number of workers. The moving interface approach was sketched out by Lynch et al (17) for a general eutectic system with idealized composition profiles. The model was set up with a mass balance at the interface, and with some manipulation the interface velocity can be found as a function of diffusivity and time such that

$$\frac{dy}{dt} = K \left[\frac{R_S \sqrt{D_S} - R_L \sqrt{D_L}}{\sqrt{t}} \right], \qquad 33.$$

where R_S and R_L are unspecified functions of time and temperature. The form of Equation 33 indicates that the profiles could be modeled as error functions to which an analytic solution can be obtained. Lynch et al did not extend this model, probably because R_S and R_L are empirical functions that are specific to the system.

Ikawa et al (18) were able to extend this approach to obtain a complete analytical solution. Their work, based on the TLP process in nickel base superalloys, modeled the isothermal solidification step and the homogenization step separately. The solidification process depends on the flux of the MPD, either phosphorous or boron, into the solid given by

$$J = -\left(\frac{D_S}{V_S}\right)\frac{dC_S}{dy}, \qquad 34.$$

where V_S is the molar volume. This flux is equal to the rate of growth of the solid phase

$$J = \left(\frac{C_{\alpha L}}{V_L} - \frac{C_{\alpha S}}{C_S}\right)\frac{dy}{dt}. \qquad 35.$$

Setting Equations 34 and 35 equal and integrating gives the concentration profile as

$$\frac{C}{V_S} = \left(\frac{C_{\alpha S}}{V_S}\right)\left[erfc\frac{\left(y - \frac{W_{max}}{2}\right)}{2\sqrt{D_S t}}\right], \qquad 36.$$

where the gradient at the interface is given by

$$\left(\frac{dC}{dy}\right)_{Y_S} = -\frac{C_{\alpha S}}{\sqrt{\pi D_S t}}. \qquad 37.$$

A substitution of Equation 36 into 37 gives an expression for the displacement of the interface as a function of time and temperature such that

$$(W_{max} - 2y) = \left(\frac{4C_{\alpha S}}{\sqrt{\pi} V_S}\right)\left(\frac{C_{\alpha L}}{V_L} - \frac{C_{\alpha S}}{V_S}\right)^{-1}\sqrt{Dt}, \qquad 38.$$

where the gradient is

$$\ln[m] = \ln\left[\frac{4C_{\alpha S}}{\sqrt{\pi} V_S}\left(\frac{C_{\alpha L}}{V_L} - \frac{C_{\alpha S}}{V_S}\right)^{-1}\right] + \frac{1}{2}\ln[D_o] - \frac{Q}{2RT}. \qquad 39.$$

When $\ln m$ is plotted against the inverse of temperature, an activation energy for diffusion of phosphorous of 247.8 kJ mol^{-1} is obtained. Nakao et al (13) later repeated the experiment and obtained a value of 284 kJ mol^{-1}. The thickness of the interlayer was determined by measuring the width of the Ni-P eutectic, W_e, then converting to the actual thickness, W_e, using the lever rule from the phase diagram.

In this work, the width of the interlayer was held constant by molybdenum wire spacers. As Nakagawa et al (15) pointed out, the solidification time is shorter for the constrained width method used here. The time required for isothermal solidification, t_3, was derived by Nakao et al (13) as

$$t_3 = \left[\left(\frac{C_{\alpha L}}{V_L} - \frac{C_{\alpha S}}{V_S}\right)\left(\frac{\sqrt{\pi} V_S}{4 C_{\alpha S}} \frac{W_{max}}{\sqrt{D_S}}\right)\right]^2. \qquad 40.$$

A straight line fit was obtained in $t_3^{1/2}$ vs $W_{max}/D^{1/2}$ plots, but the kinetics are an order of magnitude faster than would be obtained using a reasonable pre-exponential factor for the diffusivity. Tuah-Poku et al (1) derived a slightly different governing equation for the variable width case. They found that the maximum width of the interlayer could be derived from the mass balance relationship

$$W_o C \rho_l = (W_{max} - W_o) C_{\alpha L} \rho_l + W_o C_{L\alpha} \rho_S, \qquad 41.$$

where W_o is the initial interlayer width, W_{max} is the maximum interlayer width, and ρ_l, ρ_S are the densities of the interlayer and the parent metal, respectively. Upon rearranging, W_{max} was found to be 4.9 W_o, which is close to the experimentally obtained value of 5.3 W_o, the difference arising partly because the change in density during liquefaction was taken into account. Once again, the interface kinetics were modeled through the use of a general relation such that

$$Y_3 = \frac{(W_{max} - W)}{2} = 2 K_3 \sqrt{D_S t}. \qquad 42.$$

The experimental value of K_3 was found to be nearly three times that derived from the phase diagram. It was suggested that this discrepancy was the result of the solidification process, possibly involving a ledge-type migration mechanism, wherein the diffusional mobility of the interface is then controlled by the two-dimensional nucleation and growth of ledges, which involves a cooperative movement of groups of atoms. This process was said to speed up the advancement of the solid liquid interface and reduce the time needed for solidification. However, it should be a slower process than if each atom could attach at any site.

An estimate for the time required for the completion of solidification using a mass balance under the assumption that the densities were equal was given as

$$W C_{\alpha L} = W_{max} C_{\alpha L} + 2 \int_0^t D_S \frac{dC}{dy} dt. \qquad 43.$$

The error function solution for concentration derived earlier is then inserted into the integral and an expression for solidification becomes

$$t_3 = \frac{\pi W_o^2}{16 D_S} \left(\frac{C_{\beta o}}{C_{\alpha L}} \right). \qquad 44.$$

Using their experimental conditions, namely a 79-μm copper foil in silver at 820°C, Equation 44 predicts a time of 1200 hr as compared to the experimental extrapolated value of about 200 hr. This suggests that either the experimental method allowed some loss of the liquid interlayer, or some other solidification mechanism was operating. Writing Equation 44 explicitly to show the dependence of solidification time on temperature gives

$$t_3 = \frac{\pi W_o^2}{16 D_o} \left[\frac{\exp\left(\frac{Q}{RT}\right)}{(C_{\alpha S})^2} \right]. \qquad 45.$$

As the temperature increases, the diffusivity increases via the exponential term, but the concentration $C_{\alpha S}$ decreases, causing greater widening. By approximating the solidus as a straight line, a linear relationship is obtained between $C_{\alpha S}$ and temperature. Substituting this into Equation 45, then taking the first derivative set to zero, gives an expression for the minimum solidification time temperature as

$$T = [-Qn \pm \sqrt{(Qn)^2 + (8QCRn)}] \frac{1}{4Rn}, \qquad 46.$$

where $1/n$ is the slope of the solidus and C is a constant derived from the phase diagram equal to $C_{\alpha E} T_M/(T_M - T_E)$. The temperature dependence was determined for several systems, but no experimental verification of this behavior has been presented. The maximum alloy concentration at this minimum may exceed the maximum allowable design concentration, thus requiring further homogenization time. A higher bonding temperature will result in a lower final concentration, which indicates that there is a trade-off in selecting the process temperature. The total isothermal solidification time is then

$$t_3 = \frac{\pi W_o^2}{16 D_S^\alpha C_{\max}^2}, \qquad 47.$$

when starting with a pure metal interlayer.

Ramirez & Liu (19) performed TLP bonding experiments with the nickel-boron system and found that there were two distinct regimes of solidification behavior. A faster initial regime had a displacement rate ten times as great as the second regime. That the interface will slow with time is apparent from consideration of Equations 12 and 13, which show that the displacement rate varies with the inverse root of time. The two distinct regimes that were found suggest that at early times excess liquid was being squeezed out of the joint. Despite this, an activation energy of 166 kJ mol^{-1} was obtained, which is significantly lower than the values reported earlier by Nakao et al (13) and Ikawa et al (18).

STAGE IV: HOMOGENIZATION   There have been no explicit models developed to describe homogenization during TLP bonding. Homogenization of cast alloys in general, however, is a well-studied subject from which suitable models can be selected. The bonding temperature at which a specified maximum allowable concentration of the MPD is achieved in the minimum amount of time is not necessarily the same as that for the minimum in the solidification time.

EXPERIMENTAL

Interface Kinetics

Measurement of the interface migration rate has been accomplished three different ways as shown in Figure 5. The earliest approach taken by Tuah-Poku et al had a tantalum washer to maintain a minimum spacing held in place by a miniature clamp. Their results indicated that the interlayer widened in an unconstrained manner, exceeding the theoretical maximum width as given by Equation 33. This is surprising since one would expect that the action of the clamp and washer would be to fix the interlayer width at the thickness of the tantalum foil used. That it did not do so indicates that lateral dissolution of the silver was not a factor. The extrapolated isothermal solidification time obtained was considerably shorter than the calculated time. Tuah-Poku et al suggest that this is due to the complex conditions prevailing at the interface. Another possible explanation is that the liquid interlayer leaked out of the joint region.

Nakao et al (13) used a hot press to hold samples together with a molybdenum wire spacer inserted to fix the liquid width. Under load, the liquid was extruded from the joint region, which reduced the overall solidification time. The accuracy of this measurement method would be reduced if there were deformation of the solid nickel alloy around the wire. There should also be a step in the interface displacement vs time plot. The solidification kinetics will be faster at the start when the interface is

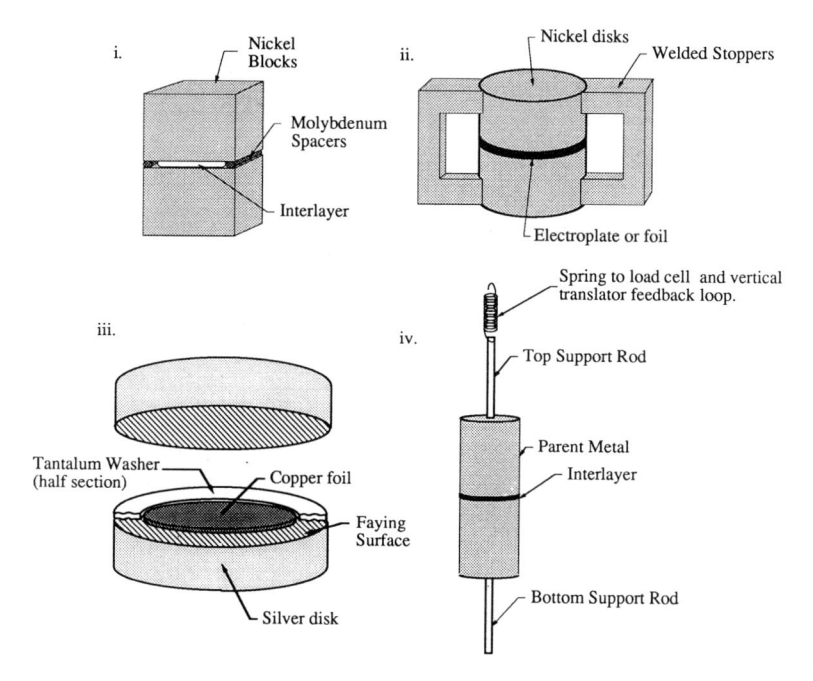

Figure 5 Comparison of the experimental samples developed for TLP bonding from (i) Ikawa et al (17); (ii) Nakao et al (12); (iii) Tuah-Poku et al (1); and (iv) MacDonald & Eagar's present setup, which allows for free expansion of the joint. Not to scale.

constrained and excess liquid is being squeezed out. At longer times, the liquid width will be less than the spacer width, and the interface will again behave as an unconstrained interlayer. Nakeo et al (13) also designed the welded stopper arrangement in order to fix the width of the interface and obtained good agreement with the calculated values. This is somewhat surprising because one would expect that the liquid interlayer would leak out since it obviously wets the solid. It is not clear from the paper whether a stop-off was used around the complete perimeter of the joint.

A new experimental technique has been developed in our laboratory that allows the interface to expand and contract without constraint. A feedback loop operating between a load cell and vertical translator maintains a constant force on the joint. By reducing the load to a few grams, the surface tension of the interlayer prevents the liquid from leaking out of the joint, thus maintaining the mass balance. Experience has shown that in the silver-copper system the interlayer liquid will wick up the side of the parent metal even if the load is minimal. To alleviate this problem, the sides of the parent metal were coated with a non-wetting agent; in our case, chromium on copper with a silver interlayer. This method also prevents Kirkendall porosity, which develops when the two parent metals are constrained at a fixed distance.

Grain Size Effect

Kokawa et al (20) were able to show a distinct effect of grain size and orientation on the solidification process. Pure nickel samples with grain sizes of 4 mm and 40 μm were TLP bonded with a 38-μm thick nickel-phosphorous interlayer at 1200 K for 2.75 hr. The interlayer width of the fine grain sample was 9 vs 18 μm for the coarse grain sample. The liquid penetration into the boundary was also shown to be a function of grain boundary orientation. A saw-toothed solid-liquid interface shape developed with cusps at each grain boundary. The cusps are a result of surface tension effects as the liquid preferentially wets the grain boundaries. It was claimed that the resulting increase in the surface area of the interface promoted faster solidification in the fine-grained sample. It is also likely that phosphorous has a higher diffusion rate along grain boundaries. The rapid grain boundary diffusion enhances the migration rate of the MPD. An effective diffusion constant could be obtained from a rule-of-mixtures approach so that

$$D_{\text{eff}} = A_l D_l + A_{gb} D_{gb}, \qquad 48.$$

where the subscripts refer to lattice and grain boundary.

Diffusion Profiles

Concentration profiles of the MPD have been obtained for several bonding systems including titanium and nickel superalloys (1, 10, 18, 21, 22). Ikawa & Nakao (23) were the first to show the concentration profile of phosphorous in a nickel joint becoming flat after a 16 hr hold at 1200°C. In general, an asymptotic approach to equilibrium is observed although the parent metal alloy composition is not reached. Even after extended homogenization, a slight variation on the composition can be seen at the original bond line. Analytical models usually predict less diffusion at early times, whereas some numerical models seem to agree more closely with the data (24).

Solidification Front

It is not yet clear how the solid front advances into the liquid interlayer. Tuah-Poku et al suggest that grain growth is by a ledge-type mechanism that enhances the solidification rate. The presence of a cellular structure is often observed in specimens quenched from stage III. Some of these cells align with grain boundaries and are manifestations of the wetting phenomenon discussed earlier. Cells not associated with grain boundaries have been observed in samples with large grains, e.g. 4 mm, which suggests that there may be another reason. Cellular growth in eutectic alloys caused by constitutional supercooling is well known (25), hence impurities in the interlayer causing solute buildup at the interface may be responsible for these cellular structures.

APPLICATIONS

Titanium

In 1955, while preparing silver-brazed joints, Tiner (21), working at North American Aviation, noticed that by bonding at high temperatures and long times silver could not be detected at the interlayer. Joints with extremely high shear strengths were formed, although the grain growth slightly impaired the base metal properties. In 1959, Lynch et al (17) repeated the process by bonding titanium, using an unspecified alloy as an interlayer. Although no specific processing conditions were given, a series of micrographs revealed the progressive dissolution of the interlayer and the eventual formation of a joint that was "effectively just a grain boundary."

The NOR-TI-BOND process reported by Wu (26), based on a patent by Wells & Mikus (27), was developed by Northrop in 1971 for bonding titanium I and T beams, channels, and other structural shapes. The two-step process used local resistance heating to melt an electrolytically deposited copper interlayer. Subsequently, a diffusion treatment at 927°C for four hr reduced the peak copper content from 70 to below 6%, which produced a strong, tough joint. Freedman (28), in an extensive account of the mechanical properties of these joints, showed that parent metal strengths could be achieved using the optimum coating thickness and processing conditions. The plane strain fracture toughness varied from 40 to 90% of that for annealed Ti-6Al-4V. No effect on stress corrosion cracking susceptibility was observed.

Liquid interface diffusion (LID) bonding (also termed Rohr bonding) was developed by Rohr industries to bond titanium honeycomb sandwich structures using nickel-copper interlayers. Schwartz (29) showed how large structures containing several components, such as a jet engine case, are formed in one step. As of 1986 over 5000 engine ducts had been produced using this technology (22). The interlayer was originally deposited, but now 25-μm alloy foils are used, which melt congruently at the bonding temperature, thus avoiding dissolution. Fixturing of the joints is provided by thermal expansion controlled tooling, which on heating places the joint in compression. The tensile properties of the joints are dependent on the surface finish, therefore a smoother finish results in stronger joints.

Nickel Superalloys

Activated diffusion bonding was described by Hoppin & Berry (30) in 1970 for joining nickel base superalloys. This is the earliest reference to the TLP bonding with a melting point depressant (MPD). Interlayer compositions were prepared with the same nominal composition as the parent metal except for the addition of an unspecified MPD, probably boron or phosphorous. Several superalloys were joined, including René 80, and subjected to stress-rupture tests. Joint strengths of 70–90% of the parent metal were obtained, but ductilities were low.

Using nickel-copper interlayers, Duvall et al (31), working at Pratt and Whitney, were able to diffusion weld the superalloy Udimet 700 without forming a brittle gamma prime precipitate at the interface. Interface-free welds were achieved using an electroplated nickel-cobalt interlayer with joint efficiencies approaching 100%. The process, patented by Owczarski et al (32), reportedly occurred in the solid state, but was more likely a TLP process. Subsequently this was clarified by Duvall et al (33) who, using the same materials, coined the term TLP. The patented process (34) was applied to turbine vane clusters used in the low pressure end of a jet engine. Joint efficiencies of 100% were achieved. Adam & Steinhauser (35) showed how this process could be applied to make and repair compound turbine vanes.

Nakao et al (36) extended the concept to include the use of a powder addition to the interlayer in order to reduce the solidification time. Termed transient liquid insert metal (TLIM) diffusion bonding, the process time was reduced by over two orders of magnitude as compared to the conventional method of TLP bonding superalloys. Suzumura et al (37) found that by using amorphous alloy interlayers close to the composition of the parent metal, solidification times were reduced. Joint efficiencies of 100% were also obtained.

Viskov et al (38) used a niobium powder interlayer, which forms a liquid phase with the nickel superalloy Kh20N80 at 1200°C. After 36 hr of annealing, only a few particles of $(Ni,Cr)_3Nb$ remained at the interface. The authors ruled out the use of phosphorous or boron as the MPD because of the adverse influence these elements have on mechanical properties, which contrasts with previous investigations.

Dissimilar Metals

A concept similar to TLP was developed by Owczarski (39) in 1961 for joining Zircaloy 2 to 304 stainless steel. At that time, no suitable braze filler metals had been found. The two materials were simply butted up to one another forming a liquid phase by a diffusion controlled eutectic reaction. The three major components, iron, chromium and nickel, formed low melting eutectic liquids with zirconium, and the nominally quaternary system 304-Zr formed a liquid at 980°C. The joints so produced had fair strength and good corrosion resistance, but lacked ductility. This latter drawback was the consequence of forming intermetallic compounds at the interface. Other dissimilar metals such as aluminum and titanium have been joined in this manner, as described by Enjo et al (40). It is not really a TLP process because at long times the joint does not solidify, but rather continues to dissolve the parent metals.

Andryushechkin & Dahkova (41) investigated isothermal solidification of titanium coatings on steel to provide a protective coating. The interlayer width was only dependent on the width of the initial paste and consisted of an α-solid solution of Fe in Ti, FeTi intermetallic, and TiC particles. Contact melting occurred rapidly, thus making it difficult to control the structure and properties of the coating. Longer hold times resulted in an increase in both the size and volume fraction of titanium carbide particles.

Eagar (42) reported joining of copper to molybdenum using copper-gold-nickel interlayer alloys. By use of a circular geometry, which creates large compressive stresses on the liquid interlayer because of the thermal expansion of the two base metals, a very thin liquid region is produced that creates a TLP bond within a few tens of seconds. In this case, the gold diffuses only into the copper base metal and not into the molybdenum.

A completely different approach was proposed by Zhang et al (43) to join dissimilar metals. Termed instantaneous liquid phase (ILP) bonding, the process rapidly forms a thin liquid film on non-contacting faying surfaces, then quickly closes the joint and rapidly cools the parts. Unlike TLP, very little diffusion occurs, thus preventing the formation of undesirable intermetallic compounds.

Semiconductors

In 1966, Bernstein (44) and Bernstein & Bartholomew (45) developed the solid-liquid inter-diffusion (SLID) process for joining semiconductor components. The process was divided into five stages, namely, wetting, alloying, liquid diffusion, gradual solidification, and solid diffusion. It was concluded that bonding occurred as a result of interdiffusion, but no quantitative analysis was performed. Evaluation of the joints consisted of determining the unbonding temperature, i.e. the temperature at which a joint would reliquefy and fail under a shear load. The scatter in their results was an indication of intermetallic formation by amalgamation, as discussed by MacDonald & Eagar (2). SLID bonding was designed to overcome the limitations imposed during device fabrication where each sequential joining operation has to be performed at a lower temperature.

Roman (46) investigated low temperature TLP bonding as a method for surface mounting of semiconductor chips. In this process, as in SLID bonding, the solidification reaction occurs as a result of intermetallic formation and not by diffusion of the MPD. Shear tests showed considerable scatter, which indicated that it is difficult to control the extent of bonding. To be an effective process, selection and control of the intermetallic formation are required.

Composites

Niemann & Garrett (10), of McDonnell-Douglas, described eutectic bonding as a reliable low pressure, low temperature method for fabricating aluminum-boron composites with a copper interlayer. Titanium to aluminum joints were also made, achieving moderate properties; however a 0.040" reaction zone of intermetallics was observed. Evaluation of the joint was by a three point interlaminar shear bending test.

Klehn (47) applied the TLP process to aluminum metal matrix composites. A marked difference in the joint structure was noted between pure metal and alloy interlayers. A zone of high particulate density was produced along the bond line as a result of excess dissolution and squeezing out of the base metal. This effect was minimized when alloy interlayers such as aluminum-silicon eutectic were used. Again, a dependence on the surface finish was noted; smooth machined surfaces provided the strongest joints.

Other Systems

A transient liquid phase was used to bond aluminum-lithium alloys as described by Ricks et al (48). A roll-clad zinc base alloy interlayer was used to overcome the tenacious surface oxide. The process was amenable to producing superplastic components. There are several other related processes that act in the same manner as TLP bonding. These are amalgamation, transient liquid phase sintering, as applied to powdered metals (49) and ceramics (50), and diffusion solidification (51). In each case, the process requires that the liquid phase disappears either by diffusion or reaction as the equilibrium solid phase grows.

TLP bonds can also form during soldering operations. In one case, an eutectic Pb-Sn solder with a melting point of 183°C was attached to two pieces of copper. During thermal stress testing of the assembly for ten hr at 200°C, the tin diffused into the base metal forming copper-tin intermetallics. The remaining solder alloy consisted of approximately 95% Pb, which melted above 300°C after the isothermal hold. This ability to produce high temperature creep-resistant bonds using low temperatures is a key feature of a number of commercial TLP bonding processes.

SUMMARY

The process of transient liquid phase bonding has ancient origins, but is only recently finding new applications for specific alloy systems including titanium, nickel, aluminum, composites, and semiconductors. It has been used successfully for manufacturing aerospace components for nearly two decades. To expand the realm of application, systems amenable to the process need to be identified, selection criteria for the interlayer composition and form and a method for determining the thinnest interlayer that can be used for a given system are required. Theoretically, it is not clear what effect intermetallics have on the kinetics of bonding or strength of the resulting joint, and experimental methods need to be developed that accurately measure the kinetics in order to determine the rate controlling steps. If these questions can be answered, it is clear that TLP bonding will find increasing application in a number of systems, particularly with advanced materials in which fusion of the base metal must be avoided.

ACKNOWLEDGMENTS

The authors wish to thank the Office of Naval Research and the National Science Foundation for support of this work.

Literature Cited

1. Tuah-Poku, I., Dollar, M., Massalski, T. B. 1988. *Metall. Trans.* 19A: 675–86
2. MacDonald, W. D., Eagar, T. W. 1992. *Proc. TMS Symp. Mater. Sci. Joining.* In press
3. Cellini, B. 1568. *"Due Trattati, uno intorno alle otto principali arti del l'oreficiera. L'altro in material dell'arte della scultura."* Florence
4. Hawthorne, J. G., Smith, C. S. 1963. *On Divers Arts, The Treatise of Theophilus.* Chicago: Univ. Chicago Press. 216 pp. (Translation)
5. Hawthorne, J. G., Smith, C. S. 1974. *Mappae Clavicula. Trans. Am. Philos. Soc.* 64: 3–128 (Translation)
6. Smith, C. S. 1981. *A Search for Structure,* pp. 92–94. Cambridge, Mass.: MIT Press. 423 pp.
7. Littledale, H. A. P. 1933. *Brit. Patent No.* 415,181
8. Wells, R. R. 1976. *Weld. J.* 55: 1: 20s–27
9. Crank, J. 1956. *The Mathematics of Diffusion,* pp. 13–14. Oxford: Clarendon. 347 pp.
10. Niemann, J. T., Garrett, R. A. 1974. *Weld. J.* 53: 4: 175s–83, 53: 8: 351s–60
11. Darken, L. S., Gurry, R. W. 1953. *Physical Chemistry of Metals,* pp. 445–57. New York: McGraw-Hill. 458 pp.
12. Shewmon, P. 1989. *Diffusion in Solids,* pp. 37–39. Warrendale, Penn.: Metall. Soc. 246 pp. 2nd ed.
13. Nakao, Y., Nishimoto, K., Shinozaki, K., Kang, C. 1989. *Trans. Jpn. Weld. Soc.* 20: 1: 60–65
14. Melwyn-Hughes, E. A. 1947. *The Kinetics of Reaction in Solution,* pp. 374–77. Oxford: Clarendon. 428 pp.
15. Nakagawa, H., Lee, C. H., North, T. H. 1991. *Metall. Trans.* 22A: 2: 543–55
16. Liu, S., Olsen, D. L., Martin, G. P., Edwards, G. R. 1991. *Weld. J.* 70: 8: 207s–15

17. Lynch, J. F., Feinstein, L., Huggins, R. A. 1959. *Weld. J.* 38: 2: 85s–89
18. Ikawa, H., Nakao, Y., Isai, T. 1979. *Trans. Jpn. Weld. Soc.* 10: 1: 24–29
19. Ramirez, J. E., Liu, S. 1990. *Proc. AWS Symp.* pp. 272–73
20. Kokawa, H., Lee, C. H., North, T. H. 1991. *Metall. Trans.* 22A: 7: 1627–31
21. Tiner, N. A. 1955. *Weld. J.* 34: 11: 846–50
22. Norris, B. 1986. *Proc. Inst. Metals Conf. Designing with Titanium*, Bristol, pp. 83–86
23. Ikawa, H., Nakao, Y. 1977. *Trans. Jpn. Weld. Soc.* 8: 1: 3–7
24. Nakao, Y., Nishimoto, K., Shinozaki, K., Kang, C. Y., Shigeta, H. 1990. *Proc. Int. Symp. Jpn. Weld. Soc. 5th Tokyo.* pp. 133–38
25. Flemings, M. C. 1974. *Solidification Processing*, pp. 58–91. New York: McGraw-Hill. 264 pp.
26. Wu, K. C. 1971. *Weld. J.* 50: 9: 386s–93
27. Wells, R. R., Mikus, E. B. 1968. *US Patent No.* 3,417,461
28. Freedman, A. H. 1971. *Weld. J.* 50: 8: 343s–56
29. Schwartz, M. M. 1978. *Weld. J.* 57: 9: 35–38
30. Hoppin, G. S. III, Berry, T. F. 1970. *Weld. J.* 49: 505s–9
31. Duvall, D. S., Owczarski, W. A., Paulonis, D. F., King, W. H. 1972. *Weld. J.* 51: 2: 41s–49
32. Owczarski, W. A., King, W. H., Duvall, D. S. 1970. *US Patent No.* 3,530,568
33. Duvall, D. S., Owczarski, W. A., Paulonis, D. F. 1974. *Weld. J.* 53: 4: 203–14
34. Paulonis, D. F., Duvall, D. S., Owczarski, W. A. 1972. *US Patent No.* 3,678,570
35. Adam, P., Steinhauser, L. 1986. *AGARD Adv. Join. Aerospace Mater.* NTIS No. n87-17059/3/HDM
36. Nakao, Y., Nishimoto, K., Shinozaki, K., Kang, C. Y., Shigeta, H. 1990. *Proc. Int. Symp. Jpn. Weld. Soc. 5th, Tokyo.* pp. 139–44
37. Suzumura, A., Onzawa, T., Tamura, H. 1985. *J. Jpn. Weld. Soc.* 3: 321–27
38. Viskov, A. S., Nesvetaeva, O. A., Solov'eva, L. N. 1988. *Metally* 5: 173–75
39. Owczarski, W. A. 1962. *Weld. J.* 42: 78s–83
40. Enjo, T., Ando, M., Hamada, K. 1985. *Trans. Jpn. Weld. Res. Inst.* 14: 93–96
41. Andryushechkin, V. I., Dashkova, I. P. 1985. *Steel in the USSR* 15: 11: 558–60
42. Eagar, T. W. 1990. *Proc. Adv. Join. Newer Struct. Mater.* pp.3–14. Oxford: Pergamon
43. Zhang, Y.-C., Nakagawa, H., Matsuda, F. 1987. *Trans. Jpn. Weld. Res. Inst.* 16: 17–29
44. Bernstein, L. 1966. *J. Electrochem. Soc.* 113: 12: 1282–88
45. Bernstein, L., Bartholomew, H. 1966. *Trans. AIME* 236: 405–12
46. Roman, J. W. 1991. *An Investigation of Low Temperature Transient Liquid Phase Bonding of Silver, Gold and Copper*. MS thesis. Mass. Inst. Technol., Cambridge. 112 pp.
47. Klehn, R. 1991. *Joining of 6061 Aluminum Matrix-Ceramic Particle Reinforced Composites*. MS thesis. Mass. Inst. Technol., Cambridge. 77 pp.
48. Ricks, R. R., Winkler, P. J., Stoklossa, H., Grimes, R. 1989. *Proc. Aluminum-Lithium Alloys*, pp. 441–49. Birmingham, England: Materials and Component Engineering
49. German, R. M., Dunlap, J. W. 1986. *Metall. Trans.* 7A: 205–13
50. Atlas, L. M. 1957. *J. Am. Ceram. Soc.* 40: 196–99
51. Langford, G., Cunningham, R. E. 1978. *Metall. Trans.* 9B: 5–19

CONTROLLED PERMEABILITY POLYMER MEMBRANES

W. J. Koros, M. R. Coleman, and D. R. B. Walker

Department of Chemical Engineering, The University of Texas at Austin, Austin, Texas 78712-1062

KEY WORDS: diffusion, solubility, selectivity, gas, separation

INTRODUCTION

The complex and interacting issues involved in effectively separating commercially important gas mixtures using membranes are summarized in Figure 1. Oxygen and nitrogen separation from air are both important goals; however, nitrogen enrichment of air is much easier to achieve than oxygen enrichment of air with the current generation of membranes (1). Nitrogen-enriched air streams can be produced at purities above 98–99% with ease, and higher purities can be achieved with the addition of catalytic units to remove traces of oxygen (2). Based on a recent analysis by Stern, no materials are currently satisfactory for economical large scale oxygen enrichment to greater than 30 mol% (3, 4), in spite of significant strides in this area of materials science over the past decade.

Hydrogen separation from various supercritical gases is important in both refinery and chemical processing. With the emphasis on cleaner air, requiring hydrotreating of oil, the concept of hydrogen management has renewed interest in opportunities for hydrogen separation from waste and other marginal streams. Good membranes for this type of application exist; however, materials with improved thermal and chemical stability are still desirable for the demanding refinery and petrochemical industrial environments where such separations are practiced.

The third major type of gas separation applications (Figure 1) pertains to the removal of acid gases such as CO_2 and H_2S or simple water vapor contaminants from natural gases to meet pipeline specifications. Within this group of applications, carbon dioxide recovery is especially attractive

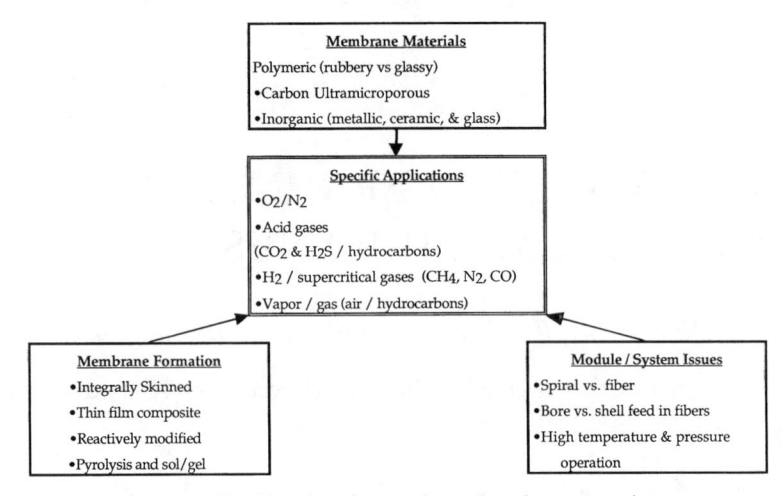

```
                    ┌─────────────────────────────────────────┐
                    │         Membrane Materials              │
                    │ Polymeric (rubbery vs glassy)           │
                    │ •Carbon Ultramicroporous                │
                    │ •Inorganic (metallic, ceramic, & glass) │
                    └─────────────────────────────────────────┘
                                       │
                                       ▼
                    ┌─────────────────────────────────────────┐
                    │         Specific Applications           │
                    │ •O2/N2                                  │
                    │ •Acid gases                             │
                    │  (CO2 & H2S / hydrocarbons)             │
                    │ •H2 / supercritical gases (CH4, N2, CO) │
                    │ •Vapor / gas (air / hydrocarbons)       │
                    └─────────────────────────────────────────┘
```

┌──────────────────────────┐   ┌─────────────────────────────────┐
│   Membrane Formation   │   │   Module / System Issues   │
│ •Integrally Skinned   │   │ •Spiral vs. fiber   │
│ •Thin film composite   │   │ •Bore vs. shell feed in fibers │
│ •Reactively modified   │   │ •High temperature & pressure   │
│ •Pyrolysis and sol/gel   │   │   operation   │
└──────────────────────────┘   └─────────────────────────────────┘

Figure 1 Considerations for membrane-based gas separations.

and represents an extremely large potential market. Carbon dioxide is an ideal material for enhancing the recovery of oil from wells that have ceased to produce under their own natural pressure. Miscibility of CO_2 with the oil and reductions in viscosity promote mobility of the oil through reservoirs. As the oil exits from the well, dissolved CO_2 and natural gas are produced so that recovery of CO_2 from the natural gas for reinjection is attractive.

Membranes have attractive features for such applications, since capacity can be added as needed, thus eliminating guess work and expending the full cost too early in the project. Such separations have turned out to be more difficult than originally envisioned, however, because of high solubility of the acid gases in many polymers used for these applications. Exposure of polymers to highly sorbing penetrants, such as CO_2, leads to a swelling of the polymer matrix and an increase in the segmental mobility. This increase in the chain mobility in the presence of highly sorbing penetrants is known as plasticization. Plasticization leads to an increase in diffusivity and therefore an increase in the permeability of all penetrants. Unfortunately, plasticization leads to a significant decrease in permselectivity for membranes used to separate gas mixtures that contain highly sorbing penetrants. Because of space limitations, issues of the complex interaction of penetrants with the polymer that alter membrane properties cannot be dealt with here in detail; however, they will be identified in the following discussion in an overview fashion.

Finally, the separation of hydrocarbon vapors from air promises to be an important application of membranes. Environmental restrictions aimed at limiting emissions to the atmosphere promise to drive the development of such systems, even when the cost of the recovered products alone cannot justify the system cost. This application is different from most of the cases mentioned above because the permeate is a vapor. Moreover, the membrane materials of choice tend to be rubbery in nature as opposed to the glassy materials used for most other cases.

The detailed issues determining success in the various applications indicated in Figure 1 make it clear that the technology of gas separation membranes is highly dependent upon basic support from polymeric materials science. Of course, membrane module construction and system design are also crucial to the successful implementation of this technology. Nevertheless, the breakthroughs in new materials for making membranes and the protocols for forming ultrathin-skinned asymmetric structures to provide high productivity have been responsible for the burgeoning interest in this topic. The formation of membranes from diverse polymers is being considered in a review (5) that complements the area of membrane materials (Figure 1), which is the focus of this review.

The most popular materials for membrane-based gas separation applications are polymeric in nature. Four specific families of polymers and representative members of these families are discussed below. The molecular structures of these materials are shown in Tables 1, 2, and 3. Discussions of the less common carbon microporous, metallic, ceramic, and glass membranes for gas separation have been presented elsewhere (6).

To establish a framework for discussion of the materials science aspects of gas separation based on polymeric membranes, several simple relations are useful. First, as indicated in Equation 1, the permeability, P_A, of a membrane for a given gas is simply a pressure and thickness normalized flux that provides a convenient measure of the ease of transporting the gas through the material.

$$P_A = (\text{flux of A})(l)/(\Delta p_A). \qquad 1.$$

Since the permeability is independent of the thickness of the membrane, it is truly a property of the polymeric material and the processing history. As shown in Equation 2, the permeability can be written as a product of two terms; a thermodynamic factor, S_A, called the solubility coefficient, and a kinetic parameter, D_A, called the diffusion coefficient.

$$P_A = (S_A)(D_A). \qquad 2.$$

For cases such as those discussed here, where the downstream pressure of the membrane is negligibly low, the solubility coefficient for component A equals the slope of the sorption isotherm, i.e. $S_A = C_A/p_A$, the con-

Table 1 Polycarbonate structures

Polymer structure	Abbreviation
	PC
	6FPC
	TMPC
	TM6FPC
	BCPC
	PCZ
	NBPC

Table 1—*continued*

Polymer structure	Abbreviation
(structure)	SBIPC

centration of A in the membrane divided by the external partial pressure of A in equilibrium with the polymeric material. The diffusion coefficient of the gas in the polymer is largely determined by the relative motion of the polymer and the penetrant inside the selective layer, which is shown as having a thickness, l, in Figure 2.

Table 2 Polysulfone structures

Polymer structure	Abbreviation
(structure)	PSF
(structure)	DMPSF
(structure)	TMPSF

Table 3 Polyimide and polypyrrolone structures

Polymer structure	Abbreviation
	6FDA-6FmDA
	6FDA-6FpDA
	6FDA-IPDA
	6FDA-ODA
	6FDA-TADPO

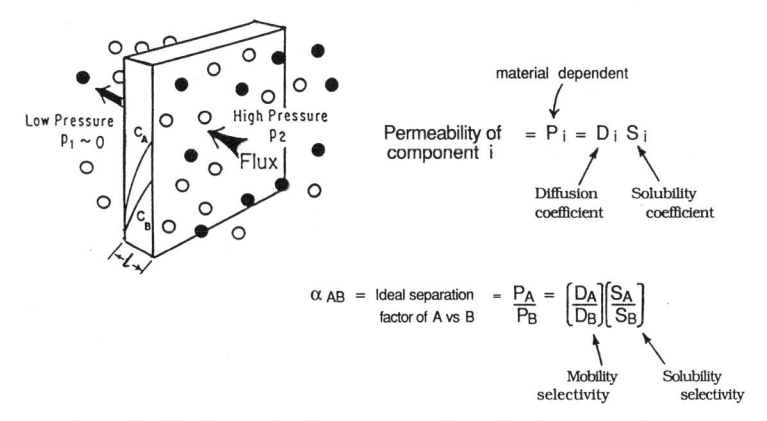

Figure 2 Fundamentals of gas transport through polymer membranes.

Permeability of component i = $P_i = D_i S_i$ (material dependent; Diffusion coefficient, Solubility coefficient)

α_{AB} = Ideal separation factor of A vs B = $\dfrac{P_A}{P_B} = \left[\dfrac{D_A}{D_B}\right]\left[\dfrac{S_A}{S_B}\right]$ (Mobility selectivity, Solubility selectivity)

A second important relation is shown in Equation 3, where the ideal selectivity of the membrane for component A relative to B is expressed as the ratio of the pure gas permeabilities of the two penetrants in the membrane material,

$$\alpha^*_{AB} = P_A/P_B. \qquad 3.$$

This factor provides a good measure of the ability of a given polymeric material to provide a permselective barrier to A relative to B. For example, in the simplest case of a 50/50 feed mixture, α^*_{AB} indicates the relative number of A molecules appearing in the downstream permeate stream compared to the number of B molecules. Therefore, an ideal selectivity of 10 in this case could produce a 90% A mixture. Other feed compositions with different penetrant ratios of A to B and with non-negligible downstream pressure can also be analyzed, but these cases are more complex and not important for the present discussion (7).

By substituting Equation 2 into Equation 3 for the two components, A and B, the selectivity can be factored into a mobility selectivity term and a solubility selectivity term as shown in Equation 4

$$(P_A/P_B) = (D_A/D_B)(S_A/S_B). \qquad 4.$$

This factoring of the thermodynamic and kinetic factors allows the effects of detailed changes in the characteristics of the polymer backbone to be analyzed.

If the selectivity factor for a given gas pair, for example, oxygen and nitrogen, is used in the absence of groups that specifically attract one of the components, the relative solubility is determined largely by the relative condensibility of the two components. The critical temperature of a component is a good indicator of condensibility; therefore, oxygen's higher critical temperature typically makes it more condensible and hence more soluble than nitrogen in most media. For low molecular weight solvents, this is illustrated in Table 4, where the range of solubility selectivities at 25°C is only from 1.35–1.89 (8, 9).

This range of solubility selectivities is also typical of what is found in polymers that do not have the ability to complex with oxygen. For instance, bisphenol-A polycarbonate (shown in Figure 3) is made by condensing bisphenol-A with phosgene to form a fairly high glass transition material (T_g = 150°C). The solubility selectivity of this material for O_2 over N_2 is 1.65, which positions it roughly in the middle of the range for all materials that have been investigated. Clearly, this is a rather limited range and is not sufficient to produce a high separation factor for this gas pair. On the other hand, the solubility selectivity is not a limitation for the separation of highly condensible vapors, e.g. Freons® from gases such as air or nitrogen (10). For most applications indicated in Figure 1, however, it

Table 4 O_2/N_2 solubility selectivities in liquids at 25°C

Medium	S_{O_2}/S_{N_2}
n-Hexane	1.38
n-Heptane	1.60
n-Octane	1.54
n-Nonane	1.65
2,2,4-Trimethylpentane	1.83
Cyclohexane	1.64
Benzene	1.83
Perfluoroheptane	1.42
Perfluorobenzene	1.35
$(C_4F_9)_3N$	1.49
CCl_4	1.85
Chlorobenzene	1.82
Methanol	1.45
Ethanol	1.61
Butanol	1.65
Isobutanol	1.74
Acetone	1.53
DMSO	1.89
Water	1.68
Acetic acid	1.50
Methyl acetate	1.49
Average	1.62±0.27

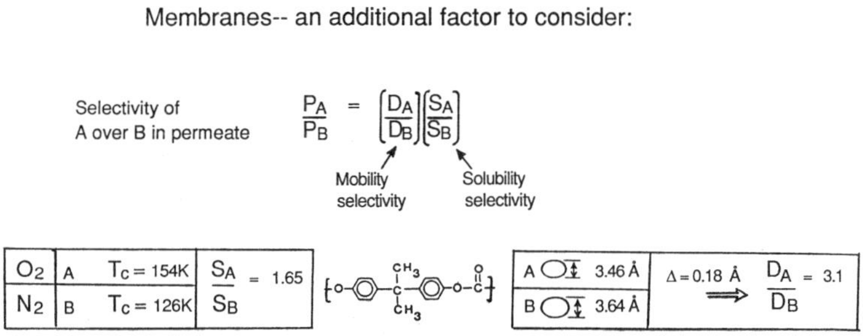

CONTROLLED GAS PERMEABILITY 55

Membranes-- an additional factor to consider:

Selectivity of A over B in permeate
$$\frac{P_A}{P_B} = \left[\frac{D_A}{D_B}\right]\left[\frac{S_A}{S_B}\right]$$

Mobility selectivity   Solubility selectivity

| O_2 | A | $T_c = 154K$ | $\frac{S_A}{S_B} = 1.65$ |
| N_2 | B | $T_c = 126K$ | |

[structure: –O–C(CH₃)₂–O–C(=O)–]

| A ⊙ 3.46 Å |
| B ⊙ 3.64 Å |

$\Delta = 0.18$ Å $\Rightarrow$ $\frac{D_A}{D_B} = 3.1$

Figure 3 Effect of penetrant characteristics on diffusion and solubility selectivities.

presents a sufficiently serious problem that precludes strictly solubility-based selection of materials for practical separations of gas pairs where both of the components are well above their critical temperatures.

The realization of the limitations imposed by relying on such purely thermodynamically based separations has motivated interest in adjusting the mobility selectivity by controlling the diffusion coefficient of gas A relative to gas B. As shown in Figure 3, in order to regulate the mobility selectivity, the minimum dimensions (sieving diameter) of the two gas molecules must be used. The sieving diameters of the molecules are the principle determinants of their relative ability to execute a diffusive jump through the spectrum of transiently opening-size-selective, molecular-scale intersegmental gaps in the polymeric matrix (11). For instance, in the case of oxygen and nitrogen, the sieving diameters have been estimated, by determination of the minimum zeolite window that will allow each to pass, to be 3.46 and 3.64 Å, respectively (12). As shown in Figure 3, this 0.18 Å difference in minimum size gives rise to a threefold difference in the diffusion coefficient of O_2 relative to N_2 in bisphenol-A polycarbonate at 35°C.

If the membrane is considered to be essentially an isotropic medium through which the penetrants execute jumps of length λ at a frequency f, the diffusion coefficient can be written as shown in Equation 5, viz.,

$$D = f\lambda^2/6, \qquad 5.$$

Since the sizes of most penetrants are very similar, the jump length term is generally thought to be similar for different penetrants. The diffusion coefficient ratio, therefore, can be ascribed essentially to the jumping frequency of O_2 relative to N_2 in the matrix. Soft, rubbery materials generally have large scale segmental motions rendering them unable to produce subtle size selective transient gaps, and the mobility selectivity for

O_2/N_2 will typically be less than 1.3 for rubbery matrices (13). Thus, in the absence of extreme solubility selectivity such as is the case for vapors over gases, rubbery matrices are typically not well suited for membrane separation involving gases.

On the other hand, highly rigid glassy materials such as those considered in the following discussion may have mobility selectivity up to 5 for the O_2/N_2 system. As was the case with solubility selectivity, the mobility selectivity of bisphenol-A polycarbonate also falls in the middle of the range of presently achievable properties. The ensuing discussion considers how one can adjust the solubility and diffusivity factors to obtain optimized membrane properties for several important gas pairs mentioned in Figure 1.

To expand our perspective beyond the O_2/N_2 system, it is useful to consider Table 5, which shows the zeolite sieving diameter of a number of different penetrants. Oxygen is shown as 3.46 and nitrogen at 3.64 Å. When the size difference beyond the 0.18 Å value for the O_2/N_2 pair to 0.5 Å is expanded, as in the case for the CO_2/CH_4 system, the mobility selectivity for CO_2 relative to CH_4 can be as high as 20 for some of the highly rigid materials. As noted earlier, it can be much lower (~ 1.3) for a rubbery matrix such as silicone rubber, which does not have the ability to produce subtle size-selective transient gaps.

By further extending the size difference to 0.91, and then 1.2 Å, as is the case for the H_2/CH_4 and He/CH_4 systems, respectively, the mobility selectivity can be as high as 2000 in highly rigid matrices. This does not imply that such materials will have an overall selectivity of 2000 for the He/CH_4 pair (13) since we have solubility factors in opposition. In fact, for the He/CH_4 pair, the much lower condensibility of helium causes its solubility to be only about one eighth to one thirtieth as large as that of methane in most liquids and polymers (9, 14). Fortunately, for high mobility-selective media, the product of these two factors can still produce an impressive overall selectivity above 100. Therefore, although moderated by the solubility selectivity, mobility is the dominant feature allowing the control of selectivity behavior, except for the prior-mentioned case of vapors relative to gases (15–54).

Table 5 Penetrant zeolite sieving diameters

Molecule	He	H_2	NO	CO_2	Ar	O_2	N_2	CO	CH_4
Kinetic diameter (Å)	2.6	2.89	3.17	3.3	3.4	3.46	3.64	3.76	3.80

BASIS FOR STRUCTURE-PERMSELECTIVITY TAILORING

A simple concept is useful for understanding much of the data relating the effects of structural changes on the resultant permeability and permselectivity properties of a given family of polymers (e.g. polycarbonates, polyimides, etc.). Hoehn suggests that it is desirable to do two things simultaneously when changing the structure within a family of polymers: inhibit intersegmental packing while simultaneously hindering the backbone mobility (55).

Inhibitions to intersegmental packing are reflected by an increase in the free fractional volume (FFV) of the polymer matrix not occupied by the electron clouds comprising the filled space. The inhibition of segmental and subsegmental mobility is reflected in increases in the glass transition and/or sub-T_g transition temperatures. Simultaneous inhibitions to segmental or subsegmental mobility and increases in the FFV of the matrix can be achieved in various ways, each with associated advantages and disadvantages.

The importance of free volume and free volume fraction has been understood for some time in the context of barrier packaging (56–58). In general, an increase in FFV leads directly to an increase in the diffusion coefficient and, therefore, the permeation rate of all penetrants through the polymer. For membranes, however, a generalized opening of the matrix tends to undermine the ability to perform mobility selectivity. The solubility of a given penetrant in such a generally open matrix tends to be higher, presumably because of energetically more facile placement into the matrix (13). The solubility selectivity, which is primarily controlled by penetrant condensibility, is not usually affected strongly by such subtle changes in the polymer matrix. Therefore, reductions in mobility selectivity are reflected directly in a corresponding drop in overall permselectivity.

An extreme example of this is seen in the case of the highly packing-inhibited glassy poly (trimethyl-silyl-propyne) (PTMSP), which has the highest diffusion coefficient recorded for a polymer, but with an essentially insignificant mobility selectivity of 1.47 (21). The solubility selectivity of the PTMSP for O_2/N_2 is 1.17 (21), which places it slightly below the normal range indicated in Table 4.

If such opening of the polymer matrix is accompanied by a generalized inhibition of subsegmental motions associated with the detailed penetrant diffusional jumps, a favorably higher jumping frequency by the smaller of the two penetrants and, therefore, a high mobility selectivity can be maintained. Various means can be used to inhibit motion, including intrasegmental steric effects, polar substituent groups, and intersegmental steric effects. The utility of these approaches for producing tradeoffs between permeability increases while maintaining or even increasing permselectivity for a particular gas pair is discussed below.

The diffusion coefficient can be represented as a temperature-dependent parameter with a preexponential factor, D_0, and an activation energy in the exponential, similar to reaction rate coefficients, e.g. (32, 59–61),

$$D = D_0 \exp(-Ed/RT). \qquad 6.$$

A simple interpretation of the activation energy in terms of the energy needed to open a cylindrical path of diameter σ and length λ in the matrix, with cohesive energy density (CED), allows interpretation of this fact (62)

$$Ed = \lambda \pi \sigma^2 CED/4. \qquad 7.$$

Owing to the exponential dependence of D on Ed, small changes in Ed can cause significant changes in diffusivity and hence diffusivity selectivity. Based on this interpretation, the squared power dependency of Ed on the minimum penetrant diameter, σ, exaggerates the importance of size differences on the mobility selectivity. The CED is a quantity defined as the energy necessary to vaporize a given volume of a substance. The greater the CED, the greater the attractive forces between the polymer chains, and thus the greater the energy expenditure to open a transient gap through which a penetrant molecule can jump.

It is useful to visualize the significance of free volume in the context of diffusion through a local region of a selective membrane, as shown schematically in Figure 4. Over an extremely short time scale, the move-

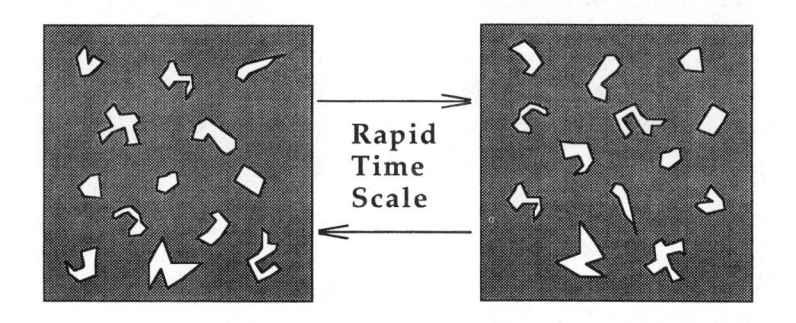

FFV = (V-Vo)/V

V = ▸ from Density

Vo = ▪ from Group Contribution

Figure 4 Free volume redistribution.

ment of the constituent segments and subsegments comprising the polymer can be thought of as a flickering picture. Exchanges between subtly different conformations occur over a characteristic time scale moderated by the intrinsic rigidity of the polymer backbone, existing intersegmental attractions, and the temperature. The summation of the open space represents the total specific free volume (57, 63, 64). The specific free volume is defined as the difference between the total specific volume, V, and the occupied volume, V_0, obtained from group contribution estimations (57). While this free space tends to be mobile, it exists as more or less disconnected packets, except as it exchanges positions during the flickering process moderated by movements of the polymer at a very subtle level. Presumably, this process is responsible for activated diffusional jumps by penetrants from one location to another. The ratio of the specific free volume to the specific volume of a polymer sample is defined as the fractional free volume (FFV).

Since such phenomena still cannot be accurately modeled in realistically complex amorphous glasses, any attempt to provide more detail about the relationship between these hypothetical flickering motions and diffusion of penetrants is schematic and cartoon-like. One can imagine, for instance, segments of the simple bisphenol-A polycarbonate glass discussed in the context of Figure 3.

The schematic in Figure 5 represents a locally nested arrangement between two repeat units in an overall amorphous tangle. A rocking motion around a mobile hinge comprised by the carbonate linkage provides a hypothetical activated penetrant scale gap through which a gas molecule can translate. A structural variation that increases the average distance between the backbones of the nested segments, as shown in Figure 5, requires a smaller rocking motion to create an activated penetrant scale gap. The larger penetrant molecules are better able to take advantage of this process, and the mobility selectivity is decreased. Therefore, corresponding inhibition of the freedom of angular displacements of moving subsegments in Figure 5 is required to maintain size and shape selection for penetrants with sub-Å differences in dimensions.

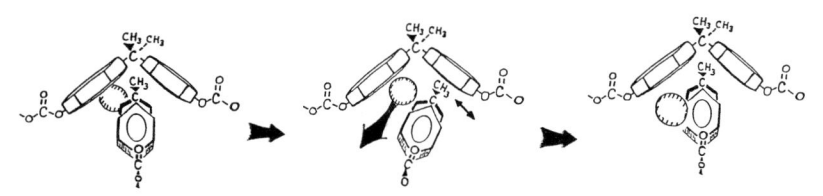

Figure 5 Diffusive jump of penetrant molecule through transient gap opening.

For polymers comprised of complex repeat units, besides the total FFV, a more complex issue of how the fractional free volume is distributed may also affect the tradeoff between permselectivity and permeability as structures are changed. One can imagine situations in which the total FFV may be identical to that represented in Figure 4; however, its distribution may be different. In such a case a different gas separation performance can result if the same amount of volume is distributed in large vs small packets. Data will be shown that suggest this may be the case and that tailoring of such a distribution may be advantageous.

Illustration of Currently Achievable Tailoring of Structures

A typical trade-off curve between permselectivity of O_2 over N_2 and the permeability of the desired permeate oxygen is presented in Figure 6. The units of permeability are cumbersome, but conventional in the field. Curves with similar qualitative shapes can be considered for the CO_2/CH_4 and He or H_2/CH_4 systems. A characteristic cross-hatched area below the solid line is associated with low free volume glasses and rubbery polymers.

As noted earlier, rubbery materials typically have a poor mobility selectivity, on the order of 1.1–1.3 for O_2/N_2 (14), since in such media, diffusion effectively occurs through a high molecular weight liquid with huge, free volume packets incapable of significant size and/or shape selectivity. Due to high rates of diffusion in such media, rubbery materials typically have high permeabilities associated with their low permselectivities. On the other hand, for the well-packed, low free volume glasses, one finds equally unattractively low permeabilities with high selectivities. The solid line represents an optimistic trade-off behavior that can be achieved if the best commercially available engineering resins are used as membranes.

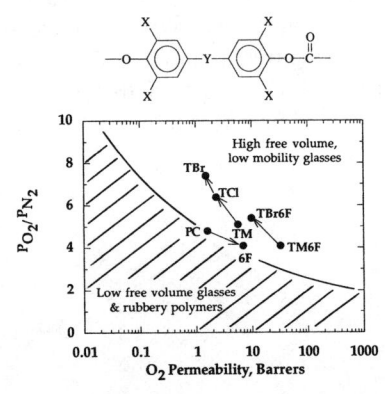

Figure 6 O_2/N_2 tradeoff at 35°C for polycarbonates.

Standard bisphenol-A polycarbonate falls on this line and represents an attractive, but not spectacular, membrane material. As a rule of thumb, given current fabrication abilities to make thin selective layers of approximately 1000 Å, adequate flux rates for commercially attractive processes can be achieved with polymers having permeabilities of at least 1 Barrer[1] (1, 3, 4, 65).

The desirable regime up and to the right of this standard line is the domain of high free volume, low segmental mobility glasses. A model compound of a polycarbonate is represented in Figure 6 with Y and X varied to illustrate the effects on transport properties of inhibiting segmental packing and motion. For example, standard polycarbonate can be represented with this model when the central Y unit is an isopropylidene and compact H is at the X positions.

Tetramethyl-substituted polycarbonate, TMPC, can be represented with methyl groups at all four X positions with no change in the isopropylidene Y. This structure is characterized by significantly inhibited motion around the carbonate linkage, thereby moving the selectivity above that for the unsubstituted standard PC. Significant inhibition of packing also attends the four methyl ring substitutions because of restriction of the number of facile nesting conformations of the phenyl rings like those shown in Figure 5. The net effect of these changes is to increase the fraction of unoccupied space and rigidity of the matrix, thus resulting in a productivity three times that of standard polycarbonate without a loss in permselectivity.

The less desirable effects of simply inhibiting packing with little change in sub-T_g segmental mobility can be seen from consideration of hexafluoropolycarbonate (6FPC). This structure corresponds to leaving the Xs as compact hydrogens, but substituting bulky -C(CF$_3$)$_2$- groups in place of the -C(CH$_3$)$_2$- at the Y position in bisphenol-A polycarbonates. Space-filling molecular models show that the -C(CF$_3$)$_2$- groups in the isopropylidene positions effectively lock the phenyl rings into orientations that greatly hinder packing. Unlike the -CH$_3$ substitutions on the ring positions mentioned above, however, the -CF$_3$ groups in the isopropylidene positions essentially do not hinder the motions around the mobile carbonate link. Consistent with expectations, this structure displays an even higher FFV and permeabilities than the TMPC with the space-filling methyl substituents. Based on the sub-T_g transition data in Table 6, the substitutions on the 6FPC do not significantly inhibit rotational mobility, thus there is some loss in selectivity.

The above-mentioned trends can be quantified by comparison of the

[1] $1 \text{ Barrer} = 10^{-10} \frac{\text{cm}^3(\text{STP}) \cdot \text{cm}}{\text{cm}^2 \cdot \text{s} \cdot \text{cmHg}} = 0.335 \frac{\text{mmol}}{\text{m} \cdot \text{s} \cdot \text{TPa}}.$

Table 6 Polycarbonate fractional free volumes and sub-T_g transitions

Polymer	$FFV = \dfrac{V - V_0}{V}$	$T\gamma$ °C
PCZ	0.156	−22
PC	0.164	−70
6FPC	0.195	−71
TMPC	0.180	+111
TM6FPC	0.220	+107

FFV, the T_g, and sub-T_g transitions of these three materials in Table 6. The FFV values were calculated using the group contribution method of Van Krevelen for estimation of occupied volume as suggested by Lee (57, 66). The use of the FFV rather than simple free volume is preferred to avoid biasing the results when heavy elements such as fluorine are present (25). Performing both substitutions on the same unit, i.e. methyls at X and hexafluoroisopropylidene (6F) at Y between phenyl rings, leads to dramatically higher (20-fold) permeabilities with some loss in selectivity.

The permeability and selectivity data, along with their factoring into diffusion and sorption coefficients and selectivities, for the four polycarbonates are described in Table 7. These data clearly show that the small loss in selectivity with the TM6FPC is not the result of lost mobility selectivity resulting from excessive packing inhibition. Indeed, the mobility selectivity of this most open structure (TM6FPC) is the highest of the four materials. The slightly lower overall permselectivity is due to a surprisingly low value of the independently measured solubility selectivity of O_2 relative to N_2. The detailed cause of this effect is not clear, but it illustrates the importance of considering sorption and diffusion data. Without the

Table 7 Permeabilities, diffusivities, and solubilities of O_2 and N_2 at 35°C and 2 atm

Polymer	P_{O_2} (Barrers)	$\dfrac{P_{O_2}}{P_{N_2}}$	$D_{O_2} \times 10^8$ (cm²/sec)	$\dfrac{D_{O_2}}{D_{N_2}}$	S_{O_2} cc (STP) / cc atm	$\dfrac{S_{O_2}}{S_{N_2}}$	FFV
PC	1.5	5.1	5.6	3.1	0.2	1.65	0.164
6FPC	6.9	4.3	10.6	2.9	0.5	1.5	0.195
TMPC	5.6	5.1	8.1	3.8	0.52	1.45	0.180
TM6FPC	32	4.1	31	4.0	0.78	1.05	0.220

factored data, it is attractive to suggest that the structure had been opened to such a degree that mobility selectivity is lost. Based on the actual data, if the solubility selectivity could be returned to even a typical value of 1.4–1.5, the resultant material would have outstanding properties.

In this respect, one might consider copolymerization of TM and TM6F monomers. Copolymerization would most likely raise the solubility selectivity to near the log average of 1.23, while giving a mobility selectivity near the log average of 3.9 of the two homopolymers. Indeed, the topic of copolymerization and blending requires more study to determine the correct rules to guide the performance trade-offs with various combinations used to make the copolymer. Studies of dissimilar blend pairs by Paul (24, 65, 67, 68) and Ranby (69) suggest that linearity of the various solubility, diffusivity and permeability coefficients on a log coefficient vs volume fraction of the components is roughly adequate. A recent work dealing with blends of isomers by Coleman & Kohn (70) also tends to support this simple mixing rule.

In addition to intrasegmental control of segmental mobility, interchain attractions can be used to inhibit motion, thereby adding another means of tailoring membrane material properties. For example, the effective occupied volume of a Cl atom and a -CH_3 group are almost identical (66); however, the Cl atom introduces significant polar attraction possibilities. Consider the trade-off trajectory shown in Figure 6 that results from the introduction of halogens in the X positions of the polycarbonate, instead of the -CH_3 groups in TMPC. Clearly, the addition of the polar Cl and Br atoms have profound effects that produce significant increases in the selectivity while causing some reductions in permeability. The TBrPC material lies well above the trade-off curve in Figure 6.

The overall trade-off behavior related to the introduction of the polar halogens is dealt with in greater detail in Table 8. The estimated CEDs of PC, TMPC, TClPC, and TBrPC from Hoy (71) are 10.05, 9.39, 9.97, and

Table 8 Permeabilities, diffusivities, and solubilities of O_2 and N_2 at 35°C and 2 atm

Polymer	P_{O_2} (Barrers)	$\dfrac{P_{O_2}}{P_{N_2}}$	$D_{O_2} \times 10^8$ (cm^2/sec)	$\dfrac{D_{O_2}}{D_{N_2}}$	S_{O_2} $\dfrac{\text{cc (STP)}}{\text{cc atm}}$	$\dfrac{S_{O_2}}{S_{N_2}}$	FFV
PC	1.5	5.1	5.61	3.10	0.2	1.65	0.164
TMPC	5.6	5.1	8.11	3.77	0.52	1.36	0.181
TClPC	2.3	6.4	2.95	4.54	0.59	1.40	0.183
TBrPC	1.36	7.5	1.69	4.97	0.61	1.50	0.180

10.09 $(cal/cc)^{1/2}$, respectively. By increasing the cohesive energy density of the material relative to TMPC, the diffusivity is lowered and the mobility selectivity is markedly increased, with little difference in solubility or solubility selectivity. These data suggest that the effective increase in the attraction between neighboring segments in the more polar TClPC and TBrPC relative to TMPC, with little or no change in the FFV, provides additional resistance to the opening of transient gaps of sufficient size to allow the execution of a given penetrant jump (72). Although data do not prove this suggestion, it is also consistent with the fact that the mobility selectivity is higher in the more polar environment.

An unusual aspect of the halogen-containing materials is their high absolute solubility level for both gases. The high solubility, coupled with the high mobility selectivity, appears inconsistent at first; however, upon inspection it can be understood in terms of a description of the glassy polymer state often referred to as the dual mode model. This model suggests that to a first approximation in glassy polymers, aside from normally densely packed regions of the glass with essentially nested natures, typical of the schematic representation in Figure 5, a small but significant second environment exists. This second population can be viewed as comprising penetrant-scale packing defects. It seems reasonable that in a highly packing inhibited matrix such as the TMPC, TClPC, and TBrPC, a significant number of such packing defects can exist. Strong evidence for this suggestion has been provided by Muruganadam et al (24). Additional repositories exist for penetrants, nevertheless, greater size and shape discrimination occurs in the polar materials. This explains the higher solubility, with relatively little change in solubility selectivity coexisting with the higher mobility selectivity in such matrices.

The above series of materials illustrates two different approaches to inhibit motion and thereby achieve higher selectivities. The first approach, relying upon intrasegmental steric inhibitions, such as the four methyl substitutions on the phenyl rings, may be more resistant to penetrant- or impurity-induced losses in selectivity. Actual data to support such a suggestion are lacking, but this is a promising area for future work. The second, the use of polar groups, is clearly powerful in its effects. This approach is most appropriately applied in cases where the sorption level of penetrants and trace condensible agents are sufficiently low to avoid undue swelling of the matrix. Swelling of the matrix can break the intersegmental attractive forces, thus causing a large drop in mobility selectivity. In the presence of a condensible agent, disruption of polar attractions may undermine the efficacy of such an approach.

An additional example of the use of large polar-inhibiting atoms to affect the trade-off between permeability and permselectivity is illustrated by the modifications of polyphenylene oxide (PPO). This topic has been considered by several researchers (36–39). Recently, it was shown that bromine substitution on the ring position, as opposed to the methyl side chain, hinders intrasegmental motion around the ether linkage (45). This modification also leads to significant packing inhibition relative to the unmodified PPO, as can be seen from the schematic in Figure 7 and from FFV measurements (45). In addition to its intrasegmental inhibition of motion and packing inhibition caused by the induced kinky nature of the backbone, which is apparent in Figure 7, the polar Br presumably increases intersegmental attractions as in the case of TBrPC and TClPC. This suggestion is consistent with the coexistence of higher mobility selectivities and enhanced solubilities for all gases in this material. As was argued for the packing-inhibited PCs, the higher solubility, without higher solubility selectivity is consistent with additional packing-inhibited regions in the brominated PPO, which can be described by the dual mode sorption model.

Combined intra- and intersegmental inhibition to rotation may be introduced using steric factors, as illustrated by considering para- and meta-isomers of some polymers. Although this is not possible with the bisphenol-A type polycarbonates because they lack the meta-isomer of standard bisphenol-A, polyesters studied by Chern et al (22) derived from meta and para di-acids illustrate these effects. An even more extreme illustration is apparent in Table 9 for polyimides derived from the 3,3' and 4,4' diamines condensed with the so-called 6F dianhydride, analogous to the 6F connector in the polycarbonates discussed above. Stern (33) has shown that polyimides containing an ether linkage in the diamine residue have markedly different transport properties for the para- vs the meta-connected

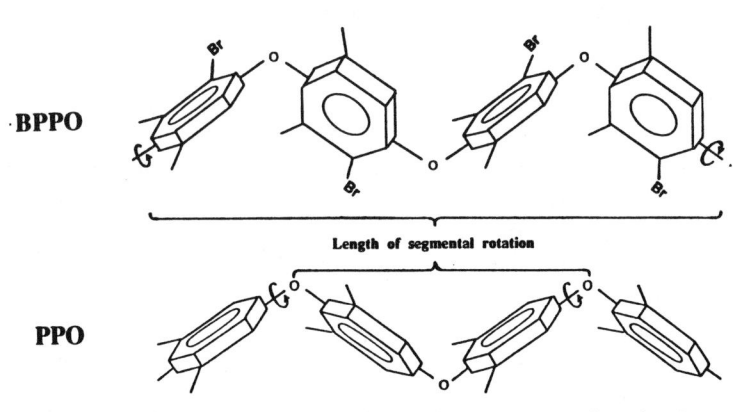

Figure 7 Effect of bromine substitution on the segmental motion of PPO.

Table 9 Permeabilities, diffusivities, and solubilities of O_2 and N_2 at 35°C and 2 atm

Polymer	P_{O_2} (Barrers)	$\dfrac{P_{O_2}}{P_{N_2}}$	$D_{O_2} \times 10^8$ (cm²/sec)	$\dfrac{D_{O_2}}{D_{N_2}}$	S_{O_2} $\dfrac{\text{cc (STP)}}{\text{cc atm}}$	$\dfrac{S_{O_2}}{S_{N_2}}$	FFV
6FDA-6FpDA	16.3	4.7	12.5	3.15	0.99	1.48	0.190
6FDA-6FmDA	1.8	6.9	2.23	3.92	0.612	1.73	0.175

materials. For example, the para-material has a CO_2 permeability that is twice that of the meta-connected isomer. Moreover, the meta-isomer shows 60% higher CO_2/CH_4 ideal separation factor at 35°C compared to the para-isomer. Even more extreme differences were discovered by Coleman & Koros (34) in comparison of the 6F dianhydride-derived polyimide based on meta- and para-isomers of the diamine connected by a 6F linkage, rather than by an ether linkage. Roughly nine and tenfold higher permeabilities were observed in the para-connected isomer for O_2 and CO_2 compared to the meta-isomer. Permselectivities for O_2/N_2 and CO_2/CH_4 also differed by as much as 30–50% between the two isomers.

The polyester and both sets of the polyimide data can be explained in terms of a combination of higher intra- and intersegmental resistance to motions in the meta-connected isomer because of the difficulty in moving the large nonsymmetrical meta-connected phenyl rings in the well packed glass. By analogy to Figure 5, the effective size of the moving element appears larger because of its noncentrosymmetric connection (33). The intersegmental contribution to higher selectivity for the halogen-containing materials results from resistance to rotation due to attractions between neighboring segments. On the other hand, for the meta isomers, the selectivity-enhancing intersegmental hindrance to motion is more purely steric in nature. As in the case of the attraction-generated resistance to rotation, the intersegmental steric resistance may be undermined by the presence of a strongly sorbing component such as CO_2, which results in losses in selectivity.

Still another structure-property issue related to intersegmental mobility resistance structures involves the effect of nonsymmetrical substitution on backbone ring structures. As is shown in Table 10, symmetrical substitution of methyl groups on the bisphenol-A phenyl rings of the polysulfones, analogous to the polycarbonates mentioned above, tends to cause corresponding increases in both permeability and selectivity relative to the unsubstituted polysulfone. On the other hand, nonsymmetrical substitution of methyl groups on the bisphenol phenyl rings tends to cause significant reductions in the permeability of the polysulfones with a simi-

Table 10 Permeabilities, diffusivities, and solubilities of O_2 and N_2 at 35°C and 2 atm

Polymer	P_{O_2} (Barrers)	$\dfrac{P_{O_2}}{P_{N_2}}$	$D_{O_2} \times 10^8$ (cm²/sec)	$\dfrac{D_{O_2}}{D_{N_2}}$	S_{O_2} $\dfrac{\text{cc (STP)}}{\text{cc atm}}$	$\dfrac{S_{O_2}}{S_{N_2}}$	FFV
PSF	1.4	5.6	4.4	3.6	0.24	1.6	0.156
DMPSF	0.64	7.0	1.7	4.2	0.29	1.7	0.149
TMPSF	5.6	5.3	8.0	3.8	0.53	1.4	0.162

larly corresponding increase in selectivity (48). Thus, while this change appears to be useful in some connections from the standpoint of membrane alteration, it, like the meta/para-isomer approach, must be used carefully to avoid excessive loss in permeability.

Dynamic mechanical spectroscopy results suggest that sub-T_g motions tend to be inhibited by such nonsymmetric substitutions, consistent with the analogy to the 3,3′ intermolecular inhibition. The methyl groups in the disubstituted polysulfone appear to fill void space between chains, which results in less FFV. Also, disubstitution leads to a significant increase in $T\gamma$, which is attributed to motion of the phenyl ring in polysulfone. Substitutions of -CH_3 on the phenyl rings increase the $T\gamma$ from −70°C for PSF to +30°C for TMPSF to +80°C for DMPSF (48). The mobility selectivity in this case also reflects a significant increase in discriminating ability, as was the case for the meta- vs the para-structural changes.

An additional example of the combination of inter- and intrasegmental effects of substitution on chain backbones is illustrated by the extensive literature on chemical modification of PPO (37, 40–42). In particular, a study by Story & Koros (44) extends the earlier modification work based on bromination of PPO, in which it was found that bromination on the methyl groups rather than on the ring, as discussed above, can lead to unsatisfactory trade-offs (73). Here methyl esters replace some of the methyl groups to form so-called MeCPPO. In this case an increase in the solubility selectivity for the CO_2/CH_4 system is consistent with the addition of carbonyl moieties that can favorably interact with CO_2 over CH_4. The mobility selectivity was also affected, and a major reduction in diffusivity was observed and explained in terms of increased intersegmental hindrance to motion in the ring-substituted PPO. Using arguments based on space-filling molecular models, it was shown that for both PPO and MeCPPO the presence of the substituent groups on the methyl side chain has negligible effect on intrasegmental rotational freedom. This is in marked distinction to the above-mentioned effects of ring substitutions with bromine. The space filling molecular models of PPO, MeCPPO, and brominated PPO (BrPPO) are shown in Figure 8.

These observations led to the conclusion that intersegmental hindrance to motion caused by the interference of the larger ester side chains with adjacent polymer segments was the principal factor in this system. The replacement of the methyl groups of PPO with -COOCH$_3$ groups decreases the diffusivity and increases the mobility selectivity in two or perhaps even three ways. First the FFV is decreased, thereby reducing the ability of gas molecules to move between polymer segments and increasing the size and shape discrimination ability of the matrix. Secondly, the additional rotational hindrance imparted by the longer MeCPPO side chains reduces the amplitude of phenyl ring motion, thereby increasing selectivity and decreasing the diffusivity. A third potential effect of the substitution is a possible introduction of dipolar attractions between neighboring ester groups, analogous to the effects of introducing polar halogens in the polycarbonates and the brominated PPO discussed above. Such modifications, therefore, have complex effects, and depending upon the per-

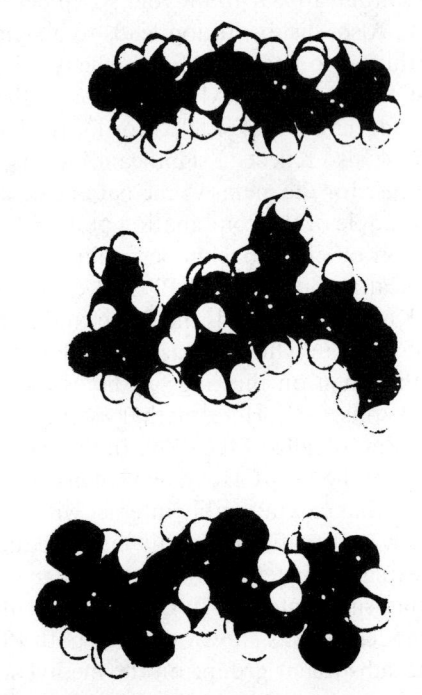

Figure 8 Molecular models of PPO (*top*), MeCPPO (*center*), and BrPPO (*bottom*). The models show 100% substitution on four repeat units.

fection of placement on the ring vs the methyl group, contradictory results can be observed. In general, it appears that one can achieve more controllable results by polymerizing specifically selected monomers rather than seeking to control the vagaries of such modification reactions.

Most of the preceding discussion has focused on polymers containing connector groups derived from an isopropylidene unit between phenyl rings. Clearly there are many other options for such connections. The polycarbonate family is again useful to illustrate the effects of such changes. Besides the interphenyl connection, the only other joint in the chain occurs at the carbonate, thereby allowing assignment of observed effects to variation in this single joint. For instance, while similar effects pertain to the polysulfones, the longer repeat unit with the relatively flexible sulfone linkage tends to dilute the effects of changes in the interphenyl connector.

The most unusual of the novel connectors is the spirobiindane unit associated with SBIPC. The structure of the SBIPC, shown in Table 1, differs from the other polycarbonate repeat units, which contain phenyl rings that are joined by a single carbon at the positions para to the hydroxyl groups. This one carbon link is, in turn, bonded to a variety of substituents. By contrast, the SBIPC repeat unit consists of two spiro-fused indane units, where the five-membered rings are attached both meta and para to the hydroxyl groups. The spiro structure locks the indane halves into a rigid conformation, with the carbonate links held out of plane to each other, thereby giving the polymer a helical twist. Molecular space-filling models show that both the repeat unit kink and the bulky methyl substituents on the indane skeleton tend to inhibit chain packing. This is evidenced by the higher FFV of SPIPC (FFV = 0.183) compared to PC (FFV = 0.164) (26).

Nevertheless, the FFV of SBIPC is somewhat lower than 6FPC and possesses more resistance to phenyl ring motions than 6FPC. From this balance of segmental packing and mobility, one might expect SBIPC to be more selective than either 6FPC or PC and to have a permeability for most gases that is intermediate between PC and 6FPC. In fact, the permeability of O_2 in SBIPC is approximately equal to the permeability in 6FPC, as shown in Table 11. Consistent with expectations, the SBIPC also has a higher permselectivity than either material for O_2/N_2, which reflects its greater mobility inhibition compared to 6FPC. The positions on the trade-off curve of SBIPC and the other polycarbonates discussed earlier for O_2/N_2 are shown in Figure 9.

The gas transport properties of the various polycarbonates with novel connectors can be discussed further in terms of the absolute penetrant solubility, diffusivity, and associated factorized selectivities for the O_2/N_2 gas pair. As with other rigid packing-disrupted polycarbonates, such as

Table 11 Permeabilities, diffusivities, and solubilities of O_2 and N_2 at 35°C and 2 atm

Polymer	P_{O_2} (Barrers)	$\dfrac{P_{O_2}}{P_{N_2}}$	$D_{O_2} \times 10^8$ (cm²/sec)	$\dfrac{D_{O_2}}{D_{N_2}}$	S_{O_2} $\dfrac{cc\ (STP)}{cc\ atm}$	$\dfrac{S_{O_2}}{S_{N_2}}$	FFV
PC	1.6	4.8	5.8	3.2	0.21	1.5	0.164
6FPC	6.9	4.1	11	3.2	0.47	1.3	0.195
NBPC	2.4	5.1	4.4	3.3	0.41	1.5	0.174
PCZ	0.6	5.7	1.8	3.8	0.25	1.5	0.103
BCPC	1.4	5.2	4.0	3.3	0.27	1.4	0.176
SBIPC	7.0	5.1	11	3.9	0.50	1.3	0.183

6FPC, the higher sorption levels for NBPC and SBIPC relative to PC may reflect an increased number of penetrant size segmental packing defects. On the other hand, based on molecular models and their lower glass transitions compared to the other nonstandard polycarbonates, both BCPC and PCZ possess greater intrasegmental flexibility. Such flexibility can promote the elimination of packing defects resulting in lower absolute sorption coefficient like that for PC.

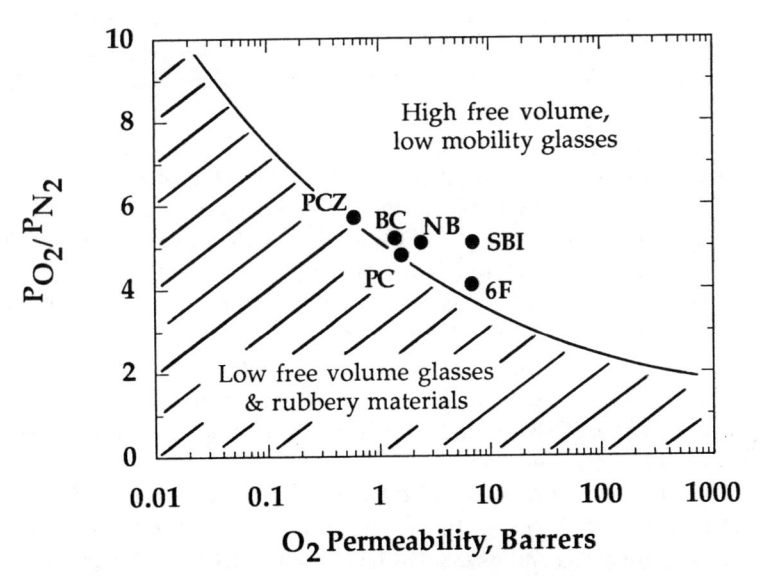

Figure 9 O_2/N_2 tradeoff at 35°C for polycarbonates listed in Table 8.

Most of these materials fail to show unusually high mobility selectivity, possibly because they fail to provide an adequate inhibition of segmental motion. PCZ achieves the necessary inhibition of segmental motion for a high mobility selectivity; however, this is due to a highly packed matrix, which results in an unattractively low permeability. SBIPC, on the other hand, fits the criteria established in the introductory section by providing a balance between inhibition of segmental mobility and packing. While the highly hindered packing and restricted segmental mobility of SBIPC provides it with an attractive combination of permeability and permselectivity, it may be possible to further optimize its properties. For example, the introduction of strong intersegmental attractions may increase selectivity; however, halogenated variants of the SBIPC monomer are not currently available to test these suggestions.

Although reference has been made to other families, most of the above discussion has been couched in terms of the diverse family of polycarbonates and polysulfones, which allow relatively clear illustrations of important principles. Limited data exist to suggest that the various principles discussed for these two families will also apply to other families, such as polyesters and polyimides. Nevertheless, the field is young, and less systematic data exists for these families.

Comparison of Different Families of Materials

This section performs some cross-cutting comparisons of members in the polycarbonate family with related members in the polyimide family to gain insights into additional structure-transport issues that, while present in the polycarbonates or any other family, might be overlooked when studying only a single family or even related families such as the polycarbonates, polysulfones, and polyesters. We call attention to the likely importance of the detailed distribution of transient free volume in a material in determining separation performance. Eventually, molecular dynamics simulations of the diffusion process may help to complement the highly qualitative arguments made here.

In addition, to further broaden the scope of the discussion beyond the O_2/N_2 case, it is useful to consider additional aspects related to CO_2/CH_4 and other systems involving He or H_2. For simplicity of arguments, only selected points will be identified on plots; however, a more complete tabulation of available data are offered in the appendices 1, 2, and 3.

Since the size difference and critical temperature difference is larger between the CO_2/CH_4 pair than the O_2/N_2 pair, more favorable mobility selectivities and solubility selectivities are typical. These combined favorable factors can lead to overall permselectivities approaching 80 as shown in Figure 10.

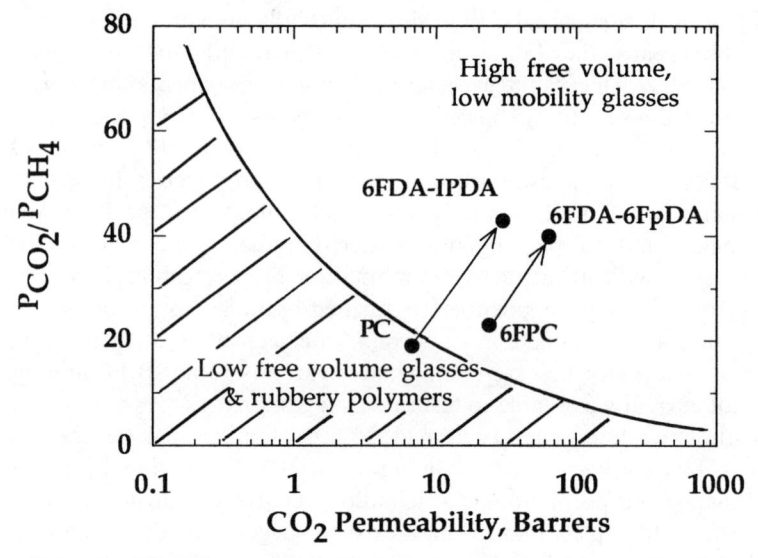

Figure 10 CO_2/CH_4 tradeoff at 35°C for polymers listed in Tables 12a and 12b.

Before considering the polyimides, it is useful to use the polycarbonates as benchmarks against which to compare them. As for most gas pairs of interest, standard polycarbonate belongs on the typical trade-off curve above the low free volume materials and below truly high free volume materials. For the substituted polycarbonates discussed in connection with the O_2/N_2 pair, we see the same trend, except the loss in selectivity associated with the 6F replacement at the isopropylidene is absent. For the less demanding CO_2/CH_4 gas pair, which differs in minimum diameter by 0.5 Å, the mobility selectivity is not subverted by opening up the structure with the 6F connector group. Indeed, the progressively more packing-inhibited structures (PC to TM6FPC) form a desirable trajectory that results in a gain of about a factor of 20 in CO_2 permeability, with a small gain in selectivity as shown in Appendix 1. These gains occur prior to the onset of plasticization-induced selectivity losses for strongly sorbing penetrants such as CO_2.

The upper right hand corner of the trade-off plot in Figure 10 shows a grouping containing polyimides. A flat dotted/underlined symbol in Table 3 emphasizes the existence of broad flat structures in two polyimides containing the analogue of the bisphenol-A and 6F-bisphenol-A structures in PC and 6FPC. While the items 1 and 2 and 3 and 4 in Tables 12a and 12b are analogues, they differ significantly in that 1 and 3 have a carbonate

Table 12a Permeabilities, diffusivities, and solubilities of CO_2 and CH_4 at 35°C and 10 atm

Polymer	P_{CO_2} (Barrers)	$\dfrac{P_{CO_2}}{P_{CH_4}}$	D_{CO_2} × 10^8 (cm²/sec)	$\dfrac{D_{CO_2}}{D_{CH_4}}$	S_{CO_2} cc (STP) cc atm	$\dfrac{S_{CO_2}}{S_{CH_4}}$	FFV
(1) PC	6.8	19	3.2	4.7	1.6	4.0	0.164
(2) 6FDA-IPDA	<u>30</u>	43	5.4	<u>12.1</u>	<u>4.2</u>	3.5	0.168
(3) 6FPC	24	23	6.7	5.9	2.7	3.9	0.195
(4) 6FDA-6FpDA	<u>64</u>	40	8.1	<u>9.6</u>	<u>6.0</u>	4.2	0.190

Table 12b Permeabilities, diffusivities, and solubilities of O_2 and N_2 at 35°C and 2 atm

Polymer	P_{O_2} (Barrers)	$\dfrac{P_{O_2}}{P_{N_2}}$	D_{O_2} × 10^8 (cm²/sec)	$\dfrac{D_{O_2}}{D_{N_2}}$	S_{O_2} cc (STP) cc atm	$\dfrac{S_{O_2}}{S_{N_2}}$	FFV
(1) PC	1.6	4.8	5.8	3.2	0.21	1.5	0.164
(2) 6FDA-IPDA	<u>7.53</u>	<u>5.6</u>	6.36	3.14	<u>0.9</u>	1.8	0.168
(3) 6FPC	6.9	4.1	11	3.2	0.47	1.3	0.195
(4) 6FDA-6FpDA	<u>16.3</u>	<u>4.7</u>	12.5	3.18	<u>0.99</u>	1.48	0.190

link, while 2 and 4 have two broad flat aromatic imide units connected by a structurally disruptive 6F unit. This 6F isopropylidene unit between the aromatic dianhydride rings serves to inhibit packing, as it did when positioned between the phenyl rings of the polycarbonates.

The effects on sorption and transport properties of incorporating the long, flat packable aromatic dianhydride rings with the structurally disruptive 6F moiety as a connector is shown in Table 12a for the CO_2/CH_4 gas pair. Interestingly, the overall FFV is unchanged within each of the two respective sets of analogues (item 1 and 2, 3 and 4). In spite of this constancy of overall FFV in each pair, the permeability of CO_2 is almost fivefold higher for the polyimide material in the first pair and threefold higher in the second pair compared to their respective polycarbonate analogues.

A significant cause for the higher permeabilities in the polyimides vs the polycarbonates are the much higher (two to fourfold) solubilities of CO_2 in the imide matrix compared to the corresponding carbonate of equivalent FFV. This point is emphasized by the underlined values in Table 12a. This difference is not the result of a specific solubility preference for CO_2, for instance, because of a specific interaction of CO_2 with the carbonyls. Clearly both CH_4 and CO_2 sorb more readily into the imides than into the carbonates with similar overall FFV, as evidenced by the similar solubility selectivities. In addition, a higher diffusion coefficient and a significantly higher mobility selectivity for CO_2/CH_4 is apparent in comparing both 1 and 2 and 3 and 4.

When materials with equivalent FFV are compared, the ability to sorb gas molecules of all types is increased in the matrix containing the long, flat packable structures. Moreover, in such a matrix, the ability of the CO_2 to jump relative to CH_4 has been significantly increased. It is impressive that the same effect occurs for both the hexafluoro analogues (3 and 4) and the more flexible analogues (1 and 2). Results are seen for the O_2/N_2 gas pairs in Table 12b, with the polyimides exhibiting greater solubilities and permeabilities than the polycarbonates with similar FFV.

It is useful to refine and try to generalize the above observations. Specifically, it appears that positioning a long, flat aromatic imide moiety between packing disruptive 6F groups increases the mobility selectivity and the solubility without any significant change in FFV. The mobility selectivity increase, coexisting with the higher absolute solubility, may reflect an improved ability to control jumping through lower FFV regions communicating between local areas of high FFV, which may easily accommodate penetrant. Such a molecularly interconnected set of environments comprising a packing disrupted and an efficiently packed environment in intimate communication represents a special form of free volume distribution. In other words, while the total FFV in the two systems may be the same, a much different distribution of this volume may exist in the two polymer families for the structures considered here.

Extending this concept to the polypyrrolone family, an even longer and more packable section of chain can be produced between the packing disruptive 6F units. In this case, the long, flat chain section has been produced by substituting a planar pyrrolone-fused ring system for the imide ring system, which contains a flexible link. This change is achieved by using a tetramine (TADPO) in place of a diamine (ODA) with the 6F dianhydride shown in Tables 13a and 13b. It is useful to compare the polyimides and this related polypyrrolone to further understand the effects of introduction of broad, flat extended regions between disruptive links such as the 6F moiety.

For the polyimide and polypyrrolone, introducing the locked flat planar pyrrolone linkage between the packing disruptive 6F unit increases the FFV and the permeability through an increase in diffusion coefficient, with a negligible change in solubility. The nearly twofold increase in diffusion coefficient of O_2, which results from the substitution of the packing-

Table 13a Permeabilities, diffusivities, and solubilities of CO$_2$ and CH$_4$ at 35°C and 10 atm

Polymer	P_{CO_2} (Barrers)	$\dfrac{P_{CO_2}}{P_{CH_4}}$	$D_{CO_2} \times 10^8$ (cm^2/sec)	$\dfrac{D_{CO_2}}{D_{CH_4}}$	S_{CO_2} cc (STP) / cc atm	$\dfrac{S_{CO_2}}{S_{CH_4}}$	FFV
6FDA-ODA	23.0	60.5	3.6	16.3	4.9	3.7	0.164
6FDA-TADPO	27.6	51.1	4.8	19.4	4.4	2.6	0.196

Table 13b Permeabilities, diffusivities, and solubilities of O$_2$ and N$_2$ at 35°C and 10 atm

Polymer	P_{O_2} (Barrers)	$\dfrac{P_{O_2}}{P_{N_2}}$	$D_{O_2} \times 10^8$ (cm^2/sec)	$\dfrac{D_{O_2}}{D_{N_2}}$	S_{O_2} cc (STP) / cc atm	$\dfrac{S_{O_2}}{S_{N_2}}$	FFV
6FDA-ODA	4.3	5.2	3.2	2.8	1.03	1.9	0.164
6FDA-TADPO	7.9	6.5	5.8	5.1	1.04	1.3	0.196

resistant pyrrolone, leads to a twofold increase in permeability as shown in Table 13a. A corresponding significant increase in mobility selectivity is achieved for the O$_2$/N$_2$ system with the substitution of the rigid fused ring system of TADPO. Note that the difficulty in achieving good mobility selectivity in the close cut (0.18 Å) size difference for O$_2$/N$_2$ with the polycarbonate and its analogous polyimide has been eliminated in the analogous pyrrolone.

The introduction of the rigid TADPO also results in a significant increase in diffusion coefficient of CO$_2$ and in the mobility selectivity of CO$_2$/CH$_4$, as shown in Table 13b. The corresponding slight decrease in solubility of CO$_2$ coupled with the diffusion increase results in an overall small increase in permeability. The reduction of carbonyl density in the CO$_2$/CH$_4$ system is reflected by a lower solubility selectivity in the pyrrolone, thereby slightly reducing its permselectivity. Nevertheless, this change may be attractive in lowering the sensitivity of the material to CO$_2$-induced plasticization. Moreover, the permselectivity of the pyrrolone is still very attractive for commercial applications.

Tables 14a and 14b give data for the polycarbonate and polyimide with essentially equivalent FFVs to the TADPO polypyrrolone material. Comparisons between the 6FDA-ODA imide and 6FDA-TADPO pyrrolone are based on both polymers containing 6F isopropylidene transphenyl and ether transphenyl links. On the other hand, comparisons with the

Table 14a Permeabilities, diffusivities, and solubilities of CO_2 and CH_4 at 35°C and 10 atm

Polymer	P_{CO_2} (Barriers)	$\dfrac{P_{CO_2}}{P_{CH_4}}$	$D_{CO_2} \times 10^8$ (cm²/sec)	$\dfrac{D_{CO_2}}{D_{CH_4}}$	S_{CO_2} cc (STP) cc atm	$\dfrac{S_{CO_2}}{S_{CH_4}}$	FFV
6FPC	24	23	6.7	5.9	2.7	3.9	0.195
6FDA-6FpDA	64	40	8.1	9.6	6.0	4.2	0.190
6FDA-TADPO	27.6	51.1	4.8	19.4	4.4	2.6	0.196

Table 14b Permeabilities, diffusivities, and solubilities of O_2 and N_2 at 35°C and 2 atm

Polymer	P_{O_2} (Barriers)	$\dfrac{P_{O_2}}{P_{N_2}}$	$D_{O_2} \times 10^8$ (cm²/sec)	$\dfrac{D_{O_2}}{D_{N_2}}$	S_{O_2} cc (STP) cc atm	$\dfrac{S_{O_2}}{S_{N_2}}$	FFV
6FPC	6.9	4.1	11.0	3.2	0.47	1.3	0.195
6FDA-6FpDA	16.3	4.7	12.5	3.18	0.99	1.48	0.190
6FDA-TADPO	7.9	6.5	5.8	5.1	1.04	1.3	0.196

polycarbonate are based on the 6F isopropylidene transphenyl links and the equivalent FFV of the TADPO pyrrolone and 6FDA-6FpDA polyimide. Clearly, large cross-family differences are apparent that shed additional light on the issues identified herein.

Comparisons of the 6FDA-6FpDA polyimide and polypyrrolone with FFV similar to the 6FPC reveal rather different behavior for the O_2/N_2 and the CO_2/CH_4 systems. For CO_2/CH_4, as shown in Table 14a, with a relatively large size difference between penetrants, the polycarbonate and polypyrrolone display similar CO_2 permeability, but the polycarbonate has much less selectivity because of its lower mobility selectivity. Equivalent permeability in the polycarbonate, compared to the polypyrrolone with the long flat units, is achieved by having a higher overall diffusion coefficient with a lower CO_2 solubility. These observations support the general suggestions made above regarding the distribution of free volume in the 6FDA-6FpDA and 6FPC. For instance, if packing is more or less uniformly disturbed in the 6FPC material, while in the imide and pyrrolone there is a combination of efficiently packed and highly disturbed packing regions, high solubility can coexist with good mobility selectivity, which is consistent with observation.

Such a complex morphology could exist if the potentially packable regions periodically achieve effective nesting on a local basis to provide low FFV barriers between regions of interconnecting high FFV, where most of the penetrant resides. The existence of such highly disturbed regions in the presence of the flat units is reasonable in cases where flat structures are not stacked, but because of kinetic limitations to organization, are significantly out-of-equilibrium nesting. In fact, inefficiently packed long flat units comprise extraordinarily disrupted packing over extended regions, more so than for a corresponding flexible unit with sufficient joints to allow relaxation of out-of-equilibrium packing defects. In considering the O_2/N_2 system, one notes that the polypyrrolone of comparable FFV to the polycarbonate exhibits markedly superior mobility selectivity relative to 6FPC. This further supports the simple explanation offered in terms of the biphasic molecular-scale interconnected morphology noted above; however, the low mobility selectivity for the polyimide in Table 14b is not consistent. The suggested morphological ideas, therefore, should be considered reasonable hypotheses, with some supporting evidence, but requiring additional critical study. The validity of such ideas could be explored by extending them to other polymer families through the use of large packable monomers between packing disruptive units such as the 6F-isopropylidene. For instance, copolymerization of a bisphenol derived from fluorene with 6F-bisphenol-A would introduce long, flat packable units between the mobile carbonates and the packing-inhibited 6Fbisphenol linkages. In such a case, marked increases in mobility selectivity might result as the more or less uniformly disturbed polycarbonate takes on some of the nature hypothesized to exist in the polyimides and polypyrrolone.

For the TADPO structure considered here, the only significantly mobile linkage in the backbone is the ether oxygen. Clearly, other members in this family can be considered by using tetraamines analogous to hindered IPDA and 6FpDA diamines used to prepare the polyimides in Table 14. Eventually, with increasing rigidity, insufficient molecular mobility may lead to brittleness; however, the TADPO material does not reflect such problems, and it is difficult to predict the onset of such problems.

Penetrant and Thermal History Behavior

A few comments are appropriate concerning the effects of elevated levels of sorbed penetrant and temperature on materials used in gas separations. The effect of sorbed penetrants is an extension of the discussion of CO_2/CH_4, while the effect of temperature is used to consider the He/N_2 system. This latter case is often taken as a surrogate for the H_2/N_2 system.

As mentioned above, transport plasticization refers to the situation where the diffusivity of a penetrant increases significantly due to the presence of other penetrants in its neighborhood. This phenomenon is generally associated with an upward inflection in the permeability vs pressure plot of a component. Plasticization is also signaled by a reduction in the selectivity of the membrane in mixed gas studies for systems containing strongly sorbing penetrants such as CO_2 at elevated pressures.

Plasticization becomes apparent for different polymers at different upstream pressures (74–77). The issue of plasticization of glassy polymers is extraordinarily complex; however, some insights derived from the preceding discussion of structure-property issues are offered. Data for CO_2 permeation and mixed gas selectivities as functions of CO_2 total feed pressure for a 50/50 CO_2/CH_4 mixed gas feed with vacuum downstream are shown in Figure 11. The TM6FPC material is the highly packing-disrupted polycarbonate material described in connection with simple structure property considerations. The ratio of pure gas permeabilities of CO_2/CH_4 as listed in Appendix 1 is 24, in good agreement with the mixed gas values below the pressure at which significant plasticization occurs. This highly packing-disrupted material lacks the long flat packable units present in the pyrrolone and has markedly higher FFV. In spite of this fact, the TM6FPC has slightly lower CO_2 solubility compared to the pyrrolone [$S = 4.1$ vs 4.4 cc(STP)/cc atm for the pyrrolone] (78). Therefore, at equivalent pressure, more CO_2 exists in the pyrrolone, and yet it shows a much lower tendency to plasticize, as indicated by a maximum loss in selectivity of 50% compared to almost 100% for the TM6FPC at 500 psia of CO_2. Moreover, the lower tendency for the CO_2 permeability of the pyrrolone to show an upswing with increasing pressure also indicates better resistance to plasticization.

These results are consistent with the morphology suggested earlier in which the pyrrolone may accommodate a significant fraction of sorbed penetrants in packing-disturbed regions between the better packed domains responsible for principal size and shape discrimination. Sorption into these regions is most likely to cause plasticization-induced losses in selectivity. At equivalent sorption levels in the 6FPC and pyrrolone, a larger fraction of penetrant is expected to reside in these packed regions in the 6FPC than in the pyrrolone, since the pyrrolone contains more packing defects (47). The low solubility of CO_2 in the packed regions may contribute to the plasticization resistance.

The composite form of the permeability coefficient represented as a product of solubility and diffusivity in Equation 2 indicates that the permeability will be affected by the result of the temperature-dependence of these two coefficients. Typically, the solubility decreases and the diffusivity increases exponentially with increasing temperature. For more condensible agents such as CO_2, the reduction in solubility tends to be more significant

CONTROLLED GAS PERMEABILITY 79

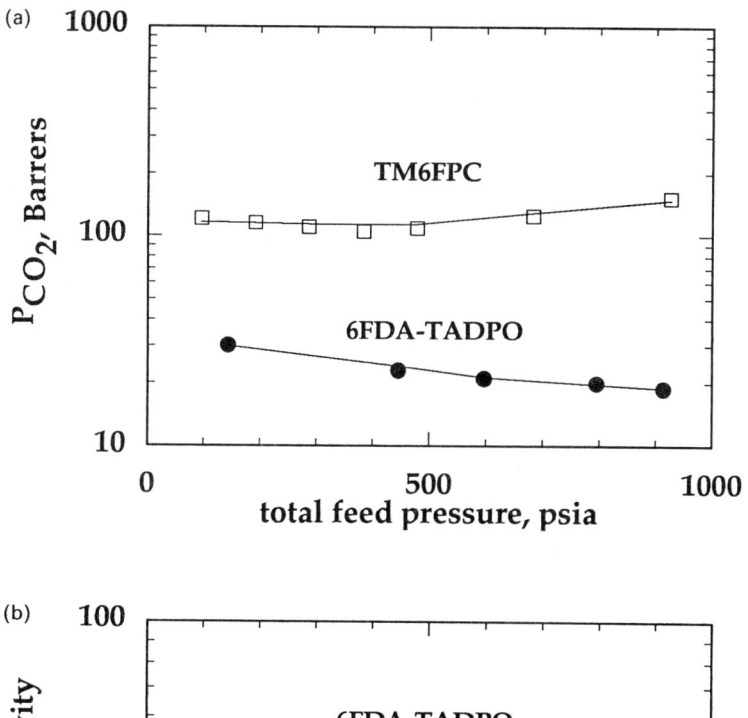

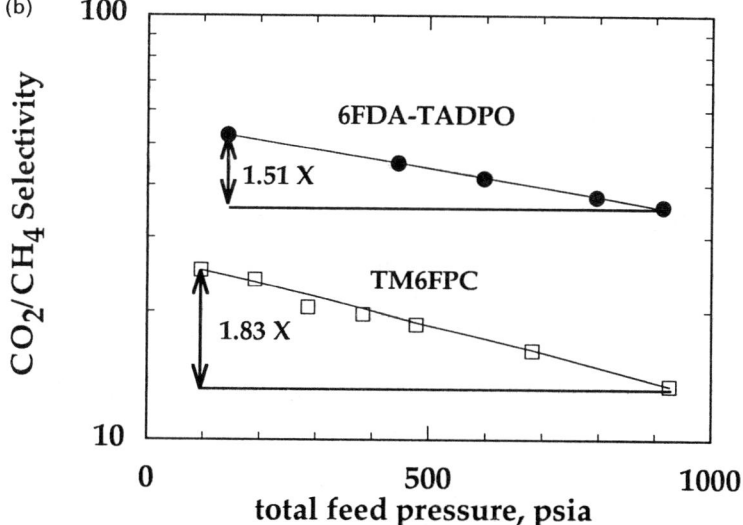

Figure 11 (*a*) Carbon dioxide permeability for a 50/50 carbon dioxide-methane feed at 35°C. (*b*) Mixed gas selectivity for a 50/50 carbon dioxide-methane feed at 35°C.

than for supercritical gases such as He, H_2, N_2, O_2, and CH_4. For these supercritical gases, changes in the diffusion coefficient with increasing temperature generally favor the larger penetrant more than the smaller one, so the mobility selectivity, and hence the overall selectivity, generally drop with increasing temperature. Factors affecting the activation energy have been discussed briefly in the context of Equation 6. Based on these relations, the tendency for the mobility selectivity to decrease most sharply with temperature for penetrant pairs with largest size differences, e.g. He/CH_4, H_2/CH_4, etc, is understandable.

It has been shown by Kim et al (32) that highly open, rigid-chained polymers tend to show lower temperature dependence of absolute permeabilities than corresponding flexible-chained materials. However, few studies over extended ranges have been studied to characterize fundamental effects in this regard. Perhaps the most extensive study has been reported by Gebben et al (46) for the triazole (TIPT) shown below.

This rigid material has a complex structure that gives rise to a strong loss in separation factor as temperature is increased for the He/N_2 and O_2/N_2 pairs. Mobility selectivity values were not reported between 50 and 200°C for this material; however, larger losses occurred in overall permselectivity for He/N_2 (76%) compared to the more similar O_2/N_2 pair (62%) with σs differing by only 0.18 Å. These relative changes are probably conservative, since losses in mobility selectivity with temperature tend to be mitigated in the overall permselectivity by solubility selectivity changes for He/N_2, but not the O_2/N_2 system. This fact reflects the tendency for N_2 and O_2 solubilities to drop at similar rates and more rapidly than that for He because temperature is increased in most polymers and liquids (9).

Based on principles considered in the preceding sections, one can hypothesize possible corrections for the adverse reductions occurring with increasing temperature as seen with the TIPT structure. The unsubstituted para-connected phenyl ring should be unusually mobile relative to the other units. Even the meta-connected phenyl should show significant intersegmental restriction to motion and the two phenyl substituents on the triazoles will hinder movements of these rings. The polytriazole is made from a polyhydrazide precursor derived from terephthaloyl chloride and isophthaloyl hydrazide, so it would be relatively straightforward to replace the terephthaloyl chloride with isophthaloyl chloride. Clearly, based on the previous work with meta- and para-isomers of polyesters and imides, this change will result in a significantly lower permeability at room temperature; however, a higher selectivity and superior maintenance of this selectivity with increasing temperature should also result. The topic of structure-property characterization as a function of temperature is badly in need of additional work, and insights gained from such studies would also be valuable in testing hypotheses regarding intersegmental resistance to motion through observation of the magnitudes of activation energies as a function of isomer types.

A final topic bears brief mention from a practical standpoint of processing advanced materials. Typically, most current generation commercial membranes have asymmetric forms with an open porous support on top of which is an integral skin formed during the casting of a solution of the polymer into a nonsolvent bath. As chain rigidity increases, it becomes more difficult to find acceptable solvents in which to dissolve the material. In fact, the TADPO-derived pyrrolone discussed above is not soluble in its final form, and a poly(amide amino acid) prepolymer must be formed, and subsequently heat-treated to close the pyrrolone ring. Moreover, although all of the polyimides discussed here are soluble in standard solvents, some of the lower free volume imides must also be cast in amic acid form followed by thermal ring closure. Some such materials, in fact, are used in commercial membranes; however, this use is clearly more complex and expensive. An additional problem with the use of exotic materials such as those represented in this review is their cost. Although it is not the topic of this discussion, it is worth noting that this problem can be potentially circumvented through the use of a technique in which the thin selective layer is deposited on an inexpensive porous support. These and related issues are the topic of another review involving the current state of the art in membrane formation (5).

APPENDIX 1 PERMEABILITIES AND PERMSELECTIVITIES AT 35°C

Polymer family and abbreviation	P_{He} Barrers 10 atm	P_{O_2} Barrers 2 atm	P_{CO_2} Barrers 10 atm	$\dfrac{P_{He}}{P_{CH_4}}$ 10 atm	$\dfrac{P_{O_2}}{P_{N_2}}$ 2 atm	$\dfrac{P_{CO_2}}{P_{CH_4}}$ 10 atm
Polycarbonates						
PC (54)	13	1.6	6.8	35	4.8	19
6FPC (54)	60	6.9	24	57	4.1	23
TMPC (54)	46	5.6	19	50	5.1	21
TM6FPC (54)	200	32	110	44	4.1	24
TBrPC (54)	18	1.5	4.2	140	7.4	34
TBr6FPC (54)	100	9.7	32	112	5.4	36
TBr/TBr6FPC (54)	49	4.9	16	110	6.2	34
TMPPC (63)	35	3.7	12	56	5.3	19.5
TMFPC (63)	18	1.8	6.2	61	5	21
SBIPC (63)	49	7	30	25	5.1	15
NBPC (27)	19	2.4	9.1	38	5.1	19
PCZ (27)	10	0.6	2.2	110	5.7	24
BCPC (27)	12	1.4	5.6	52	5.2	24
Polystyrenes						
PS (23)	22.4[a]	2.9[a]	12.4[a]	28.4[a]	5.6[a]	15.8[a]
PαMS (23)	14.5[a]	0.82[a]	3.0[a]	100[a]	5.4[a]	20.8[a]
PMS (23)	37.1[a]	7.2[a]	29.8[a]	16.9[a]	4.8[a]	13.6[a]
PAS (23)	18.6[a]	3.1[a]	16.3[a]	18.8[a]	4.8[a]	16.4[a]
PCS (23)	16.4[a]	1.2[a]	4.3[a]	62.2[a]	5.6[a]	16.4[a]
PFS (23)	34.4[a]	4.4[a]	17.2[a]	32.3[a]	5.3[a]	16.1[a]
PBS (23)	16.4[a]	1.9[a]	8.5[a]	34.2[a]	5.9[a]	17.7[a]
PMxS (23)	15.0[a]	2.6[a]	18.9[a]	16.4[a]	4.3[a]	20.7[a]
PtBS (23)	104.1[a]	35.5[a]	140.1[a]	6.2[a]	4.2[a]	8.4[a]
PHS (23)	3.4[a]	0.12[a]	-	-	8	-
Polysulfones						
PSF (48)	13	1.4	5.6	49	5.6	22
TMPSF (48)	41	5.6	21	45	5.3	22
DMPSF (48)	12	0.64	2.1	170	7	30
DMPSF-Z (48)	11	0.41	1.4	280	7.2	34
6FPSF (49)	33	3.4	12	63	5.1	22
PSF-F (49)	10	1.1	4.5	54	5.5	34
PSF-O (49)	10	1.1	4.3	56	5.6	34
TMPSF-F (50)	29	3.3	15	50	5.4	26
TM6FPSF (50)	113	18	72	38	4.5	24
Polyesters						
PPha-tere (36)	-	3.15[b]	17.2[b]	-	4.85[b]	27.3[b]
PPha-50:50 (36)	-	2.6[b]	13.5[b]	-	5.12[b]	27.1[b]
PPha-iso (36)	-	1.65[b]	7.6[b]	-	5.7[b]	28.1[b]
PAr (52)	-	-	25.5	-	-	20.9
TClPar (52)	-	-	12	-	-	34.3
TBrPPha-tere (52)	-	4.5	8.9	-	5.3	35.9
Polypyrrolone						
6FDA-TADPO (47)	89	7.9	27.6	165	6.5	51
Polyetherketones						
12H (51)	11.8	1.1	4.6	539.	5.7	20
6H6F (51)	26.4	2.4	8.6	71.4	5.4	23.2
6F6H (51)	24.8	2.3	8.9	65.3	5.5	23.4
12F (51)	42	3.7	12.9	78.2	4.8	24.1

APPENDIX 1—continued

Polymer family and abbreviation	P_{He} Barrers 10 atm	P_{O_2} Barrers 2 atm	P_{CO_2} Barrers 10 atm	$\dfrac{P_{He}}{P_{CH_4}}$ 10 atm	$\dfrac{P_{O_2}}{P_{N_2}}$ 2 atm	$\dfrac{P_{CO_2}}{P_{CH_4}}$ 10 atm
Polyphenyleneoxides						
PPO (39)	-	-	50[b]	-	-	17[b]
0.36BrPPO (39)	-	-	51[b]	-	-	19[b]
0.91BrPPO (39)	-	-	68[b]	-	-	20[b]
1.06BrPPO (39)	-	-	108[b]	-	-	17[b]
CPPO-22 (45)	-	-	22.0	-	-	19.5
MeCPPO-22 (45)	-	-	18.0	-	-	17.0
Polyimides						
PMDA-ODA (13)	8	0.61	2.71	135	6.1	46
PMDA-MDA (13)	9.4	0.98	4.03	94	4.9	43
PMDA-IPDA (13)	37.1	7.1	26.8	41	4.7	30
PMDA-DAF (13)	1.9	-	0.15	921	-	72
6FDA-ODA (13)	51.5	4.34	23	135	5.2	61
6FDA-MDA (13)	50	4.6	19.3	117	5.7	45
6FDA-IPDA (13)	71.2	7.53	30	102	5.6	43
6FDA-DAF (13)	98.5	7.85	32.2	156	6.2	51
6FDA-6FmDA (34)	48	1.8	5.1	600	6.9	64
6FDA-6FpDA (34)	137	16.3	63.9	85.6	4.7	40
6FDA-m-PDA (53)	-	2.61[c]	8.23[c]	-	7.2[c]	58[c]
6FDA-2,4-DATr (53)	-	7.44[c]	28.63[c]	-	5.7[c]	41[c]
6FDA-2,6-DATr (53)	-	11.00[c]	42.52[c]	-	5.2[c]	46[c]
6FDA-3,5-DBTF (53)	-	6.43[c]	21.64[c]	-	5.5[c]	48[c]
PMDA-3,3'-ODA (33)	-	0.13[c]	0.50[c]	-	7.2[c]	62[c]
PMDA-3BDAF (33)	-	1.4[c]	6.12[c]	-	4.8[c]	36[c]
PMDA-4BDAF (33)	-	2.9[c]	11.8[c]	-	4.4[c]	33[c]
6FDA-3,3'-ODA (33)	-	0.68[c]	22.0[c]	-	6.8[c]	64[c]
6FDA-3BDAF (33)	-	1.35[c]	6.30[c]	-	5.6[c]	48[c]
6FDA-4BDAF (33)	-	5.40[c]	19.0[c]	-	5.5[c]	37[c]
6FDA-p-PDA (33)	-	2.10[c]	11.8[c]	-	5.5[c]	65[c]
BTDA-p,p'-DAS (35)	-	0.0028[a]	0.0091[d]	-	-	-
BTDA-DAFO (35)	-	0.0030[a]	-	-	-	-
BTDA-DAF (35)	-	0.0049[a]	0.0152[d]	-	-	-
BTDA-Benzidine (35)	-	0.0088[a]	0.0310[d]	-	-	-
BTDA-p-PDA (35)	-	0.0136[a]	-	-	-	-
BTDA-DADPyS (35)	-	0.0261[a]	0.0828[d]	-	-	-
BTDA-m,m'-DABP (35)	-	0.0288[a]	0.1062[d]	-	-	-
BTDA-m,m'-MDA (35)	-	0.0310[a]	0.0447[d]	-	-	-
BTDA-m-PDA (35)	-	0.0405[a]	0.1198[d]	-	-	-
BTDA-m,p'-DABP (35)	-	0.0593[a]	0.1167[d]	-	-	-
BTDA-p,p'-DABP (35)	-	0.0594[a]	0.1733[d]	-	-	-
BTDA-m,p'-MDA (35)	-	0.0761[a]	-	-	-	-
BTDA-m,m'-DADPC (35)	-	0.0954[a]	-	-	-	-
BTDA-p,p'-ODA (35)	-	0.1219[a]	0.4692[d]	-	-	-
BTDA-DADPS (35)	-	0.1306[a]	0.4036[d]	-	-	-
BTDA-p,p'-MDA (35)	-	0.1848[a]	0.4268[d]	-	-	-
PMDA-Benzidine (35)	-	0.0059[a]	0.0104[d]	-	-	-
PMDA-m,m'-DABP (35)	-	0.0342[a]	-	-	-	-
PMDA-DADPyS (35)	-	0.0397[a]	0.1441[d]	-	-	-
PMDA-p,p'-DABP (35)	-	0.1450[a]	0.1780[d]	-	-	-
PMDA-DADPS (35)	-	0.4424[a]	-	-	-	-
PMDA-p,p'-MDA (35)	-	0.4843[a]	1.74[d]	-	-	-

[a] 1 atm.
[b] 20 atm.
[c] 1.00 psi.
[d] 0.33 atm.

Appendix 2 Solubilities and Solubility Selectivities at 35°C

Polymer family and abbreviation	S_{He} cc/cc atm 10 atm	S_{O_2} cc/cc atm 2 atm	S_{CO_2} cc/cc atm 10 atm	$\dfrac{S_{He}}{S_{CH_4}}$ 10 atm	$\dfrac{S_{O_2}}{S_{N_2}}$ 2 atm	$\dfrac{S_{CO_2}}{S_{CH_4}}$ 10 atm
Polycarbonates						
PC (54)	0.019	0.21	1.6	0.047	1.5	4
6FPC (54)	0.042	0.47	2.7	0.06	1.3	3.9
TMPC (54)	0.037	0.46	2.6	0.045	1.2	3.2
TM6FPC (54)	0.089	0.78	4.1	0.074	1.3	3.4
TBrPC (54)	0.047	0.54	2.6	0.049	1.5	2.8
TBr6FPC (54)	0.068	0.8	3.9	0.055	1.3	3.1
TBr/TBr6FPC (54)	0.049	0.78	3.1	0.044	1.5	2.8
TMPPC (63)	0.034	0.47	2.1	0.047	1.3	2.8
TMFPC (63)	0.035	0.29	1.6	0.069	1.2	3
SBIPC (63)	0.051	0.5	2.6	0.057	1.3	2.8
NBPC (27)	-	0.41	2.5	-	1.5	3.9
PCZ (27)	-	0.25	1.7	-	1.5	3.4
BCPC (27)	-	0.27	1.9	-	1.4	4.3
Polystyrenes						
PS (23)	-	0.14[a]	1.1[a]	-	1.7[a]	2.9[a]
PαMS (23)	-	0.3[a]	3.0[a]	-	1.2[a]	4.2[a]
PMS (23)	-	0.2[a]	1.6[a]	-	1.8[a]	3.9[a]
PAS (23)	-	0.17[a]	2.7[a]	-	1.9[a]	6.8[a]
PCS (23)	-	0.12[a]	1.5[a]	-	1.7[a]	4.5[a]
PFS (23)	-	0.19[a]	2.2[a]	-	1.6[a]	4.3[a]
PBS (23)	-	0.14[a]	1.8[a]	-	1.6[a]	3.8[a]
PMxS (23)	-	0.12[a]	1.8[a]	-	1.6[a]	5.2[a]
PtBS (23)	-	-	1.9[a]	-	-	2.3[a]
PHS (23)	-	-	-	-	-	-
Polysulfones						
PSF (48)	-	0.24	2.1	-	1.6	3.7
TMPSF (48)	-	0.53	2.5	-	1.4	2.7
DMPSF (48)	-	0.29	1.7	-	1.7	3.3
DMPSF-Z (48)	-	0.28	1.9	-	1.5	3.3
6FPSF (49)	-	0.41	2.5	-	1.5	3.6
PSF-F (49)	-	0.2	1.9	-	1.5	1.8
PSF-O (49)	-	0.21	1.9	-	1.5	1.7
TMPSF-F (50)	-	0.63	4.6	-	1.4	2.9
TM6FPSF (50)	-	0.91	4	-	1.4	2.7
Polyesters						
PPha-tere (36)	-	0.5[b]	2.93[b]	-	1.25[b]	3.5[b]
PPha-50:50 (36)	-	-	2.78[b]	-	-	3.3[b]
PPha-iso (36)	-	-	2.65[b]	-	-	3.2[b]
PAr (52)	-	-	2.4	-	-	3.2
TClPar (52)	-	-	3.8	-	-	3.2
TBrPPha-tere (52)	-	0.75	5.6	-	1.07	3.6
Polypyrrolone						
6FDA-TADPO (47)	0.091	1.04	4.41	0.054	1.3	2.6
Polyetherketones						
12H (51)	-	-	1.56	-	-	3.78
6H6F (51)	-	-	1.70	-	-	3.52
6F6H (51)	-	-	1.73	-	-	3.48
12F (51)	-	-	2.12	-	-	3.6

Appendix 2—continued

Polymer family and abbreviation	S_{He} cc/cc atm 10 atm	S_{O_2} cc/cc atm 2 atm	S_{CO_2} cc/cc atm 10 atm	$\dfrac{S_{He}}{S_{CH_4}}$ 10 atm	$\dfrac{S_{O_2}}{S_{N_2}}$ 2 atm	$\dfrac{S_{CO_2}}{S_{CH_4}}$ 10 atm
Polyphenyleneoxides						
PPO (39)	-	-	2.22[b]	-	-	2.06[b]
0.36BrPPO (39)	-	-	2.41[b]	-	-	2.01[b]
0.91BrPPO (39)	-	-	2.57[b]	-	-	1.93[b]
1.06BrPPO (39)	-	-	2.76[b]	-	-	1.90[b]
CPPO-22 (45)	-	-	2.74	-	-	2.36
MeCPPO-22 (45)	-	-	2.12	-	-	2.56
Polyimides						
PMDA-ODA (13)	0.053	0.49	3.65	0.056	1.7	3.8
PMDA-MDA (13)	-	-	3.4	-	-	3.6
PMDA-IPDA (13)	0.062	0.73	4.97	0.047	1.7	3.8
PMDA-DAF (13)	0.05	-	2.1	0.082	-	3.4
6FDA-ODA (13)	0.079	1.03	4.89	0.06	1.9	3.7
6FDA-MDA (13)	0.081	0.82	3.96	0.069	1.8	3.4
6FDA-IPDA (13)	0.079	0.9	4.24	0.066	1.8	3.5
6FDA-DAF (13)	0.096	1.2	5.02	0.06	1.8	3.1
6FDA-6F*m*DA (34)	0.085	0.6	2.89	0.105	1.7	3.6
6FDA-6F*p*DA (34)	0.072	1.0	5.99	0.051	1.5	4.2

[a] 1 atm.
[b] 20 atm.

Appendix 3 Diffusivities and Diffusivity Selectivities at 35°C

Polymer family and abbreviation	$D_{He} \times 10^{-10}$ cm²/s 10 atm	$D_{O_2} \times 10^{-10}$ cm²/s 2 atm	$D_{CO_2} \times 10^{-10}$ cm²/s 10 atm	$\dfrac{D_{He}}{D_{CH_4}}$ 10 atm	$\dfrac{D_{O_2}}{D_{N_2}}$ 2 atm	$\dfrac{D_{CO_2}}{D_{CH_4}}$ 10 atm
Polycarbonates						
PC (54)	50000	580	320	750	3.2	4.7
6FPC (54)	110000	1100	670	940	3.2	5.9
TMPC (54)	95000	920	540	1100	4.1	6.6
TM6FPC (54)	180000	3100	2100	590	3.2	7
TBrPC (54)	29000	210	120	2800	4.9	12.2
TBr6FPC (54)	110000	920	640	2000	4.2	11.7
TBr/TBr6FPC (54)	77000	480	410	2400	4.2	12.1
TMPPC (63)	78000	600	1200	1200	4.1	6.9
TMFPC (63)	329000	470	620	880	4.2	6.8
SBIPC (63)	73000	1100	3000	430	3.9	5.3
NBPC (27)	-	440	280	-	3.3	4.8
PCZ (27)	-	180	97	-	3.8	4.7
BCPC (27)	-	400	230	-	3.3	2.5
Polystyrenes						
PS (23)	-	1650[a]	880[a]	-	3.3[a]	5.5[a]
PαMS (23)	-	210[a]	74[a]	-	4.6[a]	4.9[a]
PMS (23)	-	2810[a]	1370[a]	-	2.7[a]	3.5[a]
PAS (23)	-	1370[a]	460[a]	-	2.5[a]	2.4[a]
PCS (23)	-	780[a]	220[a]	-	3.2[a]	3.7[a]
PFS (23)	-	1760[a]	580[a]	-	3.2[a]	3.7[a]
PBS (23)	-	1040[a]	360[a]	-	3.6[a]	4.7[a]
PMxS (23)	-	1630[a]	810[a]	-	2.6[a]	4.0[a]
PtBS (23)	-	-	5580[a]	-	-	3.7[a]
PHS (23)	-	-	-	-	-	-
Polysulfones						
PSF (48)	-	440	200	-	3.6	5.9
TMPSF (48)	-	800	640	-	3.8	8.1
DMPSF (48)	-	170	94	-	4.2	9.1
DMPSF-Z (48)	-	110	56	-	4.8	10
6FPSF (49)	-	630	360	-	3.5	5.5
PSF-F (49)	-	420	180	-	3.7	5.7
PSF-O (49)	-	410	170	-	3.7	5.6
TMPSF-F (50)	-	400	250	-	3.8	7.7
TM6FPSF (50)	-	1500	1400	-	3.3	8.9
Polyesters						
PPha-tere (36)	-	440[b]	446[b]	-	2.9[b]	7.8[b]
PPha-50:50 (36)	-	-	371[b]	-	-	8.3[b]
PPha-iso (36)	-	-	218[b]	-	-	8.8[b]
PAr (52)	-	-	450	-	-	7.5
TClPar (52)	-	-	530	-	-	15.1
TBrPPha-tere (52)	-	420	1400	-	3.8	25.5
Polypyrrolone						
6FDA-TADPO (47)	74300	577	476	3020	5.06	19.4
Polyetherketones						
12H (51)	-	-	229	-	-	5.5
6H6F (51)	-	-	358	-	-	6.5
6F6H (51)	-	-	353	-	-	6.5
12F (51)	-	-	411	-	-	6.7

Appendix 3—continued

Polymer family and abbreviation	D_{He} ×10^{-10} cm²/s 10 atm	D_{O_2} ×10^{-10} cm²/s 2 atm	D_{CO_2} ×10^{-10} cm²/s 10 atm	$\dfrac{D_{He}}{D_{CH_4}}$ 10 atm	$\dfrac{D_{O_2}}{D_{N_2}}$ 2 atm	$\dfrac{D_{CO_2}}{D_{CH_4}}$ 10 atm
Polyphenyleneoxides						
PPO (39)	-	-	1730[b]	-	-	8.2[b]
0.36BrPPO (39)	-	-	1640[b]	-	-	8.9[b]
0.91BrPPO (39)	-	-	2010[b]	-	-	10.0[b]
1.06BrPPO (39)	-	-	2990[b]	-	-	9.1[b]
CPPO-22 (45)	-	-	610	-	-	8.27
MeCPPO-22 (45)	-	-	645	-	-	6.64
Polyimides						
PMDA-ODA (13)	11500	95	56	2441	3.52	11.9
PMDA-MDA (13)	-	-	90	-	-	12
PMDA-IPDA (13)	45500	739	410	876	3.85	7.9
PMDA-DAF (13)	2860	-	5.4	11432	-	21.7
6FDA-ODA (13)	49500	320	358	2262	2.75	16.3
6FDA-MDA (13)	46900	426	370	1694	3.19	13.4
6FDA-IPDA (13)	68500	636	538	1564	3.14	12.1
6FDA-DAF (13)	78000	497	488	2608	3.47	16.3
6FDA-6FmDA (34)	-	224	134	-	3.93	17.7
6FDA-6FpDA (34)	-	1250	811	-	3.18	9.6

[a] 1 atm.
[b] 20 atm.

Literature Cited

1. Koros, W. J. 1990. *Membrane Separation Systems—A Research Development Needs Assessment*, Vol. 2, pp. 3:1–47. Washington, DC: Dept. Energy Office Energy Res.
2. *Membr. Separation Technol. News* 9: (7)9
3. Bhide, B. D., Stern, S. A. 1991. *J. Memb. Sci.* 62: 13–36
4. Bhide, B. D., Stern, S. A. 1991. *J. Memb. Sci.* 62: 37–58
5. Pinnau, I., Koros, W. J. 1991. *Polymeric Gas Separation Membranes*, ed. C. A. Martin. Boca Raton, FL: CRC. In press
6. Baker, R. W. 1990. See Ref. 1, pp. 1:1–46
7. Koros, W. J., Fleming, G., Jordan, S. M., Kim, T. H., Hoehn, H. 1988. *Progr. Polym. Sci.* 13: 339–401
8. Koros, W. J. 1985. *J. Polym. Sci. Polym. Phys.* 23: 1611–28
9. Wilhelm, E., Battino, R. 1973. *Chem. Rev.* 73: 1–9
10. Baker, R. W. 1990. See Ref. 1, pp. 2:1–37
11. Koros, W. J., Paul, D. R. 1986. *Synthetic Membranes*, ed. Ma. B. Chenoworth. New York: Michigan Molec. Inst. Symp. Ser. Vol. 5
12. Breck, D. W. 1974. *Zeolite Molecular Sieves*. New York: Wiley & Sons. 636 pp.
13. Kim, T. H. 1989. *Gas Sorption and Permeation in a Series of Aromatic Polyimides*. PhD thesis. Univ. Texas, Austin. 150 pp.
14. Bixler, H. J., Sweeting, O. J. 1971. *The Science and Technology of Polymer Films*, ed. O. J. Sweeting, 2: 1–71. New York: Wiley & Sons
15. Chern, R. T., Koros, W. J., Sander, E., Chen, S. H., Hopfenburg, H. B. 1983. *Ind. Gas Sep. ACS Sym. Ser. 223*, ed. T. E. Whyte, C. M. Yon, E. H. Wagerner, pp. 47–75. Washington, DC: Am. Chem. Soc. 292 pp.
16. Hoehn, H. H., Richter, J. W. 1980. *US Patent No. 30,351*
17. Pye, D. G., Hoehn, H. H., Panar, N. 1976. *J. Appl. Polym. Sci.* 20: 287–301
18. Pye, D. G., Hoehn, H. H., Panar, N. 1976. *J. Appl. Polym. Sci.* 20: 1921–31
19. Pilato, L., Litz, L., Hargitay, B., Osborne, R., Farnham, C., et al. 1975. *Am. Chem. Soc. Prep.* 16: 42
20. Koros, W. J., Hellums, M. W. 1989. *Gas Separation Membrane Material Selection Criteria: Differences for Weakly and Strongly Interacting Feed Components*, 5th Internat. Conf. Fluid Prop. Phase Equilib. *Fluid Phase Equilib.* 53: 339
21. Masuda, T., Iguchi, Y., Tang, B., Higashimura, T. 1988. *Polymer* 29: 2041–49
22. Chern, R. T. 1990. *Sep. Sci. Technol.* 25: 1325–38
23. Puleo, A. C., Muruganandam, N., Paul, D. R. 1989. *J. Polym. Sci. Polym. Phys.* 27: 2385–2406
24. Muruganandam, N., Koros, W. J., Paul, D. R. 1987. *J. Polym. Sci. Polym. Phys.* 27: 1999–2026
25. Hellums, M. W., Koros, W. J., Husk, G. R., Paul, D. R. 1989. *J. Memb. Sci.* 46: 93–112
26. Hellums, M. W., Koros, W. J., Schmidhauser, J. 1991. *J. Memb. Sci.* In press
27. McHattie, J. S., Koros, W. J., Paul, D. R. 1991. *J. Polym. Sci. Polym. Phys.* 29: 731–46
28. Schmidhauser, J. C., Longley, K. L. 1990. *J. Appl. Polym. Sci.* 39: 2083–96
29. Schmidhauser, J., Longley, K. 1990. *Barrier Polymer Structure*, ACS Symp. Ser., 423: 159–76
30. Hoehn, H. 1985. *Material Science of Synthetic Membranes*, ACS Symp. Ser., ed. D. R. Lloyd, 269: 81. Washington, DC: Am. Chem. Soc.
31. Kim, T. H., Koros, W. J., Husk, G. R., O'Brien, K. C. 1988. *J. Membr. Sci.* 37: 45–62
32. Kim, T. H., Koros, W. J., Husk, G. R. 1989. *J. Membr. Sci.* 46: 43–56
33. Stern, S. A., Mi, Y., Yamamoto, H., St. Clair, A. 1989. *J. Polym. Sci. Polym. Phys.* 27: 1887–1909
34. Coleman, M. R., Koros, W. J. 1990. *J. Membr. Sci.* 50: 285–97
35. Sykes, G. F., St. Clair, A. K. 1986. *J. Appl. Polym. Sci.* 32: 3725–35
36. Sheu, F. R., Chern, R. T. 1989. *J. Polym. Sci. Polym. Phys.* 27: 1121–33
37. White, D. M. 1974. *Am. Chem. Soc. Polym. Prep.* 15: 210–15
38. Chern, R. T., Brown, N. F. 1990. *Macromolecules* 28: 2370–75
39. Chern, R. T., Jia, L., Shimoda, S., Hopfenberg, H. B. 1990. *J. Membr. Sci.* 48: 333–41
40. Chalk, A. J., Hay, A. S. 1969. *J. Polym. Sci. Part A* 7: 691–705
41. Xie, S., MacKnight, W. J., Karasz, F. E. 1984. *J. Polym. Sci.* 29: 2679–82
42. Huang, Y., Cong, G., MacKnight, W. J. 1986. *Macromolecules* 19: 2267–73
43. Weinkauf, D. H., Paul, D. R. 1991. *J. Polym. Sci. Polym. Phys.* 27: 329–40

44. Story, B. J., Koros, W. J. 1991. *J. Membr. Sci.* In press
45. Story, B. J. 1989. *Sorption and Transport of CO_2 and CH_4 in Chemically Modified Poly(phenylene oxide)*. PhD thesis. Univ. Texas, Austin. 216 pp.
46. Gebbens, B., Mulder, M. H. V., Smolder, C. A. 1989. *J. Membr. Sci.* 46: 29–41
47. Walker, D. R. B., Koros, W. J. 1991. *J. Membr. Sci.* 55: 99–117
48. McHattie, J. S., Koros, W. J., Paul, D. R. 1990. *Polymer* 32: 840–50
49. McHattie, J. S., Koros, W. J., Paul, D. R. 1990. *Polymer* 32: 2618–25
50. McHattie, J. S., Koros, W. J., Paul, D. R. 1990. *Polymer*. In press
51. Mohr, J. M., Paul, D. R., Tullos, G. L., Cassidy, P. E. 1990. *Polymer* 32: 2387–94
52. Chern, R. T., Provan, C. N. 1991. *J. Membr. Sci.* 59: 293–304
53. Yamamoto, H., Mi, Y., Stern, S. A., St. Clair, A. 1990. *J. Polym. Sci. Polym. Phys.* 28: 2291–2304
54. Hellums, M. W., Koros, W. J., Husk, G. R., Paul, D. R. 1991. *J. Appl. Polym. Sci.* In press
55. Hoehn, H. H. 1974. *US Patent No. 3,822,202*
56. Koros, W. J. 1990. *Barrier Polymer Structure: Overview ACS Symp. Ser.*, ed. W. J. Koros, 423: 1. Washington, DC: Am. Chem. Soc.
57. Lee, W. M. 1980. *Polym. Eng. Sci.* 20: 65
58. Weinkauf, D. H. 1991. *Gas Transport Properties of Liquid Crystalline Polymers*. PhD thesis. Univ. Texas, Austin. 210 pp.
59. Barrer, R. M. 1937. *Nature* 140: 106
60. Van Amarongen, G. J. 1964. *Rubber Chem. Technol.* 37: 1065
61. Crank, J., Park, G. S., eds. 1971. *Diffusion in Polymers*, pp. 41–73. New York: Wiley & Sons. 452 pp.
62. Meares, P. 1965. *Polymers: Structure and Bulk Properties*, ed. D. Van Nordstrand. London: D. Van Nordstrand. 381 pp.
63. Hellums, M. W. 1990. *Gas Permeation and Sorption in a Series of Polycarbonates*. PhD thesis. Univ. Texas, Austin. 209 pp.
64. Shah, V. M., Stern, S. A., Ludovice, P. J. 1989. *Macromolecules* 22: 4660–62
65. Spillman, R. W. 1989. *Chem. Eng. Prog.* 85: 41
66. Van Krevelen, D. W., Hoftyzer, P. J. 1976. *Properties of Polymers: Their Estimation and Correlation with Chemical Structure*, pp. 51–79. New York: Elsevier. 2nd ed.
67. Paul, D. R. 1984. *J. Membr. Sci.* 18: 75–86
68. Paul, D. R., Newman, S., eds. 1978. *Polymer Blends*, Vol. 1, pp. 445–89. New York: Academic
69. Ranby, B. G. 1975. *J. Polym. Sci. Polym. Lett.* 51: 89
70. Coleman, M. R., Kohn, R. 1990. *US Patent No. 505,099*
71. Hoy, K. L. 1969. *Tables of Solubility Parameters*. Danbury, Conn: Union Carbide Corp.
72. Yee, A. F., Smith, S. A. 1981. *Macromolecules* 14: 54–64
73. Li, G. S. 1986. *US Patent No. 4,586,939*
74. Sanders, E. S. 1988. *J. Membr. Sci.* 37: 63–80
75. Jordan, S. M., Fleming, G. K., Koros, W. J. 1989. *J. Membr. Sci.* 30: 191–212
76. Jordan, S. M., Koros, W. J., Beasley, G. 1989. *J. Membr. Sci.* 43: 103–20
77. Jordan, S. M., Fleming, G. K., Koros, W. J. 1990. *J. Polym. Sci. Polym. Phys.* 28: 2305–27
78. Koros, W. J., Walker, D. R. B. 1991. *Polym. J.* 23: 481–90

DESIGN AND PROPERTIES OF GLASS-CERAMICS

G. H. Beall

Corning Incorporated, Sullivan Park FR-51, Corning, New York 14831

KEY WORDS: glass-ceramics,   microstructure,   nucleation,   composition, properties

INTRODUCTION

Glass-ceramics are microcrystalline solids produced by the controlled devitrification of glass. Glasses are melted, fabricated to shape, and then converted by heat treatment to a predominantly crystalline ceramic. The basis of controlled crystallization lies in efficient internal nucleation (1), which allows development of fine, randomly oriented grains without voids, microcracks, or other porosity.

A unique manufacturing advantage of glass-ceramics over conventional ceramics is the ability to use high-speed plastic forming processes developed in the glass industry (e.g. pressing, blowing, rolling, etc.) to create complex shapes essentially free of internal inhomogeneities. Because glass-ceramic compositions are designed to crystallize, however, they cannot be held for long periods at temperatures below the liquidus during the forming process. Therefore, the viscosity at the liquidus temperature is critical both in the choice of a forming process and in the choice of a glass composition.

The properties of glass-ceramics depend upon both composition and microstructure. The bulk chemical composition controls the ability to form a glass and its degree of workability. In order to achieve internal nucleation, suitable nucleating agents are melted into the glass. Bulk composition also directly determines the potential crystalline phase assemblage, and this in turn governs the general physical and chemical characteristics, e.g. hardness, density, acid resistance, etc. Secondly, but equally important, is the importance of microstructure. Microstructure is the key to most mechanical and optical properties; it can promote or diminish the role of the key crystals in the glass-ceramic. It is therefore necessary to characterize glass-ceramics in terms of both composition and microstructure.

COMPOSITION

The range of glass-ceramic compositions is extremely broad and requires only the ability to form a glass and to control its crystallization through internal nucleation. Phase assemblage is restricted, however, to those metastable crystals that can form from glass and to the thermodynamically stable mix of crystals governed by the laws of phase equilibria. Commercial glass-ceramics are generally based on silica-containing glasses and can be divided into three groups: silicates, aluminosilicates, and fluosilicates.

Silicate Glass-Ceramics

The silicate glass-ceramics are composed primarily of alkali and alkaline-earth silicate crystals whose properties dominate that of the glass-ceramic. Among the most important are lithium silicate, both the metasilicate (Li_2SiO_3) and the disilicate ($Li_2Si_2O_5$), magnesium metasilicate ($MgSiO_3$-enstatite), calcium-magnesium metasilicate ($CaMgSi_2O_6$-diopside), and calcium metasilicate ($CaSiO_3$-wollastonite). Compositions of some commercial silicate glass-ceramics are given in Table 1.

Table 1 Commercial silicate glass-ceramics

		Fotoform/Fotoceram Corning 8603 (wt%)		Russia Slag-sitall white (wt%)	Hungary Minelbite gray (wt%)
SiO_2		79.6	SiO_2	55.5	60.9
Al_2O_3		4.0	Al_2O_3	8.3	14.2
Li_2O		9.3	CaO	24.8	9.0
K_2O		4.1	MgO	2.2	5.7
Na_2O		1.6	Na_2O	5.4	3.2
			K_2O	0.6	1.9
Ag, Au	n	0.11, 0.001	ZnO	1.4	
			MnO	0.9	2.0
CeO_2, SnO_2	s	0.014, 0.003	Fe_2O_3 (n)	0.3	2.5
			S	0.4	0.6
Sb_2O_3	f	0.4	Crystal phases	Wollastonite $CaSiO_3$	Diopside $CaMgSi_2O_6$

[a]As analyzed at Corning Glass Works. n = nucleant; s = sensitizer; f = fining agent.

LITHIUM SILICATES Lithium silicate glass-ceramics consist of two composition groups, both of commercial importance. The first group, nucleated with P_2O_5, develops high expansion glass-ceramics, which match the thermal expansion of several nickel-based superalloys and are used in a variety of high-strength hermetic seals, connectors, and feed-throughs (2). The second group, photosensitively nucleated by colloidal silver, produces a variety of chemically machined materials that are useful as fluidic devices, cellular display screens, lens arrays, magnetic recording head pads, and charged plates for ink jet printing.

Lithium disilicate glass-ceramics nucleated with P_2O_5 are characterized by high body strength, 140–210 MPa, good fracture toughness, ~ 3 MPa $m^{1/2}$, and moderate to high thermal expansion coefficient, 80–130 $\times 10^{-7}/°C$. The compositions typically comprise 70–85 wt% SiO_2, 10–15 Li_2O, 3–10 Al_2O_3, 1–5 P_2O_5, as well as a minor amount of other modifiers including K_2O, Na_2O, CaO, and ZnO. The glasses phase separate on heat treatment and lithium orthophosphate (Li_3PO_4) precipitates as the first crystal phase. Lithium metasilicate and/or lithium disilicate then form, the latter predominating with further heat treatment. Cristobalite and β-spodumene are often auxiliary phases, and residual glass is usually present in excess of 15 vol%.

The dielectric properties of lithium disilicate glass-ceramics are surprisingly good, with dielectric constants below 6 and loss tangents below 0.01 over a wide range of temperature and frequency. They have, therefore, found application as electrical insulators under the GE trademark Re-X®.

Photosensitive lithium silicate glass-ceramics contain metals, namely 0.1% Ag and 0.001% Au, which can be precipitated thermally after ultraviolet sensitization (3). Cerous ions act as an optical sensitizer in this process: $(Au, Ag)^+ + Ce^{3+} \xrightarrow{uv} (Au, Ag)^0 + Ce^{4+}$. The metallic colloids thereby produced nucleate, a dendritic form of lithium metasilicate, which is far more easily etched in hydrofluoric acid than is the parent glass, thus allowing an irradiated pattern to be selectively removed. The resulting photo-etched glass can then be flood-exposed to ultraviolet rays and heat-treated beyond the temperature region of metastable lithium metasilicate. The stable lithium disilicate phase is then produced, and the resulting glass-ceramic is strong (~ 140 MPa), tough, and faithfully replicates the original photo-etched pattern.

CALCIUM SILICATES Calcium silicate glass-ceramics have found application because of their high hardness and low cost. For over two decades, glass-ceramics based on blast furnace slags have been produced in eastern Europe, particularly in Russia and in Hungary (4). These materials are usually rolled as sheet or cast as tiles and are used for both interior and exterior wall cladding and flooring. Referred to as slag-sitall in Russia, these glass-ceramics presently constitute the largest volume applications for crystallized glass. Melted near 1450°C, they are formed into glass sheet and subsequently heat-treated to a maximum temperature near 1000°C. Diopside, $CaMgSi_2O_6$, or wollastonite, $CaSiO_3$, are the major phases precipitated upon sulfide nuclei of zinc, manganese, or iron present in the original slag. The abrasion resistance and good chemical durability are provided by the alkaline-earth silicate crystals, which are present as fine (1–5 μm) equiaxial grains in a matrix of aluminosilicate residual glass. Recently, attractive translucent architectural panels of related wollastonite glass-ceramics have been manufactured by Nippon Electric Glass and sold in Japan and the United States under the trade name Neoparium®.

MAGNESIUM SILICATES Although no commercial applications have yet been found, glass-ceramics based on enstatite ($MgSiO_3$) (5, 6) are interesting because this phase undergoes a martensitic transformation on cooling, which produces toughening from fracture energy absorption by fine, lamellar twinning. Unfortunately, enstatite does not form a stable glass, so compositions must be diluted with other glass-forming components. Nevertheless, refractory, tough and fine-grained glass-ceramics have been produced in the SiO_2-MgO-ZrO_2 and SiO_2-MgO-Al_2O_3-Li_2O-ZrO_2 systems. These materials contain from 50–85 wt% enstatite, with auxiliary phases zircon, β-spodumene, minor tetragonal zirconia, and small amounts of glass. Representative compositions from each system are listed in Table 2 along with crystallization schedule, phase assemblage, and key properties.

The sequence of crystallization involves a key role of the nucleating agent zirconia. Phase separation occurs between 800–900°C and is rapidly followed by the crystallization of tetragonal zirconia upon which the enstatite forms above 900°C. Zirconia reacts with silica to form zircon ($ZrSiO_4$) at temperatures above 1200°C. The zirconia component is also effective in improving the glass stability, but it is not believed to play any role in the roughening mechanics. It is largely in the form of zircon in the toughest compositions.

The toughest (~ 4 MPa m$^{1/2}$) and most refractory compositions are basically two-phase enstatite zircon glass-ceramics. These have upper use temperatures approaching 1525°C, the minimum ternary eutectic in the SiO_2-MgO-ZrO_2 phase diagram.

Aluminosilicate Glass-Ceramics

Aluminosilicate glass-ceramics are of great commercial interest because they combine exceptional thermal-dimensional stability with good chemi-

Table 2 Enstatite glass-ceramics: composition and properties

	E-1	E-2
SiO_2	58.0	54.0
Al_2O_3	5.4	—
MgO	25.0	33.0
Li_2O	0.9	—
ZrO_2	10.7	13.0
Glass crystallization treatment	800°C/2 hr 1200°C/4 hr	800°C/2 hr 1400°C/4 hr
Phase assemblage	Enstatite (proto, clino), β-spodumene, tet. zirconia	Enstatite (proto, clino), zircon, minor tet. zirconia, cristobalite
Abraded M.O.R. (MPa)	193 ± 15	200 ± 15
Fracture toughness (MPa m$^{1/2}$)	3.5 ± 0.4	4.6 ± 0.6
Refractoriness	1250°C	1500°C
C.T.E. (0–1000°C)	68×10^{-7}/°C	80×10^{-7}/°C

cal durability. Strong resistance to thermal shock is based on the very low to moderate thermal expansion coefficients found in many aluminosilicate framework structures, e.g. β-quartz solid solution, β-spodumene solid solution, cordierite ($Mg_2Al_4Si_5O_{18}$), anorthite ($CaAl_2Si_2O_8$), and pollucite ($CsAlSi_2O_6$). The base system Li_2O-Al_2O_3-SiO_2 has produced glass-ceramics of the lowest thermal expansion coefficient based on either β-quartz or β-spodumene (keatite) solid solution crystal phases. The wide compositional range of these solid solutions allows one to tailor an essentially monophase composition with only minor nucleant phases and residual glass.

β-QUARTZ SOLID SOLUTION The β-quartz solid solutions are metastable, hexagonal, crystalline phases of very low coefficient of thermal expansion. The general composition is $(Li_2, R)O \cdot Al_2O_3 \cdot nSiO_2$, where n varies from 2–10 and R is a divalent cation normally Mg^{2+} or Zn^{2+}. Near the stoichiometric end-member composition, $LiAlSiO_4$, this phase is thermodynamically stable and is often referred to as β-eucryptite. In commercial glass-ceramic compositions, however, n is in the range of 6–8 because it is in this siliceous area that meltable glasses of sufficient viscosity at the liquidus to be rolled, pressed, blown, and vacuum formed are found (7).

The β-quartz solid solution structure is metastable and will break down to other phases like β-spodumene if heated above 900°C. Substitutions of MgO and ZnO for Li_2O and $AlPO_4$ for SiO_2 are useful in reducing both batch cost and liquidus temperature. The combination of TiO$_2$ and ZrO$_2$ appears most effective in nucleation of β-quartz and is used in most commercial glass-ceramics. When used at levels near 2 mol% each, ZrTiO$_4$ crystalline nuclei are observed to precipitate, and they allow very fine β-quartz crystals, less than 100 nm, to develop, which produce a transparent highly crystalline body. Both fine crystal size and low birefringence inherent in the β-quartz solid solutions allow light scattering to be minimized. The combination of transparency, near zero thermal expansion behavior, optical polishability, and strength greater than glass has generated applications such as cookware, telescope mirror blanks, woodstove windows, and infrared-transmitting range tops. Optically stable platforms and the ring-laser gyroscope are more recent applications.

Table 3 lists the compositions of three commercial β-quartz glass-ceramics from different manufacturers and their areas of application. The ingredients that compose the crystal phase are separated from those that concentrate in the residual glass, which makes up less than 10 vol% of the

Table 3 Composition of transparent glass-ceramics based on β-quartz solid solution (wt%)

		VISION® Corning	ZERODUR® [a] Schott	NARUMI® Nippon Electric
SiO$_2$	xl	68.8	55.5	65.1
Al$_2$O$_3$		19.2	25.3	22.6
Li$_2$O		2.7	3.7	4.2
MgO		1.8	1.0	0.5
ZnO		1.0	1.4	
P$_2$O$_5$			7.9	1.2
F				0.1
Na$_2$O	gl	0.2	0.5	0.6
K$_2$O		0.1		0.3
BaO		0.8		
TiO$_2$	n	2.7	2.3	2.0
ZrO$_2$		1.8	1.9	2.3
As$_2$O$_3$	f	0.8	0.5	1.1
Fe$_2$O$_3$	c	0.1	0.03	0.03
CoO		50 ppm		
Cr$_2$O$_3$		50 ppm		
		Transparent cookware	Telescope mirrors	Rangetops, stove windows

[a] As analyzed at Corning Glass Works. xl, oxides concentrated in crystal; gl, oxides concentrated in glass; n, nucleating-agent oxides; f, fixing-agent oxide; c, colorant oxides.

body. The latter, generally tramp constituents like soda and potash, form a persistent aluminosilicate glassy phase along grain boundaries. Arsenic oxide is added as a refining agent to purge gas bubbles, and various colorant transition metal oxides are present to provide, in concert with titania, a brownish tint.

β-SPODUMENE SOLID SOLUTION   Opaque glass-ceramics are readily achieved in the system $Li_2O\text{-}Al_2O_3\text{-}SiO_2$ by crystallizing heterogeneous-nucleated glasses at relatively high temperatures (1000–1200°C) and allowing development of the stable crystalline assemblage, which generally includes β-spodumene solid solution as the main phase. This tetragonal crystal, like its hexagonal predecessor, has a very low thermal expansion coefficient (8). It is a stuffed derivative of the silica polymorph keatite. Its composition varies from $Li_2O \cdot Al_2O_3 \cdot 4SiO_2$ to $Li_2O \cdot Al_2O_3 \cdot 10SiO_2$, which allows a wide range of parent glass viscosities. Figure 1 illustrates the thermal expansion behavior of these solid solutions and indicates how this parameter can be carefully adjusted by choice of composition. Considerable substitution of magnesium for lithium is permitted in the β-spodumene structure ($Mg^{2+} \to 2Li^+$), although this is less than is allowed in the metastable β-quartz precursor. The transformation from quartz to spodumene

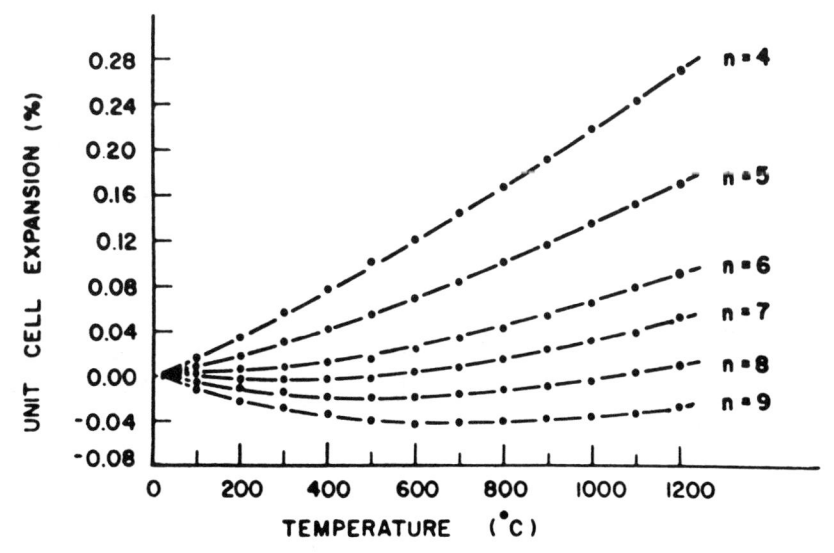

Figure 1   Thermal expansion of β-spodumene solid solutions $Li_2O \cdot Al_2O_3 \cdot nSiO_2$ (after Ostertag et al, 8).

usually occurs between 900 and 1000°C, is irreversible, and is accompanied by an increase in grain size (usually five- to tenfold). When TiO$_2$ is used as a nucleating agent, rutile development accompanies the silicate phase transformation. Because of the high refractive index and birefringence of rutile, a high degree of opacity is developed.

Typically, β-spodumene glass-ceramic grains are in the 1–2 μm range. Secondary grain growth is sluggish and generally linear with the cube root to time. The resistance to grain growth is particularly important in view of marked anisotropy in the thermal expansion of β-spodumene crystals (8).

Table 4 lists the compositions of two commercial β-spodumene glass-ceramics, one used for cookware and one for ceramic regenerators in turbine engines. The former shows a multicomponent glass containing titania as the basic nucleation agent. Magnesia partially substitutes for lithia to lower the liquidus temperature, thereby allowing sufficient viscosity for pressing, blowing, and tube drawing (see curve D, Figure 2). This glass-ceramic is crystallized with a maximum temperature near 1125°C, about 100°C below the liquidus. It is highly crystalline (>93%) and contains β-spodumene as the dominant phase with minor spinel, rutile, and glass also present. The thermal expansion coefficient is 12 × 10^{-7}/°C (0–500°C), and the abraded flexural strength is about 100 MPa.

Table 4 Composition of glass-ceramics based on β-spodumene solid solution

		CORNING WARE®		CERCOR® Corning	
		(wt%)	(mol%)	(wt%)	(mol%)
SiO$_2$	xl	69.7	73.6	72.5	75.9
Al$_2$O$_3$		17.8	11.0	22.5	13.6
Li$_2$O		2.8	5.9	5.0	10.5
MgO		2.6	4.1		
ZnO		1.0	0.8		
Na$_2$O	gl	0.4	0.4		
K$_2$O		0.2	0.1		
TiO$_2$	n	4.7	3.7		
ZrO$_2$		0.1	0.1		
Fe$_2$O$_3$	c	0.1	0.1		
As$_2$O$_3$	f	0.6	0.2		
		Cookware, hot plates		Heat exchangers, regenerators	

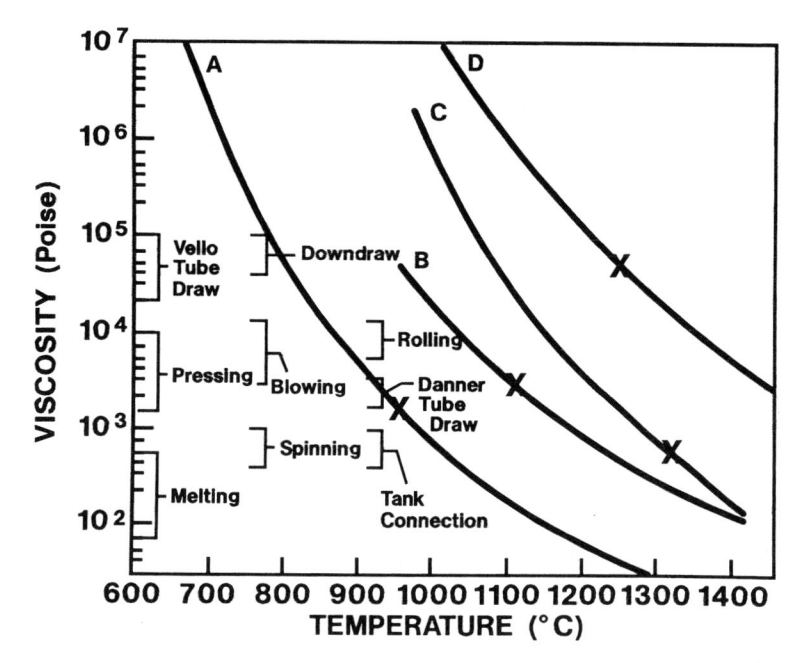

Figure 2 Temperature-viscosity curves for some commercial glass-ceramic-forming glasses (× marks the liquidus temperature).

The regenerator glass-ceramic is produced as a honeycomb product for a turbine engine heat exchanger. It is made from a powdered glass frit, which as a slurry can impregnate paper that is wound as alternately corrugated sheets on a wheel. After firing, a porous ceramic regenerator wheel is produced that allows energy to be transferred from hot exhaust gases to the cold intake air in the turbine engine. The very low thermal expansion [$\sim 5 \times 10^{-7}/°C$ (0–1000°C)], and high thermal stability (>1200°C) are important in this application.

CORDIERITE Glass-ceramics based on the hexagonal form of cordierite are strong, have excellent dielectric properties, good thermal stability, and thermal shock resistance. The commercial composition given in Table 5 is the standard glass-ceramic used for missile nose cones. It is a multiphase material nucleated with titania, but the major constituent is cordierite ($Mg_2Al_4Si_5O_{18}$) with some solid solution toward "Mg-beryl" (i.e. $Mg^{2+} + Si^{4+} \rightarrow 2Al^{3+}$). This phase is mixed with cristobalite, rutile, magnesium dititanate, and minor glass, which is isolated at grain-boundary

Table 5 Commercial cordierite glass-ceramic (Corning 9606)

Composition		wt%	mol%	Phases
SiO$_2$	xl	56.1	58.1	
Al$_2$O$_3$		19.8	12.1	Cordierite
MgO		14.7	22.6	Cristobalite
				Rutile
CaO		0.1	0.1	Mg-dititanate
TiO$_2$	n	8.9	6.9	
As$_2$O$_3$	f	0.3	0.1	
Fe$_2$O$_3$		0.1	0.1	
		Use: Radomes		

nodes. The mechanical properties of these glass-ceramics have been studied extensively (9). A Weibull plot of flexural strength data on transverse-ground bars hewn from a slab of this commercial composition is shown in Figure 3, which illustrates the narrow range and predictability of strength. Other important properties include coefficient of thermal expansion (0–700°C) $45 \times 10^{-7}/°C$; fracture toughness (K_{1C}) 2.2 MPa m$^{1/2}$; thermal conductivity 0.09 cal/s·cm·°C; Knoop hardness 700; dielectric constant; and loss tangent at 8.6 GHz: 5.5 and 0.0003, respectively.

The choice of composition in this case was based primarily on glass-forming considerations. If thermal stability and shock resistance were the only important factors, a stoichiometric cordierite base would have been chosen. However, to optimize viscosity at the liquidus, the lowest ternary eutectic in the refractory system, MgO-Al$_2$O$_3$-SiO$_2$, was approached. This ternary eutectic, cordierite-enstatite-cristobalite, has an equilibrium temperature of 1355°C. The addition of 9% TiO$_2$, which is required for internal nucleation, further decreases the liquidus to 1330°C, where sufficient viscosity is achieved for centrifugal casting or spinning (see curve C, Figure 2).

One result of the choice of the eutectic composition is that significant cristobalite had to be incorporated into the glass-ceramic, which had the adverse effect of raising its thermal expansion. This free silica incorporation, however, allows a post-ceram surface leaching treatment with hot caustic to produce a porous skin that tends to prevent initiation of flaws and further enhances strength.

OTHER ALUMINOSILICATE SYSTEMS   There are several other aluminosilicate glass-ceramics which, although they are not yet produced commercially, have potentially useful properties. Mullite glass-ceramics can be produced

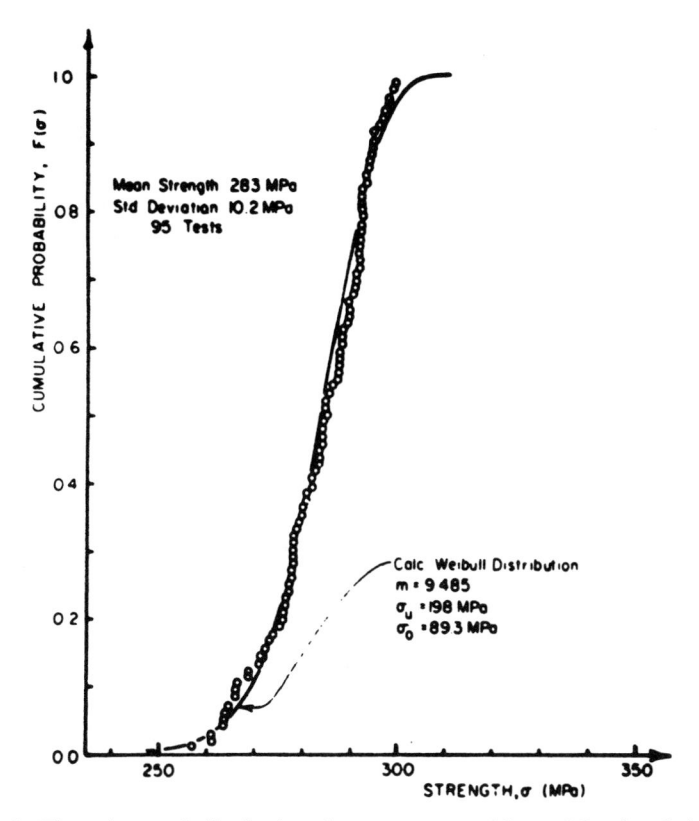

Figure 3 Flexural strength distribution of transverse ground bars of Corning Code 9606 glass-ceramic (after Lewis et al, 9).

in modified binary Al_2O_3-SiO_2 glasses. Phase separation in this system allows extremely fine self-nucleation of mullite. Transparent glass-ceramics that show efficient near-infrared luminescent characteristics when doped with Cr^{+3} can be produced (10). Solar collector and laser applications have been investigated.

Calcium aluminosilicate glass-ceramics based on anorthite can be formed from surface crystallization sintering of glass frit. They form unique glass-ceramic matrices for silicon carbide fiber-reinforced composites (11). The good thermal stability (up to 1500°C) of anorthite combined with its near match in thermal expansion with silicon carbide make this composition particularly attractive.

Glass-ceramics in the Cs_2O-Al_2O_3-SiO_2 system are highly refractory (12). Pollucite-mullite glass-ceramics crystallized at 1600°C show thermal stability from beam-bending viscosity tests that are 350° higher than fused silica glass. Unfortunately, these refractory materials must be melted near 1900°C.

Fluosilicate Glass-Ceramics

Fluosilicate glass-ceramics are characterized by unique mechanical properties dependent upon highly anisotropic crystals that take a one- or two-dimensional form. Thus mica glass-ceramics display mechanical machinability, while glass-ceramics based on chain silicates have shown extreme strength and toughness. The monovalent fluorine anion is required in both cases to stabilize the sheet and chain structures.

FLUORMICA SHEET SILICATES   Machinable glass-ceramics are based on internally nucleated fluormica crystals in glass (13). One commercial material has been marketed for twenty years under the trademark MACOR® and has found wide application in such diverse and specialty areas as precision electrical insulators, vacuum feedthroughs, windows for microwave tube parts, samples holders for field ion microscopes, seismograph bobbins, gamma-ray telescope frames, and boundary retainers on the space shuttle. The precision machinability of the MACOR material with conventional metal-working tools, combined with high dielectric strength (~ 40 kV/nm) and very low helium permeation rates, are particularly important in high-vacuum applications.

Although the MACOR glass-ceramic is based on the fluorine-phlogopite phase ($KMg_3AlSi_3O_{10}F_2$), this stoichiometry does not form a glass. The bulk composition had to be altered largely through additions of B_2O_3 and SiO_2 to form a stable although opalized glass (see Table 6). The parent glass is composed of a dispersion of aluminosilicate droplets in a magnesium-rich matrix (14). The crystallization begins near 650°C when a metastable phase chondrodite, $2Mg_2SiO_4 \cdot MgF_2$, forms in the magnesium-rich matrix at the interfaces of the aluminosilicate droplets. The chondrodite subsequently transforms to norbergite, $MgSiO_4 \cdot MgF_2$, which finally reacts with the components in the residual glass to produce fluorphlogopite mica and minor mullite. The mica grows in a preferred lateral direction because the residual glass is fluidized by the B_2O_3 flux and is also designed to be deficient in the crosslinking species potassium.

More recently another commercial material has been developed for use in DICOR® dental restorations (15). This glass-ceramic, with improved chemical durability and translucency over the MACOR material, is based on the tetrasilicic mica, $KMg_{2.5}Si_4O_{10}F_2$. Good strength (~ 150 MPa) is associated with the development of anisotropic flakes at relatively high temperatures ($>1000°C$). Translucency is achieved by roughly matching

Table 6 Commercial fluormica glass-ceramic compositions

	MACOR® (Corning) (wt%)	DICOR® (Dentsply) (wt%)
SiO_2	47.2	56–64
B_2O_3 gl	8.5	
Al_2O_3	16.7	0–2
MgO	14.5	15–20
K_2O	9.5	12–18
F	6.3	4–9
ZrO_2 gl		0–5
CeO_2		0.05
Mica type:		
MACOR	$K_{1-x}Mg_3Al_{1-x}Si_{3+x}O_{10}F_2$	
DICOR	$K_{1-x}Mg_{2.5+x/2}Si_4O_{10}F_2$	
	$x < 0.2$	

crystal and glass indices and maintaining a fine-grained ($\simeq 1$ μm) crystal size. Ceria is added to simulate the fluorescent character of natural teeth.

The unique feature of DICOR dental restorations include the close match to natural teeth in both hardness and appearance. The glass-ceramic may be accurately cast using a lost-wax technique and conventional dental laboratory investment molds. The high strength and low thermal conductivity of the material provide advantages over conventional metal-ceramic systems.

CHAIN-FLUOSILICATES In order to improve the basic body strength of glass-ceramics, polymeric crystals, in which chains of silica tetrahedra form a mineral backbone, have been grown in acicular form in glass (5). These glass-ceramics resemble natural nephrite jade, whose interlocking and acicular microstructure is responsible for its toughness. Two chain-fluosilicates have been identified as capable of producing tough glass-ceramics: potassium fluorrichterite and fluorcanasite.

Potassium fluorrichterite Glass-ceramics with the amphibole potassium fluorrichterite, $KNaCaMg_5Si_8O_{22}F_2$, as the principal crystalline phase display the strength and toughness of a random acicular microstructure in which fractures follow a tortuous path around rod-like crystals. Complex fluosilicate glasses in the system SiO_2-MgO-CaO-Na_2O-K_2O-F, with minor additions of Al_2O_3, P_2O_5, Li_2O, and BaO, have been found to form such glass-ceramics (5). Table 7 lists an optimized composition with its derivative glass-ceramic phases. It was found important to formulate a starting glass substantially richer in silica than that of stoichiometric K-F-richterite. This produces a stable glass with sufficient viscosity at the liquidus to meet high-speed pressing requirements (see curve B, Figure 2). Excess silica in the form of cristobalite also increases the final thermal expansion coefficient of the glass-ceramic, thus enabling a compressive glaze to be applied.

The early stages of nucleation in this glass-ceramic are controlled by an amorphous emulsion in which one of the two phases is evidently close to tetrasilicic fluormica taeniolite ($KMg_2LiSi_4O_{10}F_2$) in composition. Crystallization of this component is rapid near 600°C. Because fluormica incorporates the major fluxes in the system, the crystallization occurs at very high viscosity. Figure 4 shows the relationship between viscosity and phase development during all stages of an optimized 5 hr crystallization cycle of the glass R-1 (Table 7), as determined by the bending beam technique. Mica is seen to begin crystallizing well before the glass softens to 10^{11} P. Precipitation of diopside, $CaMgSi_2O_6$, follows above 700°C, and metastable mica and diopside then react above 850°C with the residual glass to produce the stable assemblage K-F-richterite plus cristobalite, the later growing above 900°C. The minimum viscosity during crystallization is well above 10^{10} P, thus ensuring little or no deformation of the original glass article.

The composition of the residual glass was calculated to be siliceous and enriched in Al_2O_3 and P_2O_5, which indicates strong chemical resistance. Combined with the chemical stability of the constituent crystals, richterite and cristobalite, this allows a glass-ceramic of exceptional chemical durability (16). The stress-corrosion coefficient n was calculated from the slope of flexural strength vs stress rate at 100°C and found to be 38. Figure 5, which plots n vs flexural strength degradation from liquid nitrogen to ambient conditions for several glassy materials, reveals that K-F-richterite glass-ceramics are superior to other materials measured, with the exception of ultra-low-expansion titanium silicate glass.

The coefficient of thermal expansion, enhanced by the presence of cristobalite, is 115×10^{-7}/°C (0–300°C). This high expansion characteristic allows the development of compressive strengths through conventional glazing. Thus the flexural strength of the body (150 ± 15 MPa) is increased to 200 ± 15 MPa. The final glazed material is high-gloss white and displays a translucency similar to bone china. This glass-ceramic is currently being manufactured by Corning, Inc. as high-performance institutional tableware and mugs for the retail CORELLE® line.

Table 7 K-F-richerite glass-ceramic; estimated phase assemblage and composition (wt%)

	Amphibole F-K-richerite KNaCaMg$_5$Si$_8$O$_{22}$F$_2$	~50% FKR	Mica taeniolite KMg$_2$LiSi$_4$O$_{10}$F$_2$	~10% F-T	Silica cristobalite 95SiO$_2$5LiAlO$_2$	~20% Crb.	Total crystals 80%	R-1 bulk (anal.)	Residual glass ~20% (calc.)
SiO$_2$	57.3	28.7	59.3	5.9	95.0	19.0	53.6	67.1	68.6
Al$_2$O$_3$	—	—	—	—	4.0	0.8	0.8	1.8	5.1
MgO$_2$	24.1	12.1	19.9	2.0	—	—	14.1	14.3	1.5
CaO	6.7	3.4	—	—	—	—	3.4	4.7	6.6
Na$_2$O	1.7	1.9	—	—	—	—	1.9	3.0	5.6
K$_2$O	5.7	2.9	11.6	1.2	—	—	4.1	4.8	3.6
Li$_2$O	—	—	3.7	0.4	1.0	0.2	0.6	0.75	0.8
BaO	—	—	—	—	—	—	—	0.3	1.5
P$_2$O	—	—	—	—	—	—	—	1.0	5.1
Sb$_2$O$_3$	—	—	—	—	—	—	—	0.2	1.0
F	4.5	2.3	9.4	1.0	—	—	3.3	3.5	1.0
Total	102.0		103.9					101.5	100.4
	0 ≈ F 2.0		0 ≈ F 3.9					0 ≈ F 1.5	0 ≈ F 0.4
	100.0		100.0					100.0	100.0

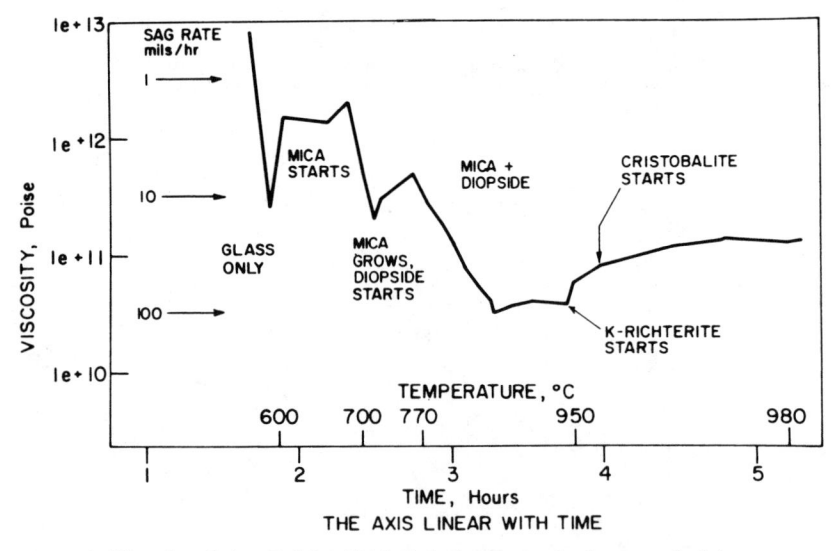

Figure 4 Viscosity of glass R-1 (see Table 7) during its standard ceram schedule.

Fluorcanasite Canasite, $Ca_5Na_4K_2Si_{12}O_{30}(OH,F)_4$, is a rare mineral found in the Kola Peninsula, Russia. The crystal structure is characterized by four silicate chains running parallel to the b-axis cross-linked to form a tubular unit (17). These quadruple chains give the basic structural unit $Si_{12}O_{30}$, a high Si/O ratio for a chain silicate, which suggests glass-forming behavior. Potassium and sodium ions are located centrally in the tubes, which are connected to one another by an edge-shared octahedral network of calcium (predominantly) and sodium oxyfluoride.

Fluorcanasite, $Ca_5K_{2-3}Na_{3-4}Si_{12}O_{30}F_4$, has been synthesized from glasses close to its stoichiometry (5, 18). Internal nucleation is achieved through precipitation of CaF_2 crystallites and sperulitic growth of canasite upon these nuclei. Addition of excess calcium fluoride to the composition allows improved nucleation and a finer-grained glass-ceramic. Thus fluorcanasite glass-ceramics can be highly crystalline and essentially monophase. Canasite glass-ceramics show typical flexural strengths near 300 MPa, a Young's modulus near 80 GPa, and a thermal expansion coefficient of about $125 \times 10^{-7}/°C$. The strain at rupture is approximately 0.35, which is high for a silicate material. The measured fracture toughness of 5.0 MPa $m^{1/2}$ infers a fracture surface energy of approximately 150 J/m^2, higher than that reported for natural jadeite.

Fluorcanasite glass is easy to melt because it has a viscosity curve some 150° lower than soda-lime silicate glass (see curve A, Figure 2). Although

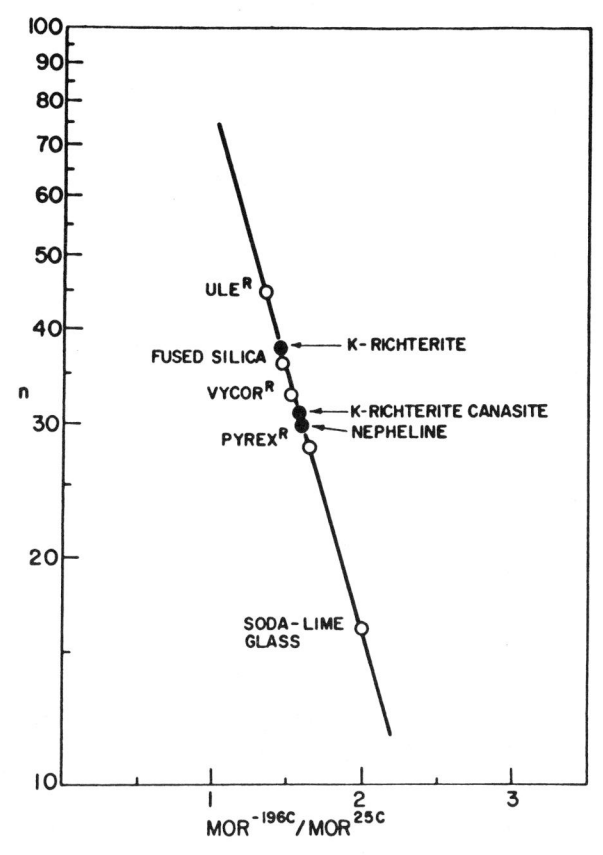

Figure 5 Stress corrosion constant vs MOR ratio for glasses and glass-ceramics (after Gulati et al, 16).

it has a relatively low viscosity at the liquidus, namely, about 2500 P at 950°C, it can still be rolled, pressed, or cast. It can be crystallized below 900°C in less than one hour. Because of its lower silica content, it is inferior in chemical durability to soda-lime glass, but nevertheless is far superior to many other building materials such as marble.

Fluorcanasite glass-ceramics are currently being manufactured by Corning, Inc. as magnetic memory disc substrates. The combination of strength, toughness, polishability, and dimensional stability is important here. This monophase material can be polished without pits, and its microstructure produces an undulating surface that allows the magnetic head to glide very close without actually touching the surface.

Other potential applications include architectural cladding and roofing, interior partitioning, thin housewares, and food packaging materials.

MICROSTRUCTURE

The design of microstructure is responsible for many of the key properties of glass-ceramics. Because crystal development is dependent on internal nucleation and growth in glass, a wide variety of textures can be produced, many of which are totally distinct from normal ceramic microstructures derived from the sintering of powders or fusion cast techniques. Some of the key glass-ceramic microstructures are (*a*) dendritic, (*b*) ultra fine-grained, (*c*) cellular membrane, (*d*) relict, (*e*) coast-and-island, (*f*) house-of-cards, (*g*) acicular interlocking, and (*h*) lamellar twinned.

Dendritic

Dentritic crystallization can result when crystal growth in a glassy medium is accelerated in certain lattice directions or planes. Although the general outline of the dendrite may mimic normal crystal morphology, the internal structure is typically skeletal and a high percent of residual glass remains within the crystallized volume. Figure 6 illustrates the dendritic micro-

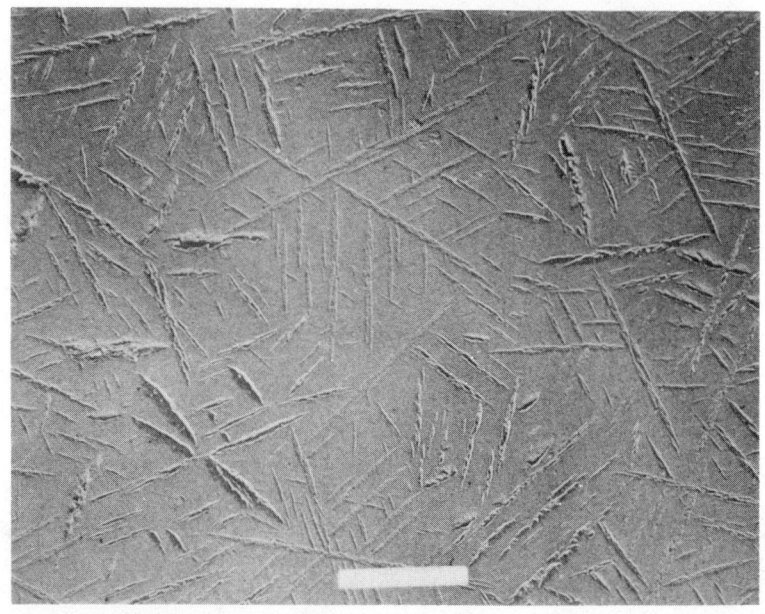

Figure 6 Microstructure of FOTOFORM glass-ceramic as revealed by r.e.m. (bar = 1 μm).

structure of lithium metasilicate crystals in the chemically etchable glass-ceramic available under the trademark FOTOFORM®. The hexagonal nature of the intersecting dendritic crystals can be observed. The dendrites form a continuous path in three dimensions and allow the glass-ceramic to be etched with hydrofluoric acid, which attacks the low-silica crystallites at a far greater rate than the durable aluminosilicate residual glass. Since the dendritic crystallization can be photosensitively nucleated, a complex etchable pattern can be transferred into the glass.

Ultra Fine-Grained

Extremely efficient nucleation is achieved in lithium aluminosilicate glasses by additions of TiO_2 and ZrO_2 in roughly equivalent molar concentrations at a total level of about 4 mol%. When such glasses are heated about 50° above the glass transition temperature, tiny zirconium titanate crystals (<100 Å in diameter) are widely precipitated throughout the glass. β-quartz solid solution is efficiently nucleated upon these oxide crystals. Figure 7 shows the development of the metastable hexagonal quartz phase upon the zirconium titanate. Figure 8 illustrates the highly crystalline microstructure developed after a heat treatment near 850°C, with the resulting particle density of the silicate phase about $5 \times 10^{-21}/m^3$. The resulting glass-ceramic is close to monophase and has a coefficient of thermal expansion of $7 \times 10^{-7}/°C$ (0–500°C), which produces outstanding thermal shock resistance. The efficient nucleation and sluggish grain growth, which allows crystals to impinge before they have grown much more than 500 Å, allows the efficient transmission of light. The low birefringence of the constituent β-quartz solid solution crystals also serves to minimize any scattering. Thus, visible light scattering cannot be perceived in a product application like integral thick-handled cookware.

Cellular Membrane

The residual glass in certain glass-ceramics can develop in a cellular membrane form. This generally occurs when the developing crystal phase is slightly lower in silica than the bulk composition, thus allowing a stable film of siliceous glass to envelop the impinging grains during crystallization.

Solid solutions of β-quartz and β-spodumene in the SiO_2-Al_2O_3-$(Li_2,Mg)O$-TiO_2 system provide a good example. Here the titanate nuclei catalyze the growth of crystals, which are lower in silica than the bulk composition, and impingment of these crystals is prevented by the increasing reluctance of the siliceous and viscous residual liquid to crystallize (Figure 7). This viscous phase adheres to the impinging polyhedral grains forming a cellular membrane network throughout the body, but generally comprises less than 10 vol%.

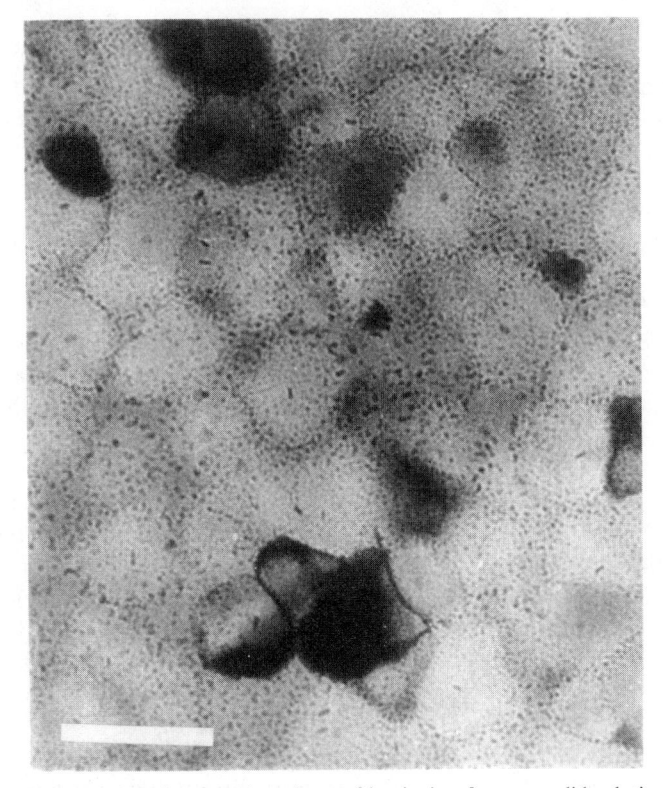

Figure 7 Transmission electron microphoto of impinging β-quartz solid solution crystals showing tetragonal $ZrTiO_4$ nuclei at the center. Note continuous residual glass and precipitation of excess $ZrTiO_4$ along crystal boundaries (white bar = 200 nm) (after Maier & Müller, 22).

This tenacious residual glass can provide some benefits in glass-ceramic properties. It acts as a barrier to diffusion of aluminum ions that control secondary grain growth in β-spodumene glass-ceramics (19). Thus these materials show isothermal grain growth dependence as the cube root of time and can be used for long periods at temperatures as high as 1200°C without the crystals approaching 5 μm, at which point thermal expansion stress anisotropy could begin to cause microcracking with repeated thermal cycling.

Reforming of β-spodumene glass-ceramics after their crystallization is also possible because of cellular residual glass. In this way, glass-ceramic sheet manufactured for laboratory benchtops can be vacuum formed or otherwise molded into sinks and other complex shapes at temperatures well below the initiation of melting, even with crystallinity well over 90%. The high creep rates at sub-solidus temperatures have been attributed to solution-precipitation phenomena involving the glassy film wetting the

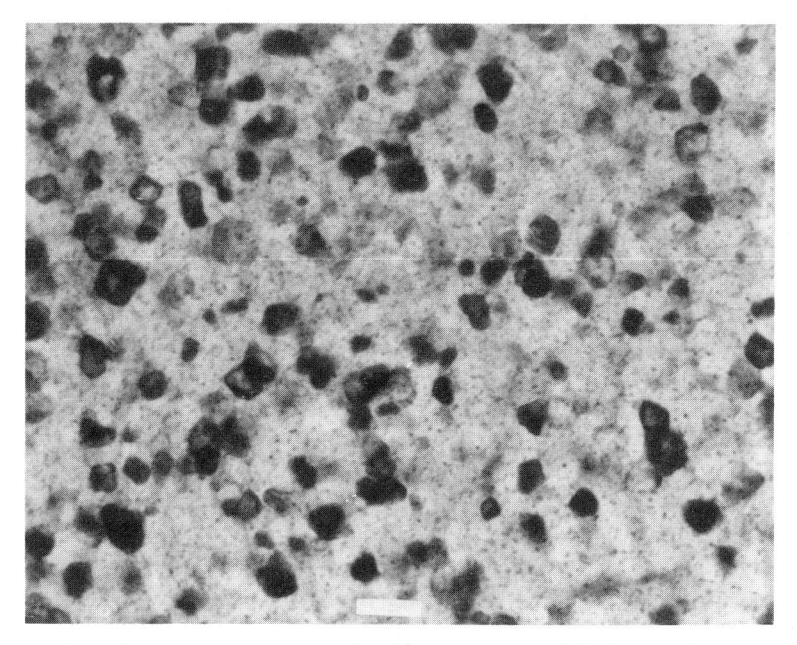

Figure 8 Microstructure of VISIONS® as revealed by TEM (bar = 100 nm).

grains (20). This model for creep involves species transport through the glass phase. The applied stress is believed to be supported at grain boundary islands where crystals meet. Grain shape elongates in the direction of tensile stress. The results of creep experiments are in agreement with interface reaction controlled creep.

Relict

Some crystalline microstructures faithfully inherit and reflect the original morphology of the parent phase-separated glass. This is true in certain mullite glass-ceramics where the original glass phase separated into aluminous droplets within a siliceous matrix (21). When heat-treated above the annealing point of the phase-separated glass, the alumina-rich droplets become more fluid than the matrix and immediately crystallize to mullite. The mullite crystals are restricted by the original droplet form and faithfully reflect this form. Because the droplets are very small (<0.1 μm), as shown in Figure 9, this glass-ceramic is transparent.

Coast-and-Island

A coast-and-island microstructure typically develops when an equilibrium crystal phase forms at the expense of a metastable assemblage of phases.

Figure 9 Relict structure in transparent mullite glass-ceramic reflecting original amorphous phase separation (white bar = 1 μm).

An example is the development of cordierite in magnesium aluminosilicate glass-ceramics at the expense of α-quartz, sapphirine, and spinel. The cordierite grows from grain boundaries and slowly becomes the predominant matrix phase. The excess silica recrystallizes as cristobalite laths, which along with titania as rutile, become partially or largely enveloped by the cordierite. This leads to an interlocking structure with good mechanical strength and toughness.

Another interesting example of coast-and-island structure involves the precipitation of pollucite ($CsAlSi_2O_6$) from previously existing mullite and glass in a partially-crystalline glass-ceramic of composition SiO_2 35, Al_2O_3 40, Cs_2O 25 wt%. The remains of the mullite-glass mixture becomes enveloped by the pollucite matrix. Since pollucite is the most refractory of these three phases, the resulting glass-ceramic (Figure 10) has a refractory crystalline matrix with mullite and glass left in islands, peninsulas, and other largely isolated forms. This accounts for the extremely high temperature for viscous deformation, i.e. 1430°C corresponding to a viscosity of 10^{12}P for this pollucite-mullite glass-ceramic, some 350°C higher than for fused silica.

House-of-Cards

The precision machinability of mica glass-ceramics such as commercial MACOR glass-ceramic with conventional metal-working tools is the direct result of the house-of-cards microstructure, as illustrated in Figure 11. Randomly oriented and flexible flakes tend to either arrest fractures or cause deflection or branching of cracks; therefore, only local damage results as tiny polyhedra of glass are dislodged. Another important property dependent on this microstructure is high dielectric strength, typically around 40 kV/nm. Such insulating qualities are possible because of the continuous interlocking form of the mica sheets. Very low helium per-

Figure 10 Pollucite-mullite glass-ceramic. Coast-and-island microstructure shows regions of pollucite (*heavily etched*) interlocking with regions of mullite and glass (white bar = 1 μm) (cerammed: 1600°C, 1 hr).

Figure 11 House-of-cards structure in machinable fluormica glass-ceramic. Note phase separated residual borosilicate glass and affinity of siliceous droplets for mica flakes (white bar = 1 μm, *right*).

meation rates, important in high vacuum applications, are also related to this microstructure. Helium can, of course, permeate most glasses, but it is severely slowed even at high temperature by mica crystals, whose basal plane is composed of an anion network of oxygens arranged in a virtual hexagonal close-packed state. The house-of-cards microstructure also contributes to the relatively high fracture toughness values in mica glass-ceramic (~ 2 MPa m$^{1/2}$), this again the result of crack blunting, branching, and deflection.

Acicular Interlocking

Interlocking rod- or blade-like crystals in a glass-ceramic matrix serve as important strengthening and toughening agents (5). A random acicular microstructure, as depicted in Figure 12, accounts for the high strength and toughness of potassium fluorrichterite glass-ceramics. The abraded modulus of rupture here is 150 ± 15 MPa. The fracture toughness, measured by the short bar technique, is 3.2 ± 0.2 MPa m$^{1/2}$. The toughening

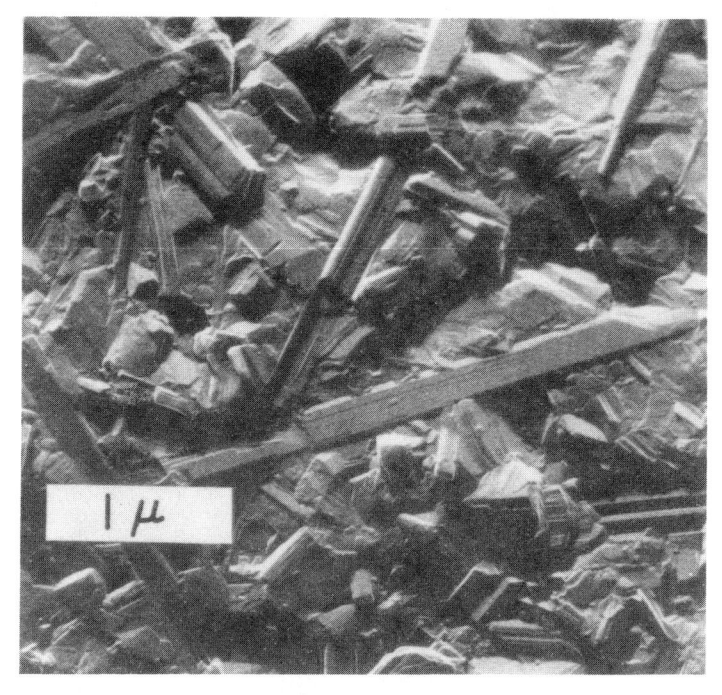

Figure 12 Fracture surface (r.e.m.) of K-F-richterite glass-ceramic showing the effects of rod reinforcement toughening.

mechanism results from rod reinforcement by the high aspect ratio K-F-richterite crystals. An even stronger and tougher microstructure (300 MPa, 5.0 MPa m$^{1/2}$), based on interpenetrating blades, is illustrated in Figure 13. Here a highly crystalline fluorcanasite monophase microstructure is shown. Cleavage splintering is observed, which causes energy absorption through crack branching and deflection. The thermal expansion anisotropy also serves to increase toughness in this material. In the chain direction, $\alpha_b = 82 \times 10^{-7}/°C$ (0–700°C). Perpendicular to the chains $\alpha_a = 159 \times 10^{-7}/°C$ and $\alpha_c = 248 \times 10^{-7}/°C$. Considerable stress is thereby developed along grain boundaries during the cooling of canasite glass-ceramics, which can produce microcracking in front of a crack tip. Specific evidence for this toughening mechanism is observed in the decrease of toughness as a function of temperature (Figure 14), which would be expected from decreasing anisotropic thermal expansion stress (5, 18). At temperatures above 600°C, the residual glass becomes plastic and begins to reverse this effect.

Figure 13 Fracture surface (r.e.m.) of fluorcanasite glass-ceramic showing interlocking blades and effects of cleavage splintering.

Lamellar Twinned

Lamellar or polysynthetic twinning is developed either during growth or upon cooling of certain silicate crystals. An example of the latter occurs in enstatite glass-ceramics where protoenstatite undergoes a martensitic transformation to clinoenstatite on cooling. The result is fine lamellar twinning on the 100 crystallographic plane as illustrated in Figure 15. Cleavage steps are also observed orthogonal to 100. This combination of fine twinning and cleavage influences fracture propagation, which causes the splintering that is easily observed on fracture surfaces (Figure 15). This energy-absorbing mechanism is believed to account for the high fracture toughness (up to 4.6 MPa m$^{1/2}$) observed in these glass-ceramics.

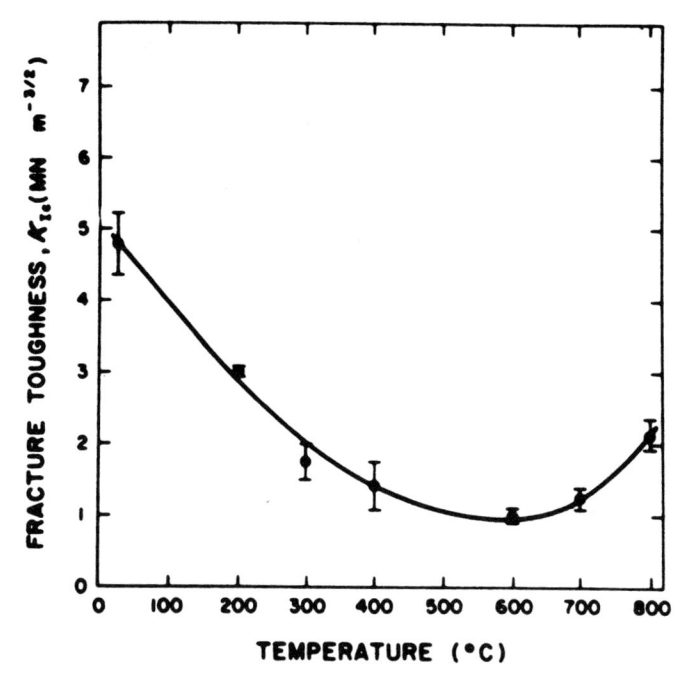

Figure 14 Fracture surface (r.e.m.) of enstatite-zircon glass-ceramic E-2 (see Table 2) showing interlocking twinned enstatite grains and nodular zircon. Note the step splintering effect of the intersection of cleavage and twinning.

CONCLUSION

Both composition and microstructure play key roles in determining the properties of glass-ceramics. Bulk composition predetermines both the nucleation and phase development sequence as well as the final phase assemblage. The latter is responsible for the physical and chemical properties of the material. The choice of composition may be limited by the desired glass forming process, which requires a glass of specific stability towards devitrification, as indicated by its liquidus-viscosity relationship.

Microstructure, which includes the geometric arrangement of the crystalline phases and the distribution of the residual glass, can be made to complement or take advantage of the properties of particular phases. Complete transparency in polycrystalline silicate ceramics depends on a microstructure with particles well below the wavelength of light, and thus far this has only been achieved in highly efficiently nucleated glass-

Figure 15 Decrease of fracture toughness with temperature in a canasite glass-ceramic (after Beall et al, 18).

ceramics. Improved or unusual mechanical properties such as high flexural strength, toughness, and machinability can be achieved through unique microstructures produced by random growth of anisotropic chain and sheet silicates.

Literature Cited

1. Stookey, S. D. 1959. *Ind. Eng. Chem.* 51(7): 805
2. Headley, T. J., Loehman, R. E. 1984. *J. Am. Ceram. Soc.* 9: 620
3. Stookey, S. D. 1953. *Ind. Eng. Chem.* 45(1): 115
4. Berezhnoi, A. I. 1970. In *Glass-Ceramics and Photo-Sitalls*, ed. A. G. Pincus. New York: Plenum. 444 pp.
5. Beall, G. H. 1991. *J. Non-Crystalline Solids* 129: 163
6. Lee, W. E., Heuer, A. H. 1987. *J. Am. Ceram. Soc.* 70(5): 349
7. Beall, G. H. 1986. Glass-Ceramics. In *Advances in Ceramics*, 18: 157–72. Am. Ceram. Soc.
8. Ostertag, W., Fischer, G. R., Williams, J. P. 1968. *J. Am. Ceram. Soc.* 51: 651

9. Lewis, D. III. 1982. *Am. Ceram. Soc. Bull.* 61(11): 1208
10. Andrews, L. J., Beall, G. H., Lempicki, A. 1986. *J. Luminescence* 36: 35
11. Cooper, R. F., Chyung, K. 1987. *J. Mater. Sci.* 22: 3148
12. Beall, G. H., Rittler, H. L. 1982. In *Advances in Ceramics*, 4: 301. Am. Ceram. Soc.
13. Beall, G. H. 1971. In *Advances in Nucleation and Crystallization in Glasses*, 5: 251. Am. Ceram. Soc.
14. Chyung, C. K., Beall, G. H., Grossman, D. G. 1974. *10th Int. Congr. Glass.* 14: 33
15. Malament, K. A., Grossman, D. G. 1987. *J. Prosthetic Dent.* 57(6): 62
16. Gulati, S. T. 1986. *Collected Papers, XIV Int. Congr. Glass, New Delhi*
17. Shigarov, M. I., Mamedov, K. S., Belov, N. V. 1969. *Dokl. Akad. Nauk SSSR* 185: 672
18. Beall, G. H., Chyung, C. K., Stewart, R. L. 1986. *J. Mater. Sci.* 21: 2365
19. Chyung, C. K. 1969. *J. Am. Ceram. Soc.* 52: 342
20. Raj, R., Chyung, C. K. 1981. *Acta Metall.* 29: 159
21. MacDowell, J. F., Beall, G. H. 1969. *J. Am. Ceram. Soc.* 52(1): 17
22. Maier, V., Müller, G. 1989. *J. Am. Ceram. Soc.* 70: C-176–78

X-RAY TOMOGRAPHIC MICROSCOPY (XTM) USING SYNCHROTRON RADIATION[1]

John H. Kinney

Chemistry and Materials Sciences Department, Lawrence Livermore National Laboratory, Livermore, California 94551

Monte C. Nichols

Materials Department, Sandia National Laboratories, Livermore, California 94551

KEY WORDS: tomography, microtomography, X-ray imaging, reconstruction from projections, X-ray microscopy

INTRODUCTION

The response of a material to an applied load is inherently three-dimensional. Heterogeneous networks of second phases, reinforcing particles, and fibers act to redistribute applied loads into complex, three-dimensional distributions of stresses and strains. Mechanical properties measurements, however, traditionally rely upon one- and two-dimensional analytical techniques to explain what are, in reality, three-dimensional responses. A load-displacement curve is an example of a one-dimensional measurement, which is often used to measure global deformation in a sample, and to infer micromechanical mechanisms that might be operating during deformation. Scanning electron microscopy (SEM) and transmission electron microscopy (TEM) are examples of essentially two-dimensional techniques that can be used to provide evidence for the origin of failure and whether

---

[1] The US government has the right to retain a nonexclusive, royalty-free license in and to any copyright covering this paper.

or not any dissipating forces, such as crack bridging or deflection, have been operative during crack advance. However, there are few good ways to three-dimensionally analyze a material's microstructure short of time-consuming serial sectioning—a technique that destroys the sample and is acknowledged to be artifact prone.

It has long been known that the linear X-ray attenuation coefficient is a sensitive measure of atomic composition and density. If the X-ray attenuation coefficient could be measured as a function of position in a sample nondestructively, then a three-dimensional image of a sample's microstructure could be obtained without time-consuming and difficult sectioning. Subtle microstructural changes from one position to the next would appear as differences in the attenuation coefficient. As long as the spatial resolution could be made small with respect to the microstructural features of interest, the three-dimensional X-ray images obtained from these measurements would provide valuable structural information.

The technique for obtaining volumetric measurements of the X-ray attenuation coefficient is known as computed tomography (CT). CT is most frequently used in medical diagnostic radiology, where the spatial resolution is of the order of 500 micrometers. CT is slowly developing for industrial applications as well. Industrial CT systems specifically designed to examine small samples typically have a spatial resolution between 50–100 micrometers. Though the resolution of industrial CT units is sufficient to detect many critically sized flaws, it is not sufficient to image microstructural features such as second phase precipitates, reinforcing particulates, or fibers in composite structures.

The ability to make precise X-ray attenuation measurements on ever smaller volume elements is a developing technology. This technology is known as X-ray tomographic microscopy (XTM) to delineate the method as a form of X-ray microscopy that uses tomographic reconstruction techniques to build three-dimensional images of microstructures. In essence, XTM carries the industrial computed microtomography techniques to their extreme limit in spatial resolution and contrast sensitivity. The highest spatial resolution (2–3 μm) is obtained using monochromatic (i.e. single energy) synchrotron radiation and is limited to inspection of small (usually less than 1 cm diameter) specimens.

This review paper describes the technique of XTM and discusses recent applications of the technique to materials studies. Because of the relative newness of the technique, most XTM studies to date have been proof-of-concept. We expect this to change as the availability of XTM increases and as technical improvements are made both to the quality of the images and to the size and type of samples that can be studied. We expect that the nondestructive, three-dimensional attributes of XTM will significantly impact our understanding of materials performance and failure—especially in composite microstructures.

BACKGROUND

Computed Tomography

XTM is based on the same principles used in medical CT. The X-ray attenuation coefficient, μ, at a point $r_{x,y,z}$ in a material is determined from a finite set of X-ray attenuation measurements (projection data) taken at different angles. The projection datum (I) is the transmitted X-ray intensity reaching a position-sensitive detector after passing through the sample. This datum, which is directly related to the materials microstructure, is given by

$$I = \int S(E) \left[\exp - \int \mu(x, y, z, E) \, dl \right] dE, \qquad 1.$$

where $S(E)$ is the energy spectrum of the X-ray source and $\mu(x, y, z, E)$ is the energy dependent attenuation coefficient at a single point in the sample. The line integral is taken along a straight path, dl, through the sample. If the radiation can be made nearly monochromatic with photon energy E_o, the energy spectrum can be approximated by a delta function and Equation 1 reduces to the familiar form of the Radon transform:

$$\ln\left(\frac{I_o}{I}\right) = \int \mu(x, y, z, E_o) \, dl. \qquad 2.$$

Measurements of the attenuation through the sample as a function of angle are used to numerically invert Equation 2 to solve for $\mu(x, y, z, E_o)$. From a priori knowledge of the phases present in the sample, μ can be used to provide a microstructural mapping of the distributions of precipitates and reinforcing phases.

In order to obtain high resolution, three-dimensional images using tomographic techniques, it is necessary to have a sufficient number of photons for good measurement statistics, and a detector that has sufficient spatial resolution to discriminate between closely separated photon ray paths. High spatial resolution also depends upon the ability to collimate the source, since source divergence leads to image blur. Because source collimation reduces X-ray intensity, the problems of spatial resolution and contrast sensitivity are interrelated.

The problem of photon statistics is of major concern for microscopic X-ray imaging. Grodzins (1) provided an estimate for the time, T, required to image a volume element to a specified accuracy provided an incident X-ray flux, N_o (photons per unit area per second):

$$T = \frac{DB e^{\mu D}}{w^5 \mu^2 (\sigma/\mu)^2 N_o},\qquad 3.$$

where D is the sample diameter, w is the edge of the cube defining the volume element, μ is the linear attenuation coefficient, $(\sigma/\mu)^2$ is the dimensionless variance, and B is a dimensionless parameter that depends on the reconstruction algorithm (usually of order 2). As an example, if 1 sec is required to image a volume element that is 100 micrometers on a side to a given level of statistical confidence, then 10^5 sec would be required to measure a volume element that is 10 micrometers on a side in the same sample. The strong dependence of the integration time on the size of the volume elements and, therefore, the spatial resolution, requires that high intensity X-ray sources be used for microscopic imaging. Using conventional X-ray sources, extremely long integration times would be required to achieve both high spatial resolution and good contrast sensitivity. Long integration times place much stronger constraints on both the X-ray source and mechanical stability of the CT system.

Industrial CT scanners, which use conventional sources of radiation, attempt to overcome their source intensity limitations either by sacrificing spatial resolution through increasing the size of the volume elements, or by reducing the statistical quality of the data. The reduced statistics can allow high resolution imaging, but only of high contrast objects such as cracks or pores. Subtle attenuation differences, which might suggest diffusion gradients or second phase precipitates, are noticeably absent in many industrial CT images because they require extremely low noise levels for detection.

XTM overcomes the statistical problems encountered with conventional sources of radiation by using synchrotron radiation sources. A synchrotron radiation source is ideal for high resolution CT imaging. Unlike X-ray tube sources, which emit only a small fraction of their dissipated power as X-rays, electron storage rings are extremely efficient at producing X-rays. Furthermore, synchrotron radiation is emitted into a small solid angle directed tangentially to the electron orbit, thereby increasing the source brightness (photons per unit solid angle) dramatically. Finally, whereas tube sources produce their highest brightness at discrete, characteristic energy levels, synchrotron radiation is continuously distributed in energy from a few eV to several tens of keV.

In a synchrotron radiation source, X-rays are produced when the electrons in the storage ring are accelerated by a magnetic field. The energy spectrum of the synchrotron radiation emitted from the storage ring depends on the electron energy E and the magnetic field B, which produces the acceleration. The energy spectrum can be approximated by (2)

$$\phi = 3.461 \times 10^3 I H_2(y)\gamma^2, \qquad 4.$$

where I is the electron current in mA in the ring, and ϕ is the X-ray flux in units of photons $\cdot$ s^{-1} $\cdot$ mrad^{-2} per 0.1% energy bandwidth. In Equation 4, the parameters are given by

$$H_2(y) = y^2 \left[K_{2/3}\left(\frac{y}{2}\right) \right]^2,$$

$$\gamma = \frac{E_o}{m_o c^2} = 1957 E(\text{GeV}),$$

$$y = \lambda_c/\lambda,$$

and

$$\lambda_c = \frac{18.64}{B(T) E^2(\text{GeV})},$$

where the $K_{2/3}(y/2)$ are modified Bessel functions of non-integer order, and λ_c is the critical wavelength where half of the photons are of longer wavelength and half are of shorter wavelength.

Figure 1 compares the intensity of a synchrotron source with a microfocus X-ray anode typically encountered in industrial CT systems. The synchrotron spectrum is calculated for the wiggler beamline 10–2 operating at the Stanford Synchrotron Radiation Laboratory, and the microfocus anode spectrum has been calculated (3) for a tube voltage of 100 kV and tube current of 5 mA incident on a tungsten anode. The units of intensity are photons per second per milliradian squared per 0.1% bandwidth [bandwidths ($\Delta E/E$) of 0.1% are not atypical of single crystal monochromators]. The most significant observation in Figure 1 is that the conventional X-ray source intensity must be multiplied by 10^9 in order to plot the two spectra on the same scale. Furthermore, the absence of characteristic peaks in the synchrotron spectrum makes it practical to continuously tune the energy of the X-rays using single crystal monochromators over the entire energy range of the source. Because of this, the optimum X-ray energy can be selected for tomographic imaging.

The ability to select the optimum X-ray energy for tomographic measurements greatly enhances both the signal-to-noise and the image information content. Differentiation of Equation 3 defines the maximum in the signal-to-noise with the relationship $\mu D = 2$. That an optimum

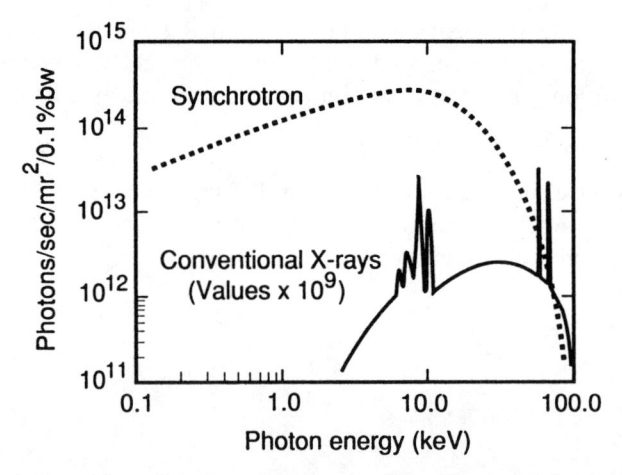

Figure 1 A comparison of the intensity per unit solid angle per unit bandwidth between a conventional microfocus X-ray source and the 31-pole wiggler beam line at Stanford Synchrotron Radiation Laboratory. The difference in intensity per unit solid angle between the two sources illustrates the advantages of using synchrotron radiation for imaging applications. Furthermore, the absence of characteristic peaks in the synchrotron spectrum makes continuous tuning of the energy practical. Because of this, the optimum energy can be selected from a synchrotron source for maximizing the signal-to-noise ratio in the tomographic image.

energy exists is not hard to visualize. Clearly, if the X-rays are too low in energy to penetrate the sample, then the statistics will be zero. Likewise, if the X-ray energy is so high that very few photons are attenuated, then the sample becomes transparent, and the information content in the attenuation image is lost. In Figure 2 we have plotted the optimal sample thickness vs X-ray energy for some materials of engineering interest calculated from tabulated attenuation data (4). The greatest problem with using synchrotron radiation sources at present is that it is difficult to obtain energies higher than 50 keV. The sample size, therefore, is limited for most materials to a few millimeters or less. The next generation of synchrotron sources, such as the Advanced Photon Source being constructed at Argonne National Laboratory, will provide much higher energy photons than can be obtained using most currently available synchrotron sources.

In addition to the signal-to-noise problems associated with using X-ray energies that are not optimized to the sample thickness, a more serious problem is encountered when using conventional sources. Because the intensity from conventional radiation sources is low, the full (or slightly filtered) polychromatic X-ray spectrum is used for imaging. A broad energy

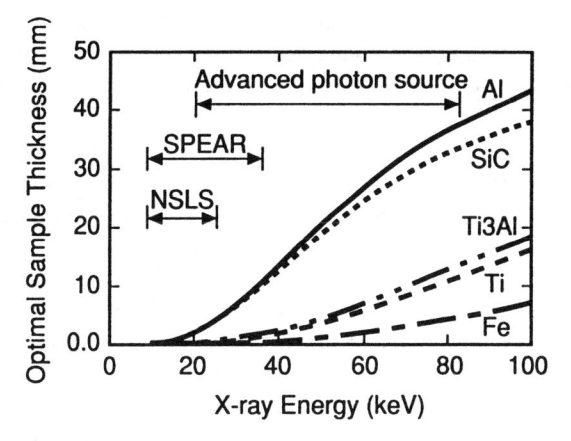

Figure 2 The optimal sample thickness as a function of X-ray energy for some materials of engineering interest. Note that the strong dependence of the photoelectric cross-section with Z makes it difficult to image large samples with X-ray energies below 100 keV. Also given are the approximate energy ranges that can reasonably be spanned at the National Synchrotron Light Source (NSLS), Stanford Synchrotron Radiation Laboratory (SPEAR), and the proposed Advanced Photon Source.

distribution in the X-rays leads to beam hardening artifacts in the resulting tomography image. Beam hardening is a term expressing the preferential absorption of low energy photons as the radiation is transported through the sample. Because of this preferential absorption, the average energy of the X-rays is higher after passing through thicker portions of the sample; hence, the measured spectrum averaged attenuation coefficient will be lower in the interior of the sample. This beam hardening leads to a pronounced contrast gradient in reconstructed images as well as great difficulty in differentiating microstructural variations in the material as compared to artifactual variations (5, 6).

Spatial resolution is determined by the X-ray source as well as the detector. Because of the natural collimation of synchrotron sources, it is possible to utilize a nearly parallel beam geometry for tomographic imaging (7). Source divergence can easily be made as low as 10^{-4} radians at the sample, which leads to penumbral blurring of only a micron over a cm of distance. Because of this, parallel beam reconstruction techniques can be applied effectively, and there is no need to consider fan- or cone-beam geometries. However, if the excellent properties of synchrotron radiation are to be utilized, it is necessary to develop extremely high resolution, linear, low-noise detectors for measuring the attenuation along parallel ray paths through the sample. In addition, to take advantage of the high intensity from synchrotron radiation sources for making three-dimensional images, it is advantageous to develop either a very fast detector or one that takes many measurements of different ray paths in parallel.

Instrumentation

Conventional CT measurements involve collecting absorption information for a single cross-sectional slice through a material. Spatial resolution is achieved either by collimating the incident beam using a pinhole (and then rastering the beam across the sample for every angular setting), or by using a position-sensitive detector to measure all of the projection data for a single angular view in parallel. Aside from its relative simplicity, the advantage of the pinhole technique is that the resolution, to first order, is determined only by the size of the collimator. Elliott & Dover (8) have successfully used pinhole scanning with a standard X-ray generator to perform tomography on mineralized tissues and composites. An energy-dispersive detector is used to count photons of only a single energy, thereby satisfying the requirement for nearly monochromatic radiation. Resolution has been sufficient to detect the 33-μm graphite cores in SiC fibers in an aluminum/SiC aligned-fiber composite. However, the sensitivity has not yet proved sufficient to detect the much larger, but lower contrast, SiC sheaths. Interfacial microcracking has not been clearly identified using the scanning method (9).

The primary disadvantage using the pinhole is that most of the incident radiation is lost in the collimation. For this reason, the rastering technique is extremely time-consuming. Acquiring the data for the reconstruction of a single cross-section of a sample using a conventional X-ray source takes upwards of 12 or more hr depending on the size of the pinhole collimator and the sample. Spanne & Rivers (10) reported an order of magnitude increase in the speed of pinhole scanning when synchrotron radiation at the National Synchrotron Light Source was used. The volume elements for the reported measurement, however, were quite large compared to those used by Elliott & Dover, and a limited number of projected ray paths (177) and angular increments (163) were sampled. Even so, if Spanne & Rivers were to have acquired the data for the 177 slices required to produce a volumetric image, it would have been necessary to irradiate the sample for over 170 hr. It is doubtful that the pinhole scanning approach for three-dimensional analysis and any real-time studies will prove useful.

Linear photodiode arrays have been used in a number of CT devices designed to operate using conventional X-ray sources (11, 12). The widespread application of linear photodiode arrays results from both their ease of use and ability to acquire upwards of a thousand projected rays simultaneously. The parallel acquisition of data improves the speed of the measurements nearly a thousandfold, and the accumulation of enough data to reconstruct a single slice becomes measured in minutes rather than days. In spite of these advantages, however, there are drawbacks to the use of the photodiode array for ultrahigh resolution characterization. The first of these drawbacks is that the photodiode array is noisy and is subject to nonlinearities. The noise limits the dynamic range and therefore the maximum contrast that can be studied in a sample. The nonlinearities introduce ring-like artifacts in the reconstructions that can further reduce the usefulness of the information obtained. Finally, even with reducing the data acquisition times for a single slice from days to minutes, it still requires days to obtain enough information for three-dimensional sample visualization.

Clearly, a two-dimensional array that records projection data for many contiguous slices simultaneously is advantageous for practical three-dimensional imaging. Feldkamp used a vidicon array as a two-dimensional detector (13). Because a vidicon is continuously read-out at video rates, the integration times are too short to detect an X-ray image using a laboratory X-ray source. Therefore, Feldkamp relies on an image intensifier to convert the X-ray photons into visible photons and then amplify the light signal by two orders of magnitude. Because image intensifiers have relatively low spatial resolution, it is necessary to use a microfocus source in a magnifying geometry, and a cone-beam algorithm is necessary to reconstruct the three-dimensional image (14). CT systems operated in this manner are limited in spatial resolution by the source spot size (~ 20–25 μm) and in sensitivity by the photon statistics and linearity of the image intensifier. Good photon statistics are difficult to obtain because signal averaging with vidicons is limited by excessive read-out noise. Nevertheless, the image intensifier with vidicon detector has been adopted by others (15), and images of 20–30 μm resolution have been demonstrated. However, in many of these images the noise is quite high, and the ability to detect other than high contrast features is not clear.

In the mid-1980s, groups from Livermore and Dortmund developed a detector system for performing XTM (16, 17), which has now been adopted by others (18, 19). The system utilizes a single crystal Si or Ge monochromator, a high resolution single-crystal scintillator, and a charge-coupled device (CCD) detector. The single-crystal monochromator allows the selection of the desired X-ray energy, and it has to remain distortionless in the high heat loads generated by the primary synchrotron radiation beam. This is accomplished by water cooling the first crystal in a double-crystal monochromator configuration.

For reasons based on photon statistics, spatial resolution, and efficiency, it was decided early on in the development of XTM that the X-rays should be converted to visible light using a high resolution scintillator screen (20). The development of these high resolution screens has been the most difficult task in achieving high resolution XTM images. The present scintillator screen consists of a single crystal of $CdWO_4$, which is highly polished to at least a 10/5 (scratch/dig) specification and is coated with an antireflective compound to reduce blurring that might be caused by reflections from the surface (21). The visible light is projected onto the CCD using variable magnification optical lenses. These lenses were especially designed for a large field of view and high spatial resolution with minimal distortion. Standard commercial lenses do not have the spatial resolution required for XTM over large fields of view (greater than 1 cm).

Finally, a two-dimensional CCD array detector was selected for converting the visible light into position-sensitive electronic signals required for tomographic reconstruction. A CCD array was chosen for the detector because it allows the simultaneous collection of both multiple ray paths and multiple slices for a given orientation of the sample—thereby greatly speeding up the tomographic measurements. Most industrial CT, and even some high resolution synchrotron CT, is still performed using linear photodiode array detectors (22). The CCD, however, is not the simple CCD found in many video cameras. Because low noise and extreme linearity of the detector are essential for tomographic imaging, the CCD is cooled either thermoelectrically or with liquid nitrogen to temperatures below $-40°C$ to reduce dark current and read-out noise. The optimal temperature for operation is CCD-dependent and must be established by making careful measurements of resolution and charge transfer efficiency. Furthermore, the CCDs are read out either as 12 or 14 bit data in a slow scan manner. Attempting to read out the CCD at video rate greatly increases the noise and significantly reduces the dynamic range of the image (G. Simms, personal communication).

Figure 3 is an artist's depiction of the XTM apparatus. In practice, a sample is positioned on a goniometer, which is mounted to a rotating stage. The goniometer allows the sample to be oriented such that upon rotation it does not precess out of the detector's field of view. The sample is initially translated out of the X-ray path, and an image is obtained of the incident X-ray beam. This reference image, taken without the sample, provides the values for I_o in Equation 2. Next, the sample is placed between the X-ray path and the scintillator, and another image, the projection image, is acquired. The projection image provides the values for I at a given angular orientation for Equation 2. The ratios of the logarithms of the reference image and the projection image provide values of the integrated attenuation along the individual ray paths. By rotating the sample by discrete angular increments through 180 degrees, enough data can be obtained to deconvolve the two-dimensional projection images into a

TOMOGRAPHIC MICROSCOPY 131

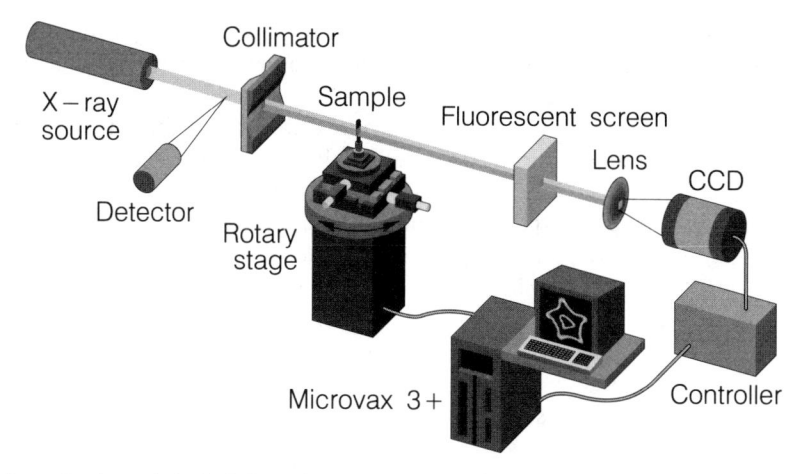

Figure 3 An artist's depiction of the X-ray tomographic microscope apparatus. X-ray photons pass through a sample that is positioned on a rotating stage. The transmitted photons are converted to visible light on a scintillator screen, and then imaged using a two-dimensional, charge-coupled device detector. The ratio of the transmitted to incident intensities as a function of spatial position and angular orientation of the sample is used to reconstruct three-dimensional images of the material's microstructure.

three-dimensional image of the attenuation coefficients. This deconvolution procedure, known as reconstruction from projections, is usually performed by a technique of Fourier-filtered back-projection (5). Using the present XTM, a 10 cubic millimeter volume can be imaged with 5 micrometer volume elements in anywhere from ten minutes to upwards of several hours, depending on the noise levels that be tolerated in the reconstructed images.

Though the XTM has successfully operated using conventional X-ray sources, the highest resolution is obtained using monochromatic synchrotron radiation. The CCD camera is read out using a MicroVax 3800 computer, which has four Gbytes of hard disk capacity. Since the present CCD is a Kodak KAF 1400 CCD, a single angular increment can represent nearly three Mbytes of data. Coupled with the need to obtain a minimum of 360 unique angular views in projection, a single volumetric image of a sample requires over one Gbyte of data. Thus, large disk capacity is required for XTM imaging purposes.

The X-ray tomographic microscope is being upgraded to run a new Kodak 2048 × 2048 CCD. Because the data requirements are four times greater for this larger format CCD, it will become necessary to improve the computation speed for performing reconstructions and image analysis. In anticipation of this, we are instrumenting the X-ray tomographic microscope with a parallel computer which, in the initial configuration, will increase processing speeds by at least a factor of 50 over that which is presently being obtained.

The stage positioning hardware currently consists of three translation stages, a rotary stage, and the stage controllers and driver electronics. All stages have stepper drive motors, incremental position encoders, and an origin signal. The stage controller provides both user and computer interfaces for up to four separate stages. The computer interface, as it is used here, is an IEEE-488 port, through which the controller can accept commands from a computer telling it how and where to position the four axes and through which it can report back to the computer its success or failure at executing the commands.

The controller and stages comprise an open-loop system. The verification that an axis is where it was commanded to go is left to the computer. Since relative encoders are used, care must be taken to preserve the fidelity of the pulse stream between them and the controller's position displays (i.e. the encoder's pulse counters) so that the position information remains accurate. It is conceivable with this type of system to have bad position information and yet not be aware of it; however, this has not been a problem to date.

The 360-degree rotational stage has a minimum step size of $0.01°$ and a maximum velocity of $20.0°/sec$. Important features for this stage are good relative accuracy (better than $0.01°$) and minimum wobble (less than 0.05 milliradians). The rotation stage is mounted on translation stages that have ± 25 mm of travel perpendicular to the optical axis, with a minimum step size of 1.0 μm and a maximum velocity of 2 mm/sec. Important features required of these translation stages are quick movement and the ability to repeatedly return the sample to the same in-beam position (within 0.1 μm).

APPLICATIONS

XTM is a relatively recent development; hence, applications of XTM have been mostly proof-of-principle (21, 25–28). This review focuses on three recent applications of XTM. The first application involves visualizing matrix consolidation during chemical vapor infiltration of a continuous fiber ceramic matrix composite. The second example describes the application of XTM to noninvasive imaging of damage development and evolution during tensile loading in metal matrix composites. The third example describes the application of XTM towards imaging hard mineral tissues. These applications represent only three examples of the application of XTM to noninvasive imaging of material microstructures. Other areas where XTM might be applied, such as in semiconductor packaging and biomaterials, can also be envisioned.

Application of XTM to Processing of Ceramic Matrix Composites

Much of the manufacture of ceramic matrix composites (CMCs) is based on textile technology. Small (10–15 μm diameter) ceramic fibers are formed into bundles (tows), which are subsequently woven into cloth. The cloth is layered over a mandrel to near net shape, and then a chemical vapor is used to grow a polycrystalline matrix around the fibers. The architecture of the cloth (i.e. the weave geometry) can be as simple as a square, two-dimensional weave (plain weave), as is shown in Figure 4, or it can be as complicated as a three-dimensional weave where stitches are used to hold the cloth layers together. The fiber architecture influences the directionality of the mechanical properties as well as the processing behavior of the composite.

Before CMCs can be used in structural applications it will be necessary to reduce the amount of porosity that remains after processing. Complete densification of sintered and chemical vapor-infiltrated (CVI) composites has yet to be attained because of the enclosure of pores by growth of the

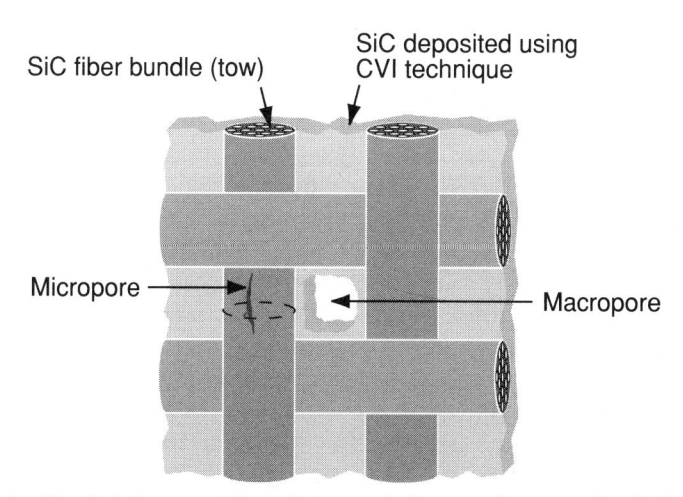

Figure 4 The cloth for a ceramic matrix composite is woven from tows (bundles) of small ceramic fibers. This figure shows a plain, or biaxial, weave architecture where the warp and the weft are orthogonal to each other. The SiC matrix is grown around the fibers by the infiltration of methyltriclorosilane gas at high temperature. The microstructural features of interest to process modeling are the amount of matrix deposition as a function of spatial position and the morphology of the micro- and macro-porosity. The gray regions represent matrix growth.

matrix on surrounding fibers. If the residual porosity cannot be significantly reduced and controlled during processing, much of the anticipated fracture toughness of CMCs may be lost.

Densification mechanisms in CMCs must be better understood if this porosity is to be controlled and minimized. A number of efforts have been made to study the evolution of the composite microstructure during densification (29, 30). These efforts have attempted to relate matrix growth rates to transport properties such as gas permeability and effective diffusion coefficient. These studies have been handicapped by the lack of direct experimental observation of the evolving microstructure. Scanning electron microscopy of interrupted CVI runs, for example, gives only a two-dimensional surface view at one particular time. It is not feasible to obtain local deposition rates or to identify three-dimensional interconnections between the pores using such methods.

Because XTM is ideally suited for imaging material microstructures in three dimensions, the microporosity between individual filaments in the fiber bundles, the channel porosity between individual cloth layers, and the interconnectedness of the large through-ply holes that remain after processing can all be examined. Current XTM research is focused on monitoring the processing of silicon carbide (SiC) matrix/Nicalon fiber (amorphous SiC) composites. These are high temperature CMCs being considered for use in engines and heat management applications. Approximately 500 Nicalon fibers are packed into the tows that compose the cloth. The matrix is grown by the decomposition of methyltriclorosilane gas in a process known as chemical vapor infiltration (CVI). Two variations for chemical vapor infiltration are being studied with XTM. The first is an isothermal process where the vapor diffuses into the fiber preform. The second is a forced flow, thermal gradient method where the vapor permeates the preform under an applied pressure gradient.

Ideally, the matrix grows uniformly in both processes, and pores remain surface connected until full density is achieved. In reality, however, each type of pore is believed to densify at a different rate and to have different effects on gas transfer through the composite during infiltration. Micropores between individual fibers become closed off early on in CVI, and macroporosity in the interstices of the cloth begins to dominate the vapor transport. The ability to follow the spatial variation of pore sizes and their interconnectedness using XTM will allow local variations of matrix deposition and pore tortuosity (directly related to vapor transport) to be measured directly. In addition, the matrix growth rate as a function of position in the preform must be related to differences in the fiber architecture of the composite. All of these microstructural measures will benefit from the three-dimensional, noninvasive imaging capability of XTM.

Preliminary experiments conducted at the Cornell High Energy Synchrotron Source (CHESS) and at Stanford Synchrotron Radiation Laboratory (SSRL) assessed the capabilities of XTM to image the microstructures of two- and three-dimensional Nicalon fiber preforms (the composite before densification) and fully infiltrated composites (31). The purpose of these studies was to determine if the spatial resolution and the contrast sensitivity of XTM were sufficient to observe both the microporosity and macroporosity in the preform and fully infiltrated states of the composites. The major uncertainty was whether the X-ray absorption contrast between the amorphous SiC in the Nicalon fibers and the crystalline SiC matrix could be contrasted using XTM. The expected contrast between the matrix and the fibers (30%) is much greater than the reconstructed image noise (< 5%). However, the small size of the Nicalon fibers (10–15 μm diameter), combined with their high packing density (500 fibers per tow), places severe constraints on the noise level in the images that will still allow fiber detection. This is related to the concept of contrast-detail-dose (the CDD concept familiar to medical imaging), which states that for a given contrast, small features are less likely to be detected by the human visual system than larger features.

Figure 5a is an XTM cross-section of a fully reacted preform. This cross-section demonstrates that the fiber tows can be imaged with excellent visual perception. Individual fibers are resolved near the tow peripheries, and it is only in the densest regions of the tows, where the fibers are touching, that the ability to completely separate individual fibers is lacking. Nevertheless, microporosity is readily apparent between the fibers. A macropore lying between tows can also be seen with exceptional contrast. Because it is possible to image the fibers, the position of the macropore within the weave can be precisely located. In this sample, the macropore is associated with both an interstitial location in the weave (similar to that shown in the example in Figure 4) and an interlamelar position.

Figure 5b is an optical micrograph of a polished section taken subsequent to XTM imaging. The close correlation between the micrograph and the cross-section in Figure 5a indicates that all porosity larger than a few micrometers in size can be detected by XTM. Subtle differences between the micrograph and XTM section are due to polishing artifacts and to the different orientation of the planes between the two sections. Close examination of the contiguous XTM sections above and below the plane shown in Figure 5a indicate that the polishing is non-uniform. The edges of the composite have evidently separated from the potting compound and have undergone preferential erosion. Furthermore, some of the fibers lying in the polishing plane have been pulled out, since they are absent in the polished section, but present in the XTM section. Because

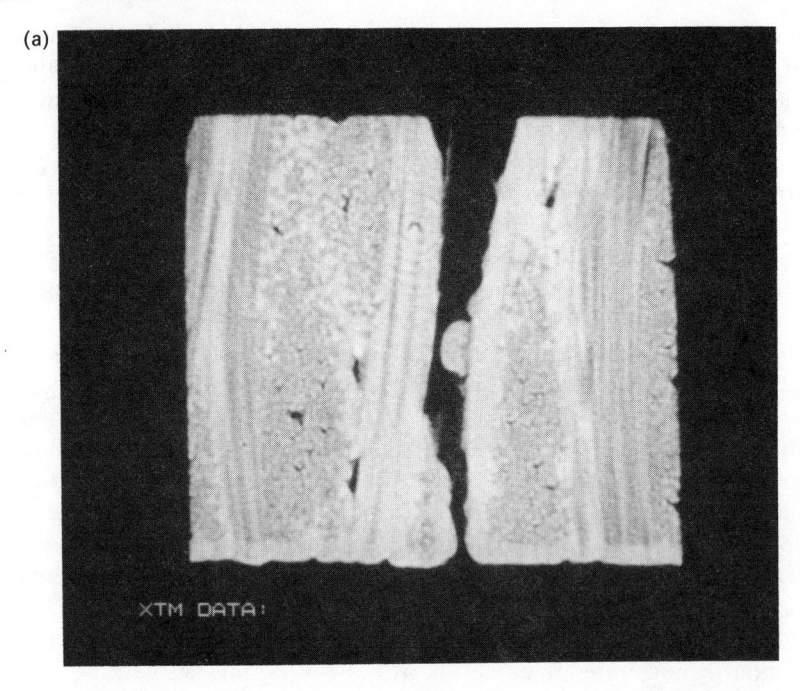

(a)

(b)

TOMOGRAPHIC MICROSCOPY 137

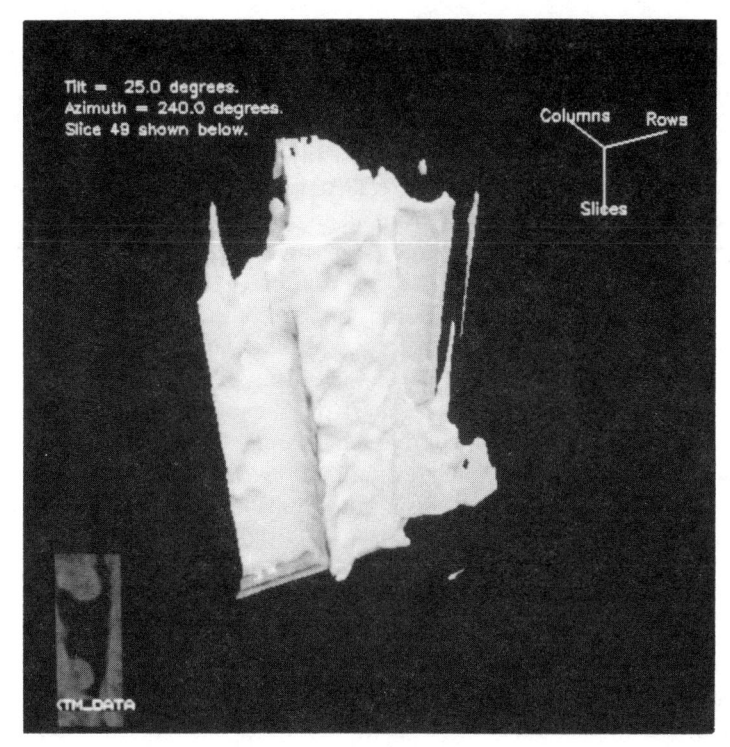

Figure 6 A three-dimensional rendering from XTM data of the macropore shown in Figure 5a,b. The matrix material has been made transparent such that only the pore can be seen. The shape of the macropore indicates that it is located in the square interstice of a single cloth layer and continues interlammelarly for a significant distance. Micropores can be seen in this rendering surrounding the macropore. The micropores are not surface connected and are confined to small regions within the tow. The shape of the micropores is generally rod-like, and the radius is usually less than a few fiber diameters (30 μm).

of this, obtaining quantitative, sectioned micrographs on these materials is difficult, and quantitative three-dimensional sectioning at the level of several micrometers is probably not practical except by using XTM.

Figure 6 is a three-dimensional rendering of the porosity in the fully infiltrated composite shown in Figure 5. The three-dimensional rendering

Figure 5 (*a*) XTM cross-section of a fully infiltrated SiC matrix/Nicalon fiber composite showing micro- and macro-porosity in addition to the individual Nicalon fibers. (*b*) Optical micrograph of a polished section near the same slice plane shown in 5a. Though great care has been taken during the polishing, fiber pullouts in the sectioned plane are unavoidable. Other artifacts from the polishing include preferential sample erosion near the edges of the sample and in some of the micropores.

of the pore has been produced from XTM data by setting all low values of the measured attenuation coefficient (representing air and very low density materials) to a high value and inverting the high attenuation values (representing matrix material) to zero. A constant contour is established, and the contour surface is rendered visually. The three-dimensional renderings show off the true power of XTM; namely, the ability to allow the microstructures to be observed from all angles.

The spatial resolution and contrast sensitivity of XTM are sufficient to image many of the important microstructural features in complicated woven fiber architectures in ceramic matrix composites. The ability of XTM to resolve individual fibers as well as micro and macroporosity has been demonstrated. Using XTM, the densification of CMCs can now be followed by interrupting the processing from fiber preform through various stages of CVI. The efficiency of different thermal gradients and gas pressures in CVI densification can be examined, as well as the effects of various fiber architectures on the processing time and the final pore density and permeability.

Application of XTM to Metal Matrix Composite Failure

Failure in continuous fiber metal matrix composites has been extensively studied over the past several years because these materials might be of use in aerospace applications such as National Aerospace Plane and the High Speed Civil Transport. Nevertheless, many fundamental questions remain regarding the mechanical response of these materials to applied loads. This is particularly true in fatigue and creep studies, where the complex interplay between matrix and fibers in these heterogeneous structures greatly complicates analytical modeling.

In a recent proof-of-concept experiment, a monotonic load sequence in an aluminum matrix/SiC SCS-8 fiber composite was examined using XTM (32, 33). The goal of this experiment was to determine whether XTM had the spatial resolution and contrast sensitivity to detect the microstructural damage that might develop under loading at stress levels greater than the yield strength of the aluminum matrix. Figure 7 shows the types of damage that might occur during loading. The important microstructural features include interfacial disbonding, fiber breaking, and matrix microcracking. Because these features might not lie in any single plane, a three-dimensional volumetric image is advantageous for assessing the nature of the damage expected in these materials.

Figure 8 shows the XTM image of the as-received microstructure in the composite. Clearly seen are the carbon cores of the SCS-8 fibers (approximately 30-μm diameter), and the contrast is sufficient to distinguish between the SiC layer and the aluminum 6061 matrix, even though

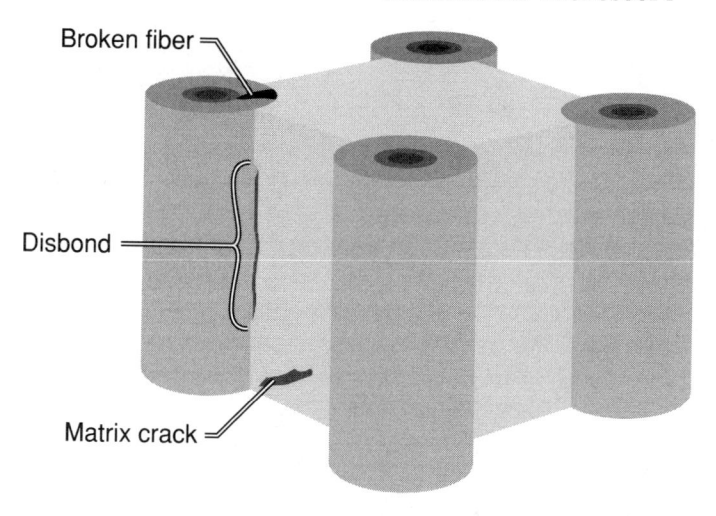

Figure 7 The important microstructural features in an aligned fiber-metal matrix composite that can be imaged using XTM. The microstructural features include fiber-matrix disbonding, broken and cracked fibers, and matrix microcracking. Fiber architecture and variations in fiber orientation can be imaged with XTM as well.

the calculated contrast difference is less than 10%. If ten or more pixels in any region are averaged to reduce statistical noise, then the measured μ values agree with the calculated μ values within a few percent, even over two orders of magnitude in value (when the contrast in the image is stretched, the carbon cores can be seen with respect to the background!) (34). The measured and calculated values of μ are listed in Table 1. The excellent agreement between the values measured using XTM and those calculated from the tabulated attenuation coefficients is another good indication that XTM is an imaging tool and, to a limited extent, an analytical tool as well.

Table 1 Calculated X-ray attenuation coefficients compared to XTM measurements in a MMC

Material	cm^{-1} μ_{obs}	cm^{-1} μ_{calc}	$\dfrac{\mu_{calc}}{\mu_{obs}}$
Crack	0.07 ± 0.04	0	—
Graphite	0.88 ± 0.44	0.84	0.95
Aluminum (pure)	8.09 ± 0.21	7.93	0.98
Aluminum (6061)	8.15 ± 0.18	8.26	1.01
Silicon carbide	8.70 ± 0.73	8.89	1.02

Coefficients calculated using X-ray energy of 21 keV and ρ (SiC) = 3.22 and ρ (graphite) = 2.2 g/cc.

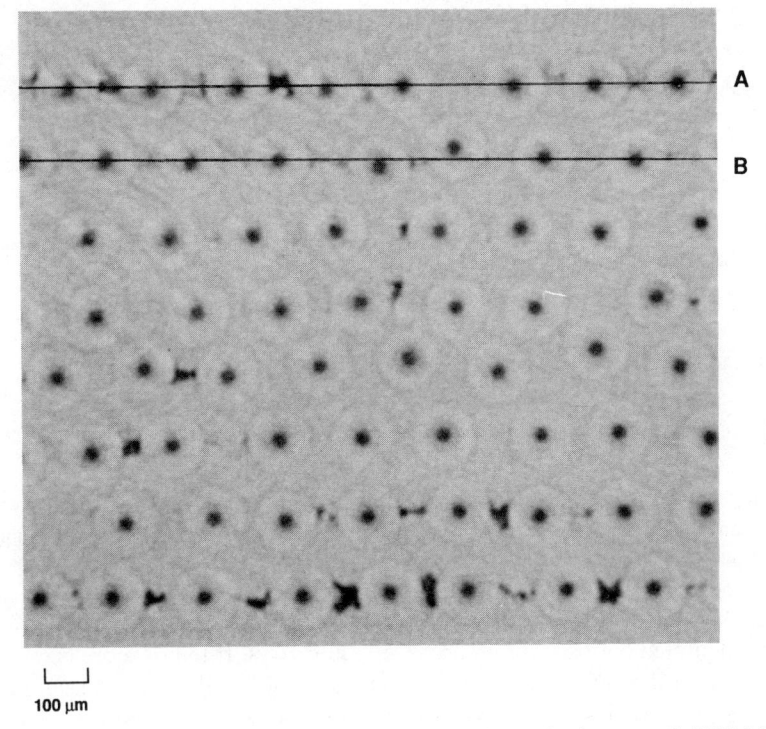

Figure 8 XTM image of the as-received microstructure in an aluminum matrix/SCS-8 SiC fiber composite. The preexisting matrix microcracking is likely caused by incomplete matrix consolidation during fabrication.

Figure 8 clearly shows significant matrix cracking, particularly in the outer plies of the composite. These cracks exist in the as-received material, and are likely the result of incomplete consolidation of the matrix during fabrication. Tensile samples were fabricated from the composite shown in Figure 8. These tensile samples were stressed in a monotonic load series extending up to the ultimate tensile strength of the composite. Several interesting observations were made (33). At stress levels greater than the yield strength of the matrix, the fibers appeared to rearrange themselves into a more closely packed configuration compared to their original distribution in the as-received panel. Also, at stress levels near half of the ultimate tensile strength of the composite, the carbon cores were seen to begin breaking. They continued to break with ever increasing frequency up to loads that ultimately produced composite failure. The core failure does not immediately seem to affect the fiber integrity, since no catastrophic fiber failures were observed after core breaking. However, this lack of affect on the fiber might be explained by the low modulus values for the carbon monofilament core reported in the literature; that is, the carbon core does not carry much of the fiber load. Nevertheless, the separation of the core exposes the interior wall of the SiC fiber and may open up defects that eventually lead to failure from the inside of the fiber.

Bhatt & Hull (35) recently reported results from single-fiber failure studies on SCS-6 and SCS-9 fibers. Identification of the primary fracture surface or the critical flaw responsible for fiber failure was difficult because of the tendency of the fibers to shatter. Nevertheless, fractographic analysis using SEM revealed that the as-received fibers failed from an unidentified flaw at the core/sheath boundary. Because XTM allows imaging of the fibers while they are still in the matrix, the fiber remains contained, and the fracture can be analyzed at the point of failure. The evidence of core/sheath failure in single fibers by Bhatt & Hull is consistent with XTM observations of core failure prior to total fiber failure in the fabricated composite. Hence, in the absence of interfacial reactions between the fiber and the matrix, it can be argued that ultimate fiber failure in the SCS fibers initiates at the fiber core.

Figure 9a shows an XTM cross-section taken through the fracture surface of a composite tensile specimen, which was loaded to failure. The cross-section is taken through the fracture surface, which has a very irregular and rough topology. An XTM section taken vertically—that is, perpendicular to the fracture surface—is shown in Figure 9b. The vertical cross-section cuts through the line marked A in Figure 8. The fracture surface is rough, and much subsurface damage (fiber breaks) can be seen. The observation of multiple fiber fractures located beneath the fracture surface is consistent with numerous observations of failed aligned-fiber tensile samples.

Figure 10 is a three-dimensional XTM rendering of a small portion of the fracture surface that shows a region of fiber pullout. The image is lower in resolution than that obtained using an SEM. Some of the lost resolution is due to the rendering algorithm, which smooths the data somewhat to reduce memory requirements. XTM is not intended to compete with the SEM in spatial resolution; rather, the power of XTM lies in its ability to image beneath the surface layers. This is demonstrated in Figure 11, where the fiber pullout has been rendered three-dimensionally in such a manner to show the multiple fiber fractures beneath the surface. Subsurface imaging of this kind is impossible with the SEM.

Figure 11 is a three-dimensional rendering of the multiple fiber fractures lying beneath the pulled-out fiber shown in Figure 10. As in the image of the pore in Figure 6, the materials data have been inverted to show space and very low density regions as being opaque, and matrix and fiber material as being transparent. Figure 11 is an image of the pulled-out region, the carbon core, and the fracture surfaces in the fiber. What is interesting in

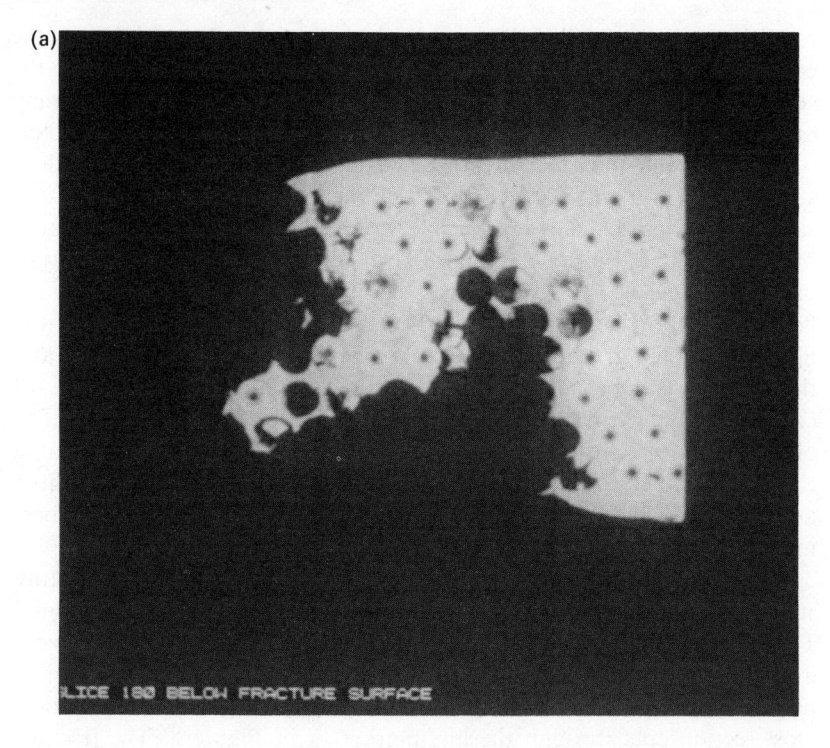

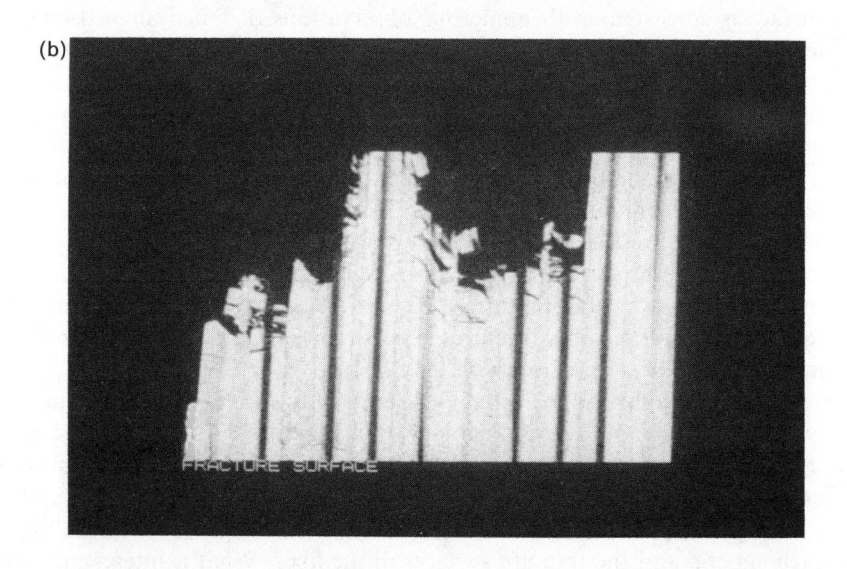

(a)

(b)

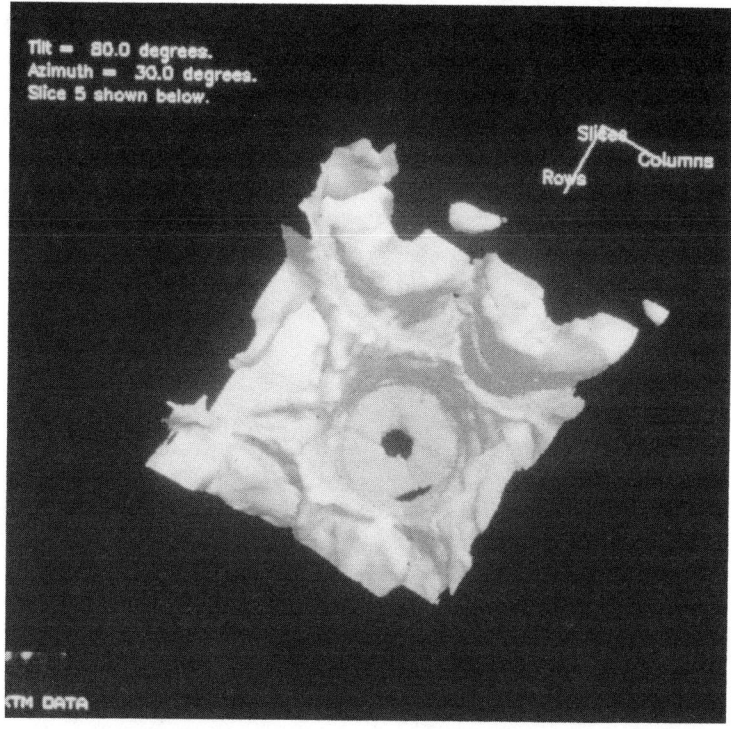

Figure 10 A three-dimensional rendering of the XTM image of a small portion of the fracture surface showing a fiber pullout. Though the XTM image is reduced in resolution by inherent smoothing in the surface rendering algorithm used to display the data, it is still possible to see the scalloped fracture markings on the broken fiber.

this figure is the observation that the fracture is at approximately 45 degrees to the load axis. That the fracture surface does not appear to lie in plane with the applied load indicates that failure, at least in these fibers, may have been caused by torsion or some unexplained features in the fiber microstructure. If torsion can be ruled out, it may be possible to explain shear-like failure in terms of a heterogeneous fiber microstructure. Three-dimensional finite element models predict this shear-like mode of failure if the fiber interior is softer than the exterior, and if the fiber is in a large state of residual compressive stress (E. Zywicz, personal communication;

Figure 9 (a) An XTM cross-section taken through the fracture surface of the composite tensile specimen loaded to failure. The cross-section is perpendicular to the load axis. Note the fiber fractures, which do not lie in the plane of the applied load. (b) An XTM section taken vertically to line *A* in 8. The fracture surface is rough, and much subsurface damage can be seen.

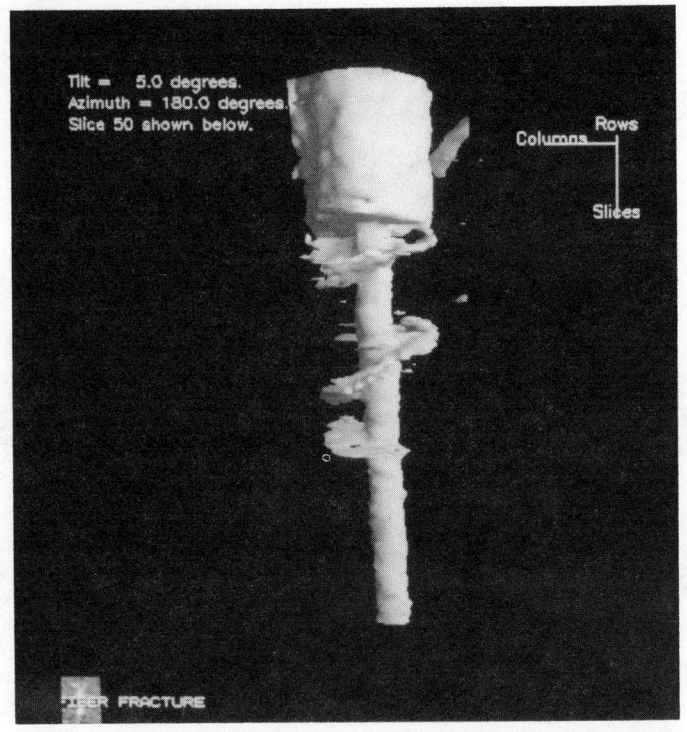

Figure 11 A three-dimensional rendering of the pulled out fiber shown in Figure 10. The materials data have been inverted to show space as opaque, and the matrix and fiber material as transparent. The image shows the carbon core and the fractures in the fiber. For the first time, fractures can be observed in-situ, and the shapes, positions, and separations can be recorded for comparison to models.

36). From what has been observed of the microstructure to date, however, this is probably unlikely (37, 38).

Application of XTM to Imaging Hard Mineral Tissues

The microchemical and microstructural transformations occurring in the hard mineral tissues in teeth as a result of carious attack are traditionally studied using microradiography and optical microscopy (39, 40). Though these studies are performed using thin sections, the thinnest sections that allow sample integrity (50–100 μm) are still too thick to prevent image blurring from overlapping microstructural features in the tissue. Furthermore, optical microscopy and microradiography are difficult to make quantitative.

In spite of the drawbacks imposed by traditional two-dimensional imaging techniques, our understanding of caries-like lesion development in enamel (the hard mineral shell surrounding the tooth) is quite good (41). However, our understanding of caries attack in dentin, the tissue lying between the enamel and the pulp, is not as thorough as it is in enamel. This is because the microstructure of dentin is much more complicated than that of enamel, and because dentin responds dynamically to infection since it is in communication with living cells in the pulpal cavity of the tooth.

Dentin, like enamel, is composed of hydroxylapatite (HAP) (a calcium phosphate mineral) and organic matter. Unlike enamel, which is 95% by weight mineral phase, dentin contains approximately 65% by weight mineral intertwined in a collagen matrix (42). The structure is composed of tubules (1–5 μm diameter) that run between the pulpal cavity and the dentin/enamel interface [the so-called dentino-enamel junction (DEJ)]. These tubules can be likened to a circulatory system for the tooth. Surrounding each tubule is a hypermineralized region ($\sim$1–3 μm thick) that is more highly mineralized than the surrounding intertubular dentin. The tubule density ranges from between 10–20,000 tubules per mm^2 near the DEJ to upwards of 70,000 per mm^2 near the pulpal cavity. It is this complex microstructure relative to enamel that leads to difficulties in modeling caries growth in dentin.

The complex, overlapping microstructure of dentin limits the usefulness of microradiography as a probe of caries processes. For example, in the formation of sclerotic dentin, the tubules become filled by mineralization; it would be difficult to observe these mineralized tubules through a thick section containing hundreds of tubules. Because of this, back-scattered scanning electron microscopy (BSEM) has recently been applied to studies of the mineralization processes occurring near carious lesions (43). BSEM offers very high spatial resolution and avoids the need to prepare the thin sections that are required for transmission studies. Nevertheless, two drawbacks prevent BSEM from realizing its full potential in studies on hard mineral tissues. First, though BSEM is sensitive to changes in average atomic composition, the backscatter yield does not provide a quantitative measure of the mineral density because there are as yet no acceptable reference standards for mineralized tissues. Second, BSEM is destructive in the sense that the specimen has to be polished prior to imaging. Therefore, time histories of carious attack are difficult to evaluate and correlate to preexisting microstructures in the tissues.

In the hard tissues of teeth, hydroxylapatite accounts for nearly all of the X-ray attenuation at energies between 15–25 keV. Therefore, an X-ray attenuation map obtained using XTM should provide a quantitative measure of the mineral density in the vicinity of caries. Furthermore, the three-dimensional attribute of XTM should facilitate the observation of those dentinal tubules that have become occluded during the formation of sclerotic dentin. XTM was used successfully to image the mineral density variations in and about a naturally occurring carious lesion in a human tooth (44). The same sample was then examined using BSEM and energy dispersive spectroscopy (EDS), using an SEM, to correlate XTM data to electron microscopy results. Preliminary data analysis from the XTM study indicates that the three-dimensional mineral density variations surrounding the lesion can be imaged with a spatial resolution nearly equal to that obtained in the EDS analysis. The mineral density in the uninfected dentin was measured by XTM to be 1.3 g/cc based upon the tabulated X-ray attenuation coefficients for HAP. This density is in excellent agreement with published estimates for the mineral density of dentin (1.4–1.5 g/cc). Furthermore, the mineral density in the XTM data varies from 2.1 g/cc in the remineralized dentin to as low as 0.7 g/cc in the infected demineralized tissue.

Figure 12 is one of several hundred contiguous XTM sections taken through a natural caries lesion in a human molar. The slice thickness is approximately 5 μm. The lesion has progressed sufficiently such that the

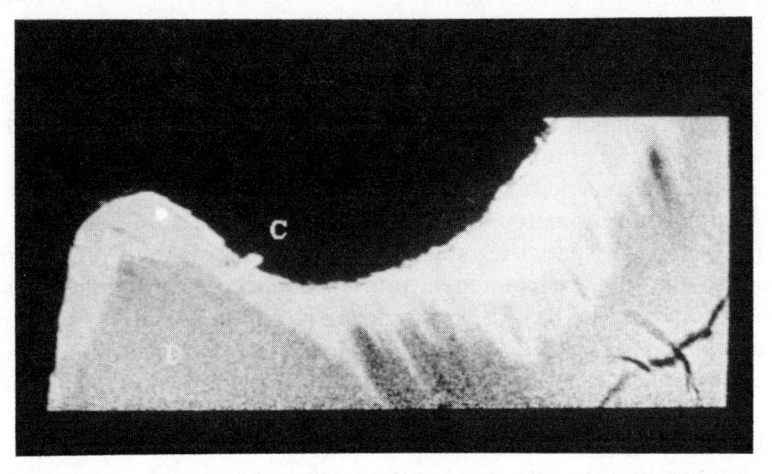

Figure 12 One of several hundred contiguous 5-μm thick XTM sections taken through a natural caries lesion in a human molar. The upper surface is tissue, which has been exposed to saliva for a considerable length of time. The protective enamel coating has been removed through demineralization resulting from bacterial activity. Small layers of calculus, C, have deposited on the surface. Normal dentin, D, has a higher mineral density than the demineralized dentin (darker gray regions) and much lower mineral density than the remineralized dentin (brighter regions in the image). Occluded dentinal tubules can be seen running perpendicular to the remineralization front.

enamel layer has been removed through complete demineralization. The exposed surface of the dentin has remained relatively free of calculus, *C*. The light gray regions of the image, *D*, contain the unaltered dentin of mineral density 1.3 g/cc. The dark gray to black regions of the image contain demineralized dentin, which has only half of the mineral density of unaltered dentin. The brighter regions near the tooth surface contain remineralized dentin that is 50% greater in mineral density than normal dentin. This indicates that the remineralized dentin most likely does not retain the structure of the natural dentin. Most important in the image of the lesion is the observation of occluded tubules, *OT*. Several of these tiny structures can be seen in the image. That these are indeed mineralized tubules has been confirmed by post-mortem sectioning and examination with an electron microscope.

The spatial resolution, coupled with the quantitative values for the mineral density in three dimensions, makes XTM a potentially powerful tool for imaging mineralized tissues. More important, however, is that by not having to prepare thin sections, it should be possible to image the evolution of the microstructure during carious attack.

FUTURE DIRECTIONS

The state-of-the-art spatial resolution of XTM is between 2 to 3 micrometers using X-ray energies from 5–35 keV. Isolated cracks as small as 0.5 to 1 micrometer can be detected as long as their length spans several detector elements (20–30 μm). No XTM experiments using synchrotron radiation in excess of 35 keV have been reported. Though the spatial resolution and crack detectability are adequate for many studies, even more applications for XTM can be envisioned if the spatial resolution could be increased to 1 micrometer, and if the energy range could be increased beyond 50 keV.

Though it is possible to design diffraction-limited optics for imaging the scintillator, and hence allow micrometer resolution in principle, it is difficult to design a scintillator that preserves spatial resolution and still manages to stop a measurable number of X-ray photons. Because of this, some researchers are developing X-ray optics for the next generation of tomographic microscopes to magnify the X-ray image with minimal distortion (25, 26, 45). Distortionless X-ray optics will relax the resolving requirements of the scintillator. Furthermore, because X-ray optics can be designed to pass only the directly transmitted radiation, blurring caused by X-ray scatter can be greatly reduced.

Asymmetrically-cut single crystals of silicon have frequently been used to magnify X-ray beams in radiographic imaging. An asymmetric crystal is one in which the Bragg diffraction planes are not parallel to the cut surface of the crystal. In this case, the incident angle that the X-rays make to the crystal surface does not equal the outgoing angle. The magnification of a parallel incident beam, M, is in only one dimension, and is given by $M = \sin(\phi_{out}/\phi_{in})$, where ϕ_{out} and ϕ_{in} are the angles between the crystal surface and the outgoing and incident beams, respectively. Magnification in two dimensions can be achieved by orienting separate asymmetric crystals at right angles to each other. Kuriyama et al (46) reported that a pair of such asymmetrically-cut silicon crystals had been used to achieve 1 μm spatial resolution at 12.25 keV in two dimensions. Magnification of 10 and more has been achieved with asymmetric crystals in tomography experiments (25, 26).

The largest obstacle to more widespread use of asymmetrically-cut crystal magnifiers in XTM is the lack of adequate experimental access to synchrotron radiation sources for imaging experiments. First, the alignment of two mutually perpendicular crystals to the precision required for tomographic imaging is time consuming. Second, once in place, the narrow angular acceptance of asymmetrically-cut single crystals decreases the incident photon flux and increases the time required for acquiring the data more than the geometric magnification arguments would suggest. Therefore, many materials experiments sacrifice spatial resolution for the sake of experimental efficiency and the ability to image more samples during the allotted beam time.

For most imaging studies, it is not necessary to use single crystals with extremely narrow bandpass. Single-crystal monochromators and magnifiers sacrifice X-ray intensity for high energy resolution. Except for those instances where it is necessary to perform chemical or elemental imaging using absorption edge differencing techniques, a larger bandpass can be tolerated without introducing beam hardening artifacts (47). Advances over the past decade in fabricating synthetic multilayers offer the promise of using monochromators and magnifying optics with much larger bandpasses and, therefore, increased photon flux (48). Recent efforts towards reducing the period spacing of multilayers to 1 nm have achieved nearly 60% peak reflectivity with 3% bandpass up to energies in excess of 40 keV (49). Furthermore, because multilayers can be grown for curved or flat substrates, it is possible to construct diffracting mirror magnifiers similar to that originally designed by Kirkpatrick & Baez (K-B microscope) in the late 1940s (50). Because the multilayers offer versatility and higher reflecting (diffracting) angles than reflective mirror surfaces, it is possible to greatly reduce the anamorphotic distortion (unequal magnifications in the two orthogonal planes) that plagued the original K-B microscopes.

The access to synchrotron radiation for imaging experiments will improve dramatically over the next several years. At least two imaging beamlines are being proposed for the Advanced Photon Source at Argonne National Laboratory, which is now under construction. In addition, imaging beamlines are being proposed for the Advanced Light Source at Lawrence Berkeley Laboratory (now nearing completion) and the European Synchrotron Radiation Facility in Grenoble, France. The presence of dedicated imaging beamlines at the new synchrotron facilities will lead to major improvements in both spatial resolution and efficiency of XTM and other X-ray imaging methods.

In the future, we can envision better access to advanced synchrotron radiation sources, improved X-ray optics, larger format detectors for imaging bigger samples, and more powerful computers for visualizing and analyzing three-dimensional data. These improvements to the existing capabilities should increase the spatial resolution, the overall size, and the type of samples that can be imaged. Furthermore, work is continuing on improved load frames and other in situ observation cells that will allow imaging of samples while they are under load or undergoing chemical transformations. All of these developments will increase the usefulness of XTM for microstructural imaging.

SUMMARY

This review has discussed the application of X-ray tomographic microscopy to materials science. An X-ray tomographic microscope has been described that provides three-dimensional images of a material's microstructure with a spatial resolution of a few micrometers. Though XTM is similar in many respects to industrial CT, the applications and resolving powers are different. Whereas industrial CT is used to certify finished components for the absence of macrostructural flaws, XTM is useful for three-dimensionally imaging the microstructure of materials with much higher resolution and sensitivity.

The three examples described in this review were chosen to illustrate how XTM can be applied to a variety of problems in materials science. The example of the ceramic matrix composite demonstrates how complex networks of porosity can be mapped out and visualized in three dimensions. The metal matrix composite study is illustrative of how the noninvasive nature of XTM can be used to study dynamic response in materials—the time sequence of fiber failure during tensile loading, for example. Finally, imaging the mineral density variations caused by a caries lesion in a human tooth illustrates how fragile, nonconductive samples can be imaged quantitatively with little or no preparation.

XTM technology is in its infancy. Even so, XTM can now be used in many applications. The resolution and sensitivity are sufficient to image many of the important microstructural features that control the properties of advanced composite materials. Coupled with its demonstrated ability to image materials during testing, XTM promises to provide valuable information as to the response of materials to creep and fatigue.

XTM is not meant to replace quantitative metallography or electron microscopies; rather, we see the technique as augmenting existing characterization methods. The ability to nondestructively and three-dimensionally image microstructures prior to sectioning and polishing will add to the effectiveness of traditional methods for analyzing materials by indicating where important features are located within the sample and may even eliminate the need for destructive analytical methods in many cases.

ACKNOWLEDGMENTS

Much of the work described in this review has been performed under the auspices of the United States Department of Energy by Lawrence Livermore National Laboratory (LLNL) under contract W-7405-Eng-48 and by Sandia National Laboratory under contract AT-(29-1)-789. The development and application of XTM to fracture-resistant composites has been provided through a grant from the Department of Energy's Advanced Industrial Concepts Materials Program (Charles Sorrell, program manager). The work on hard mineral tissues has been supported in part by National Institutes of Health grant PO1-DE-09859. The authors acknowledge the support of the staff of the Cornell High Energy Synchrotron Source (CHESS), funded by grants to B. W. Batterman from the National Science Foundation, and to the staff of the Stanford Synchrotron Radiation Laboratory (SSRL).

The authors acknowledge the computational efforts of R. A. Saroyan (LLNL), the electronics engineering support of W. N. Massey (LLNL), the mechanical engineering support of J. Celeste (LLNL), and the guidance of C. Logan (LLNL) during various phases of the development of XTM. We also appreciate and acknowledge the support and efforts of U. Bonse and R. Nusshardt (University of Dortmund) for work in all aspects of XTM design and development and particularly for their work in design and fabrication of imaging monochromators. Appreciation is also extended to R. Pahl and F. Busch (both from University of Dortmund) for their assistance in carrying out synchrotron radiation experiments. And finally, the authors acknowledge the hard work and efforts of T. M. Breunig (Georgia Institute of Technology), whose thesis work in applying XTM to imaging metal matrix composites has gone beyond the proof-of-concept phase.

Literature Cited

1. Grodzins, L. 1983. *Nucl. Instrum. Meth.* 206: 541
2. Krinsky, S., Perlman, M. L., Watson, R. E. 1983. In *Handbook on Synchrotron Radiation*, ed. D. E. Eastman, Y. Farge, 1A: 65. New York: North-Holland
3. Tao, G. A., Pella, P. A., Rousseaue, R. M. 1984. *CALCO NBSGCS: X-ray Fluorescence Quantitative Analysis Prog.*
4. Plechaty, E. F. 1981. In *Tables and Graphs of Photon-Interaction Cross Sections*, UCRL-50400, Vol. 6, Rev. 3. Livermore, Calif: Lawrence Livermore Natl. Lab.
5. Herman, G. T. 1980. *Image Reconstruction from Projections: The Fundamentals of Computerized Tomography.* New York: Academic
6. Kinney, J. H., Saroyan, R. A., Massey, W. N., Nichols, M. C., Bonse, U., Nusshardt, R. 1991. In *Review of Progress in Quantitative Nondestructive Evaluation*, ed. D. O. Thompson, D. E. Chimenti, 10A: 427. New York: Plenum
7. Bonse, U., Johnson, Q., Nichols, M., Nusshardt, R., Krasnicki, S., Kinney, J. 1986. *Nucl. Instrum. Meth.* A246: 43
8. Elliott, J. C., Dover, S. D. 1982. *J. Microsc.* 126: 211
9. Stock, S. R., Guvenilir, A., Elliott, J. C., Anderson, P., Dover, S. D., Bowen, D. K. 1989. In *Advanced Characterization Techniques for Ceramics*, 5: 161. Westerville, Ohio: Am. Ceram. Soc.
10. Spanne, P., Rivers, M. L. 1987. *Nucl. Instrum. Meth. Phys. Res.* B24/25: 1063
11. Seguin, F. H., Burstein, P., Bjorkholm, P. J., Homburger, F., Adams, R. A. 1985. *Appl. Opt.* 24: 4117
12. Cueman, M. K., Thomas, L. J., Trzaskos, C., Greskovich, C. 1989. In *Review of Progress in Quantitative Nondestructive Evaluation*, ed. D. O. Thompson, D. E. Chimenti, 8A: 431. New York: Plenum
13. Feldkamp, L. A., Jesion, G., Kubinski, D. J. 1989. See Ref. 12, p. 381
14. Feldkamp, L. A., Davis, L. C., Kress, J. W. 1984. *J. Opt. Soc.* A1: 612
15. Martz, H. E., Azevedo, S. G., Brase, J. M., Waltjen, K. E., Schneberk, D. J. 1990. *Int. J. Radiat. Appl. Instrum.* 41: 943
16. Johnson, Q. C., Kinney, J. H., Bonse, U., Nichols, M. C. 1986. In *Proc. Mater. Res. Soc. Symp.* 69: 203. Pittsburgh: Mater. Res. Soc.
17. Kinney, J. H., Johnson, Q. C., Bonse, U., Nusshardt, R., Nichols, M. C. 1986. *SPIE* 691: 43
18. Flannery, B. P., Deckman, H., Roberge, W., D'Amico, K. 1987. *Science* 237: 1439
19. Deckman, H. W., D'Amico, K. L., Dunsmuir, J. H., Flannery, B. P., Gruner, S. M. 1989. In *Advances in X-ray Analysis*, ed. C. S. Barrett, J. V. Gilfrich, R. Jenkins, T. C. Huang, P. K. Predecki, 32: 641. New York: Plenum
20. Kinney, J. H., Johnson, Q. C., Saroyan, R. A., Nichols, M. C., Bonse, U., et al. 1988. *Rev. Sci. Instrum.* 59: 196
21. Kinney, J. H., Johnson, Q. C., Saroyan, R. A., Nichols, M. C., Bonse, U., et al. 1989. *Rev. Sci. Instrum.* 60: 2475
22. Suzuki, Y., Hirano, T., Usami, K. 1990. In *X-ray Microscopy in Biology and Medicine*, ed. K. Shinohara, p. 179. Berlin: Springer-Verlag
23. Deleted in proof
24. Deleted in proof
25. Bonse, U., Nusshardt, R., Busch, F., Pahl, R., Kinney, J. H., et al. 1991. *J. Mater Sci.* 26: 4076
26. Hirano, T., Usami, K., Sakamoto, K. 1989. *Rev. Sci. Instrum.* 60: 2482
27. Hirano, T., Funaki, M., Nagata, T., Taguchi, I., Hamada, H., et al. 1990. *Proc. Natl. Inst. Polar Res. Symp. Antarctic Meteorites*, 3: 270
28. Hirano, T., Eguchi, S., Usami, K. 1989. *Jpn. J. Appl. Phys.* 28: 135
29. Starr, T. L. 1988. *Ceram. Eng. Sci. Proc.* 9: 803
30. Besmann, T. M., Sheldon, B. W., Lowden, R. A., Stinton, D. P. 1991. *Science* 253: 1104
31. Kinney, J. H., Nichols, M. C., Bonse, U., Stock, S. R., Breunig, T. M., et al. 1991. In *Advanced Tomographic Imaging Methods for the Analysis of Materials*, 217: 81. Pittsburgh: Mater. Res. Soc.
32. Kinney, J. H., Stock, S. R., Nichols, M. C., Bonse, U., Breunig, T. M., et al. 1990. *J. Mater. Res.* 5: 1123
33. Breunig, T. M., Stock, S. R., Kinney, J. H., Guvenilir, A., Nichols, M. C. 1991. See Ref. 31, p. 135
34. Breunig, T. M. 1991. PhD thesis. Georgia Inst. Technol., Atlanta
35. Bhatt, R. T., Hull, D. R. 1991. *NASA Tech. Mem. 103772/AVSCOM Tech. Rep. 91-C-014, 15th Ann. Conf. Composites Advanced Ceramics, Am. Ceram. Soc., Cocoa Beach, Fla.*
36. Deleted in proof
37. Sattler, M. L., Kinney, J. H., Nichols, M. C., Alani, R. 1992. *The Microstructures of SCS-6 and SCS-8 SiC Reinforcing Fibers, 16th Ann. Conf. Com-

posites Advanced Ceramics, Am. Ceram. Soc., Cocoa Beach, Fla.
38. Ning, X. J., Pirouz, P. 1991. *J. Mater. Res.* 6: 2234
39. Groeneveld, A., Theuns, H. M., Kalter, P. G. E. 1978. *Arch. Oral Biol.* 23: 75
40. Langdon, D. J., Elliott, J. C., Fearnhead, R. W. 1980. *Caries Res.* 14: 359
41. van Dijk, J. W. E., Borggreven, J. M. P. M., Driessens, F. C. M. 1979. *Caries Res.* 13: 169
42. Baskar, S. N. 1980. *Orban's Oral Histology and Embryology*, p. 107. St. Louis: Mosby
43. Marshall, G. W., Staninec, M., Torii, Y., Marshall, S. J. 1989. *Scanning Microsc.* 3: 1043
44. Kinney, J. H., Marshall, G. W., Marshall, S. J. 1992. *Ann. Meet. Am. Assoc. Dental Res. Abstr.*
45. Nusshardt, R., Bonse, U., Busch, F., Kinney, J. H., Saroyan, R. A., Nichols, M. C. 1991. *Synchrotron Radiat. News* 4: 21
46. Kuriyama, M., Dobbyn, R. C., Burdette, H. E., Black, D. R. 1990. *J. Res. Natl. Inst. Stand. Technol.* 95: 559
47. Kinney, J. H., Johnson, Q. C., Nichols, M. C., Bonse, U., Nusshardt, R. 1986. *Appl. Optics* 25: 4583
48. Barbee, T. W. Jr. 1990. *Physica Scripta* T31: 147
49. Baron, A. Q. R., Barbee, T. W. Jr., Brown, G. S. 1990. *SPIE* 1343: 84
50. Kirkpatrick, P., Baez, A. V. 1948. *J. Opt. Soc. Am.* 38: 766

MICROWAVE SINTERING OF CERAMICS[1]

Joel D. Katz

Materials Science and Technology Division, Los Alamos National Laboratory, Los Alamos, New Mexico 87545

KEY WORDS: processing, joining, powder synthesis

INTRODUCTION

Researchers claimed to have invented microwave processing of ceramics as recently as 1984 (1), however, the real originator is generally acknowledged to be Tinga (2), whose pioneering experiments date back to 1968. Activity in this field began to accelerate in the mid-1970s because of a shortage of natural gas. During this period microwaves were investigated by Sutton (3) and others (4) for drying and firing of castable alumina ceramics. In the late 1970s and 1980s, the microwave heating and sintering of uranium oxide (5), barium titanites (6), ferrites (7), aluminas (8), and glass-ceramics (9), among others, were investigated. A review of this area has been completed by Sutton (10). In addition, proceedings from three symposiums on microwave processing of materials have been published (3, 11, 12).

Microwave is the name given to electromagnetic radiation 1 m to 1 mm in wavelength that corresponds to a frequency of about 1 to 300 GHz. This frequency range falls just above radio waves and just below visible light on the electromagnetic spectrum. Most of the microwave frequency band is used for communications and radar, and consequently it is regulated by the Federal Communications Commission. The FCC has allocated 915 MHz, and 2.45, 5.85, and 20.2–21.2 GHz for industrial, scientific, and medical use. Special exemptions may be obtained from the FCC to use

---

[1] The US government has the right to retain a nonexclusive, royalty-free license in and to any copyright covering this paper.

some of the frequencies normally allotted to radar or communications for scientific or industrial purposes.

Present industrial and medical uses are based on heating water, so of the four frequencies permitted for use because of their suitability for this purpose, only 915 MHz and 2.45 GHz see significant application. The practical implication of this is that these are the only frequencies easily available at useful power levels for the microwave processing of ceramics. A few other frequencies are available on a limited basis, i.e. 28, 60, and 140 GHz, and 500 MHz, because of their use as power sources in accelerators, plasma fusion devices, and commercial broadcasting.

The first exposure standard for microwave radiation, based on the amount of radiation necessary to heat human tissue one degree Celsius, was established in 1966. The actual standard of 10 mW/cm^2 is a tenth of the energy level required to provide one degree heating. The exposure standard was revised in 1982 to take into account the fact that microwave energy is not absorbed with the same efficiency at all frequencies. The revised standard is based on an exposure level of 0.4 W/kg, which corresponds to 0.5 mW/cm^2 at 2.45 GHz. The health effects of microwave radiation below this level are controversial. A notorious example of misuse is the bombardment of the US Embassy in Moscow with low level microwave radiation.

MICROWAVE-MATERIAL INTERACTIONS

Dielectric Properties

Material dielectric properties are usually discussed in terms of the dielectric constant and the loss tangent. The dielectric constant can be thought of as a measure of the polarizability of a material in an electric field. While the loss tangent is a measure of the absorption of microwaves by the material, the dielectric constant can be defined through the complex permittivity, which is given by the following formula:

$$\varepsilon^* = \varepsilon' - i\varepsilon'' = \varepsilon^0(\varepsilon_{r'} - i\varepsilon_{\text{eff}''}), \qquad 1.$$

where ε^* = complex permittivity, ε' = dielectric constant, ε'' = dielectric loss factor, ε^0 = permittivity in free space, $\varepsilon_{r'}$ = relative dielectric constant, $\varepsilon_{\text{eff}''}$ = effective relative dielectric loss factor, and $i = (-1)^{1/2}$. The loss tangent is defined as follows:

$$\tan \delta = \varepsilon_{\text{eff}''}/\varepsilon_{r'}, \qquad 2.$$

where $\tan \delta$ = loss tangent.

Both the dielectric constant and the loss tangent are functions of temperature as shown in Figures 1 and 2, respectively. These measurements for high purity, 99.995% Sumitomo Alumina, were made by H. Frost (personal communication). The loss is seen to increase exponentially with temperature. This behavior can be explained by extrinsic defects in the linear region of the curve and the formation of intrinsic Schottky-type defects in the exponential region (14).

Materials can be divided into three categories according to their microwave properties as shown in Figure 3. Materials with very low loss tangents allow microwaves to pass through with very little absorption and are said to be transparent to microwaves. Materials with extremely high loss tangents, i.e. metals, reflect microwaves and are said to be opaque. Materials with intermediate loss tangents will absorb microwaves. The amount of absorption is quantified by definition of the skin depth. The skin depth, which is the distance into the material at which the electric field falls to 1/e or 37% of its initial value, is defined by the following formula:

$$SD = 1/(\pi f \mu \sigma)^{1/2}, \qquad 3.$$

where SD = skin depth, f = frequency, μ = magnetic permittivity, and σ = conductivity.

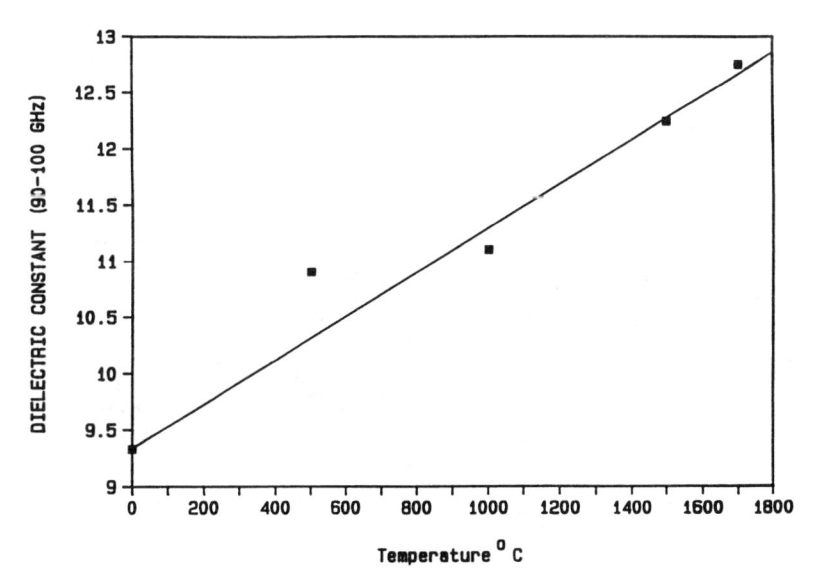

Figure 1 The dielectric constant (measured at 90–100 GHz) vs temperature for Sumitomo grade AKP-50 alumina. Note linear increase with temperature.

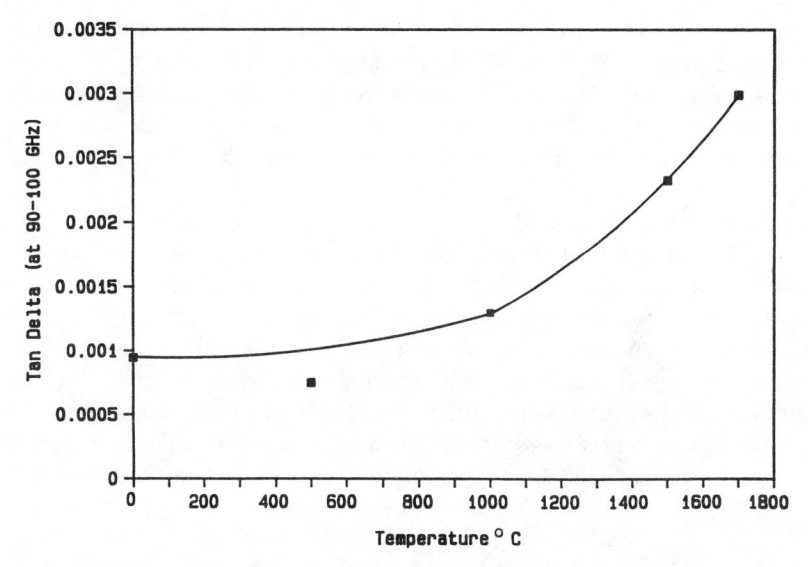

Figure 2 The loss tangent (measured at 90–100 GHz) vs temperature for Sumitomo grade AKP-50 alumina. Note sharp increase in the loss tangent above 1000°C.

Loss Mechanisms

Electromagnetic energy can be dissipated in a crystalline dielectric through several loss mechanisms. These mechanisms include electronic polarization, ionic vibration, ion jump relaxation, conduction, and interfacial polarization.

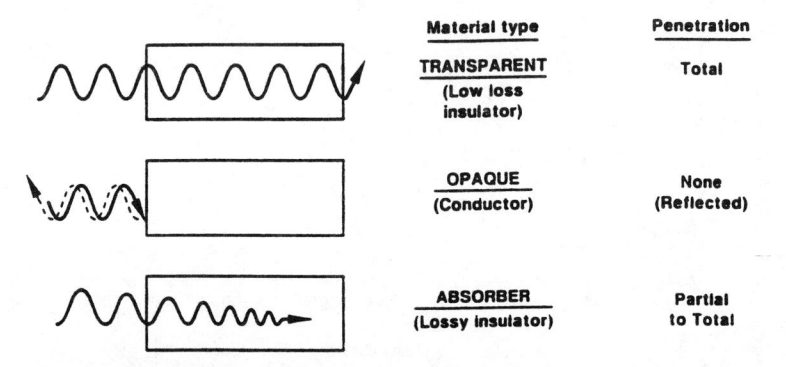

Figure 3 Materials can be divided into three categories, transparent, opaque, and absorber, depending on how they interact with microwaves.

Electronic polarization and ionic vibration are resonance phenomena and, if operable during microwave heating, have the potential to directly change the ion jump frequency. Electronic polarization would alter the energy barrier to be overcome by the jumping ion, and ionic vibration would alter the frequency of attempted jumps. Microwave frequencies are in the range of 10^9 to 10^{10} Hz, and the ion jump frequency is of the order of the Debye frequency, which is about 10^{13} Hz. Application of resonance theory shows that resonance, and hence an ionic vibration loss mechanism, is not possible for any reasonable damping force since these two frequencies differ by three to four orders of magnitude. This has been explained in detail by Kenkre (15). Since electronic polarization occurs at even higher frequencies than ionic vibration, this loss mechanism is not thought to be operable at microwave frequencies either. Thus it can be concluded that electronic polarization and ionic vibration are not important loss mechanisms for microwave heating.

Ion jump relaxation in a crystalline ceramic occurs when an aliovalent ion and vacancy form an associated pair. (An aliovalent ion is an impurity cation or anion with a valence different from that of its host sublattice.) An aliovalent ion-vacancy pair has a dipole moment associated with it that responds to the applied electric field. The vacancy is thought to jump around the aliovalent ion to align its dipole moment with the electric field (16). Interfacial polarization occurs at a structural inhomogeneity such as a grain boundary, dislocation, or vacancy cluster. In an ionic lattice there will be a localized disruption in electroneutrality at such a structural inhomogeneity with a net dipole moment that will align itself with the applied field.

Conduction is mostly of interest at low frequencies. This type of loss mechanism occurs when vacancies are not associated with other defects, and hence are not localized. Unassociated vacancies are much more mobile than associated pairs and migrate in response to the electric field.

Dielectric Heating

The power dissipated per unit volume in a dielectric undergoing microwave heating is given by the following formula:

$$P = (2\pi f \varepsilon')(E^2/2) \tan \delta \qquad 4.$$

where P = power dissipated and E = electric field.

The rate of power dissipation and the heat capacity of the material determine the rate of temperature rise. Modeling of the rate of temperature rise is not as straight forward as it might seem because of difficulties in calculating the power dissipated. The rate of power dissipation is difficult to calculate for two reasons. First, frequently the dielectric properties of the material being heated are not known as a function of temperature or density, and second, detailed information concerning the electric field in the microwave cavity and in the material being heated is not available. Nevertheless, efforts to model the heating of a dielectric material in a single-mode cavity have been made by several workers (17–19). Iskander et al (17) developed a finite-element-type model for heating alumina in a TE_{101} single-mode cavity that yields realistic results.

Microwave Cavities

Microwave cavities can be either single or multimode, tuned or untuned. When microwaves propagate into a cavity, they establish a standing wave pattern. By properly choosing the dimensions of the cavity, a fundamental standing wave pattern can be produced. Such a cavity is said to be single-moded. A multimode cavity is a cavity in which several fundamental standing waves or modes are superimposed to produce a standing wave pattern that consists of several modes. Single and multimode cavities may both be tuned to produce a desired mode pattern. Tuning is usually accomplished with the aide of a sliding short, which changes the dimensions of the cavity.

ANTICIPATED ADVANTAGES OF MICROWAVE SINTERING

Enhanced Diffusion

The observation of more rapid reaction and/or sintering during microwave processing of ceramics had led to speculation that microwave processing results in enhanced diffusion. These observations fall into two categories. The first consists of indirect evidence of diffusion such as enhanced sintering, grain growth, or reaction zones (20–22). These reactions do not provide clear evidence of enhanced diffusion since interpretation is difficult for complex processes such as sintering, grain growth, and chemical reaction.

The second body of evidence is much smaller and comprises direct measurements of diffusivities and activation energies for diffusion during microwave heating (23–25). Fathi & Clark (23) report an increase in width of the reaction zone of approximately three times for the ion exchange of potassium into a sodium-aluminum-silicate glass. These data indicate an increase in the diffusivity of approximately one order of magnitude. M. Patterson & J. McCallum (personal communication) have implanted titanium in sapphire and studied the diffusion after conventional and 2.45 GHz microwave annealing at 1000°C. Preliminary results from this study indicate that there is no difference in the concentration profiles between the conventionally and microwave-annealed samples. Janney & Kimrey

(25) report activation energies 40% lower for the tracer diffusion of oxygen in alumina annealed using 28 GHz microwaves. They obtained activation energies of 710 and 410 kJ/mol for the conventional and microwave samples, respectively, which is equivalent to an increase of approximately two orders of magnitude in the diffusivity for the temperatures studied. Katz (26) reported no enhancement for the diffusion of chromium ions in alumina as a result of microwave heating.

Katz (26) examined possible mechanisms by which microwave heating could cause an enhancement in lattice diffusion. With the exception of a possible change in the correlation coefficient for diffusion in the event that ion jump relaxation was an operable loss mechanism, no other method of altering the diffusion coefficient could be identified. Reports of an increased rate of diffusion during microwave sintering should be viewed with extreme skepticism for the following reasons. First, there is no theoretical basis for a lower activation energy for lattice diffusion in a microwave field. If it were shown that a microwave field did affect the correlation factor, the alteration to the diffusivity would be in the pre-exponential factor and not in the activation energy. Second, the enhancements in diffusion, sintering, and grain growth have with few exceptions been reported by workers in one laboratory and have not been confirmed by other laboratories. The workers at the laboratory reporting enhancements measure temperature using thermometry, almost all of the other workers in this field measure temperature using pyrometry.

Lower Processing Costs

The assumption of lower processing costs is predicated on the unique volumetric heating of microwaves. Since microwaves penetrate the dielectric with absorption and heat generation throughout the work piece, the heating process is much more efficient, possibly as high as 80–90%. This is in stark contrast to conventional heating techniques, where the surface of the work piece must first be heated by a combination of convection and radiation, and then the interior is heated by thermal conduction.

Direct volumetric heating of the work piece is expected to result in heating of a lower thermal mass because of the anticipated use of less insulation and the absence of furnace parts such as heating elements. Additional anticipated savings arise from the speed of microwave heating, which is as much as a hundred times faster than conventional heating. While cool-down times are still the same, substantial energy savings should result from the extremely rapid heat-up rates. Rapid heat-up rates should lead to significantly lower processing times and larger through-put rates, and consequently, more efficient usage of capital equipment.

Given all of the expected energy savings outlined above, one cannot help but wonder why microwave processing of alumina-based ceramics is not already standard industrial practice, especially since companies such as Special Metals (3) and General Motors (4) examined the process several years ago. Insight into why commercialization of microwave processing of alumina-based ceramics has not occurred can be gained by looking at the overall efficiency. Conversion of a fossil fuel to electricity is about 30% efficient, conversion of electricity to microwaves is about 50% efficient, and the conversion of microwaves into heat is probably about 80% efficient. The overall efficiency of the conversion of a fossil fuel to microwaves is only about 12%. The conversion of a fossil fuel to heat during a typical sintering operation is about 40% efficient.

There is little potential for energy cost savings by replacing a fossil fuel fired sintering operation with microwave sintering since the conversion of fossil fuels, especially natural gas, is so efficient. Das & Curlee (27) have performed an economic analysis of microwave sintering in which they concluded that "The adoption of microwave sintering solely on the basis of saving energy cost does not appear to be promising." Many current industrial sintering processes are carried out in electric-fired furnaces with or without controlled atmospheres. Das & Curlee did not consider the replacement of these sintering processes with microwave techniques. The potential for significant energy cost savings still exists for sintering processes currently using electricity, but is greatly reduced for fossil fuel fired processes.

At least two studies report actual numbers for power usage during microwave sintering. Katz (28) reports total power consumption to microwave sinter a 250-gm batch of alumina pellets at 4.8 kWhr/kg of alumina sintered. At $0.08/kWhr, the energy costs to sinter a kg of alumina is $0.40. Also reported are power consumption to microwave sinter alumina-5^v/o silicon carbide samples at 4.0 kWhr/kg of alumina-5^v/o silicon carbide composite sintered. At $0.08/kWhr, the energy costs to sinter a kg of the composite material is $0.32.

Patterson et al (29) have also microwave-sintered multiple samples of alumina. They calculated the net power consumption (input minus reflected power) to be 3.8 kWhr/kg of alumina. Taking into consideration the differences between the two studies and on the basis of calculated power consumption, the value of 3.8 kWhr/kg of alumina determined by Patterson et al (29) is in good agreement with the value of 4.8 kWhr/kg calculated by Katz.

Das & Curlee (27) estimate the cost to microwave sinter a kg of average ceramic at $0.155. Das & Curlee's estimate appears to be within a factor of two or three of the actual costs calculated from this study. Closer examination of their estimate indicates the seemingly good agreement is fortuitous.

Das & Curlee (27), using data from Whittemore (30), have estimated the energy needed to sinter an average ceramic to be 1860 kJ/kg. Estimating the cost of microwave energy at approximately $0.075/MJ ($0.27/kWhr), and assuming an efficiency of 90%, they arrive at a cost to microwave sinter a kg of average ceramic of $0.155. The actual power consumption is about an order of magnitude greater than that estimated for an average ceramic by Das & Curlee (27). This difference is not too surprising for two reasons. First, inefficient small scale laboratory sintering experiments were performed. Power consumption levels will be greatly lowered by scale-up to a more efficient industrial sized batch or continuous process. Second, the estimate by Das & Curlee (27) is for an average ceramic, not alumina. Another major difference is in the cost of electrical energy; Das & Curlee arrive at $0.27/kWhr for the cost to produce microwaves. This much higher number apparently arises from adding capital and depreciation costs.

Das & Curlee (27) compare the cost of microwave sintering to conventional sintering using natural gas. Such a comparison is not very valuable since it merely illustrates that natural gas is a lower cost fuel than electricity. A more valuable cost comparison was performed by Patterson et al (29), who compared the costs of microwave and conventional electric furnace sintering. These workers concluded that microwave sintering results in an energy saving of as much as 90% over conventional electric furnacing techniques.

Improved Mechanical Properties

The assumption of improved mechanical properties through microwave sintering is mostly based on (*a*) uniform volumetric heating, and (*b*) extremely rapid heat-up rates. Uniform heating ought to result in a more uniform microstructure with equiaxed grains. In addition uniform volumetric heating should minimize thermal heat-up stresses and may allow for the sintering of larger and/or more complex green shapes. Unfortunately, the assumption of more uniform heating via microwaves has not been experimentally verified. In fact, a theoretical model of microwave heating developed by Watters et al (31) for uninsulated samples indicates quite the contrary. This model predicts that microwave heating may be highly non-uniform, with temperature gradients as high as 20°C/mm.

Harmer & Brook (32) have theorized that rapid sintering will result in a finer grained microstructure. This theory applies to all rapid sintering techniques, one of which is microwave sintering. The theory of rapid sintering is based on the simple assumption that both densification and grain growth are thermally activated processes and that the activation energy for grain growth is lower than that for densification. Figure 4 is an Arrhenius plot in which the processes of densification and grain growth are depicted. In the low temperature region, grain growth is the dominant process and occurs at a faster rate than densification. In the high temperature region, the reverse is true, densification is the dominant process and occurs at a faster rate than grain growth. Thus by heating a specimen quickly to the high temperature region, grain growth, a dominant process at low temperatures, is minimized. The validity of this theory has been established through conventional rapid heating experiments.

The rapid sintering theory has been verified by Patterson et al (29) for two types of aluminas, and by Eastman et al (33) for the microwave

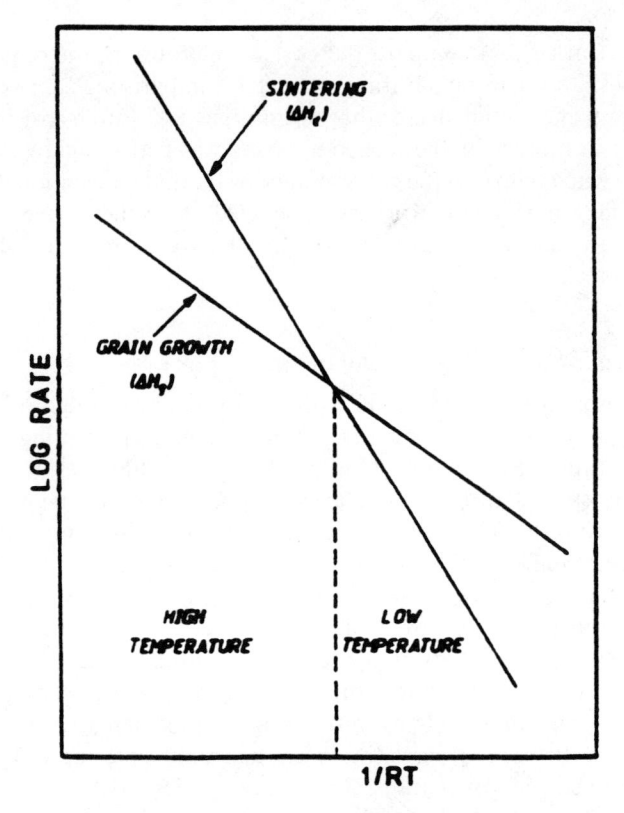

Figure 4 Arrhenius plot of the rates of densification and grain growth. The theory of rapid sintering predicts that a finer, more uniform grain structure will result by rapid heating to the high temperature region.

sintering of nanocrystalline titanium dioxide. Contrary to expectations from rapid sintering theory, however, several cases of enhanced grain growth for microwave compared to conventional sintering have been reported. Apparently enhanced grain growth results when the microwaves couple directly to the grain boundary phase. This phenomenon occurs only when the grain boundary phase has a higher microwave loss than the matrix phase and should not be seen as contradicting the rapid sintering theory.

The only demonstrated mechanical enhancements have been reported by Patterson et al (29), Tiegs et al (34), and Holcombe et al (35). Patterson reports slightly increased toughness for alumina, which is attributed to the smaller grain size attained by microwave sintering. Tiegs et al report better creep resistance for silicon nitride, which is attributed to the α-to-β silicon transformation and the formation of the acicular β grains. Holcombe et al observed greater thermal shock resistance in the form of a higher fracture stress for microwave sintered Y_2O_3-2 wt% ZrO_2. These workers theorize that there is localized melting at the surface of pores that glazes the pore surface and puts the region surrounding the pore in localized compression.

CERAMIC SINTERING

Typically sintering research is performed using 2.45 GHz microwaves, although a few studies report use of 60 (8) and 28 (21) GHz microwaves. The advantages of the higher frequencies are (a) the smaller wavelength allows for the design of a cavity with a more uniform electric field, and (b) the power dissipated in a dielectric, all other factors being equal, as seen in Equation 4, is much greater. Unfortunately there are also disadvantages to using higher frequencies. High frequency apparatus is available on a very limited basis and at an extremely high cost. A 2.45 GHz, 20 kW experimental apparatus can be purchased for well below $100,000, while a 28 or 60 GHz, 20 kW apparatus would cost $1,000,000. Another disadvantage of the higher frequencies is a thinner skin depth. Equation 3 shows that the skin depth is inversely proportional to the square root of the frequency.

Oxides

ALUMINA The vast majority of the experimental effort in microwave sintering has been directed towards oxides. Of all the oxides, alumina has received the most attention (3, 4, 23, 25, 26, 28, 29, 36–39; I. Balbaa, personal communication). This has included sintering experiments at 2.45, 28, and 60 GHz. Kimery et al (25) claim an enhancement in diffusion and densification for alumina heated by 28 GHz microwave as discussed above. Reports of enhancements in the densification kinetics for aluminas processed using 2.45 GHz microwaves are conflicting.

High purity alumina is a low loss material and as such is difficult to couple to in a 2.45 GHz multimode cavity. There is some evidence to suggest that coupling to low loss materials is easier in tuned single-mode cavities because of the high electric field strength that can be achieved locally (36). To overcome the difficulties of coupling to low loss materials, most workers use susceptors. High purity alumina bodies as large as 920 gms have been sintered using this approach (37). Workers at University of Florida (39) and Ontario Hydro (I. Balbaa, personal communication) have taken a different approach. These workers have constructed hybrid conventional-microwave furnaces that pre-heat the alumina part to achieve microwave coupling.

MIXED OXIDES Janney et al (41) report a microwave enhancement of 180°C for the sintering of ZrO_2-8 mol% Y_2O_3 to full density while using 2.45 GHz microwaves. Kimrey et al (43) have reported on the microwave sintering of Al_2O_3-10 to 70 wt% ZrO_2 using 2.45 and 28 GHz microwaves. SiC susceptors were used to produce heating at 2.45 GHz. These workers were able to achieve sintered densities in the high 90% range, which was equal to the densities produced using conventional sintering. Additionally, these workers reported a microwave effect in which they were able to achieve full density at 200 to 300°C lower temperatures, using 28 GHz microwave, than those used for conventional sintering. A similar effect was reported for 2.45 GHz, however, the microwave effect was smaller, about 100°C. Patil et al also report sintering Al_2O_3-15 to 25% ZrO_2 powders to high density, using a tunable 2.45 GHz single-mode cylindrical cavity in the TM_{012} mode.

A variety of electronic ceramics have been processed using microwaves. Harrison et al (44) demonstrated the suitability of microwave drying, calcining, binder burn-off, and sintering for PZT and PLZT. Titanites have been sintered by Humphrey (6) and Aliouat et al (45). McMahon et al (46) examined the production of ZnO-based varistors.

$YBa_2Cu_3O_{7-x}$ has been prepared using microwaves by Ahmad et al (47). Aliouat et al (45) and Cozzi et al (48) have demonstrated microwave sintering of $YBa_2Cu_3O_{7-x}$.

Non-oxides

In contrast to the large amount of work done on oxides, very little work has been performed on the non-oxides, which are either reflective, as is the case for the carbides and borides, or very low loss in the case of the nitrides.

B_4C, SiC, Si_3N_4, and TiB_2 are among the few non-oxides that have been successfully microwave sintered.

Because of the high loss of the borides and carbides, the skin depth at 2.45 GHz is very shallow, of the order of millimeters, so microwave heating of these materials is a surface phenomena and, in fact, is similar to inductive heating of a material like graphite. Clearly with such a shallow microwave penetration, many of the envisioned benefits of microwave processing will not be realized.

BORON CARBIDE   Holcombe (49) was the first to demonstrate that 2.45 GHz microwaves were readily absorbed by boron carbide. After making this observation, Holcombe (50) suggested the use of boron carbide as a coupling agent to allow the heating of materials that are not easily heated by microwaves. Katz et al (51) were the first to actually report microwave sintering of pure boron carbide, with densities as high as 97%.

A later study (52) of microwave sintering of boron carbide containing various sintering aids (carbon, Mo, CrB_2, TiB_2, and $MoSi_2$) is reported by Holcombe. Holcombe found that using 2.5% carbon produced the best densification in boron carbide. Improvements in densification of as much as 17% are reported by Holcombe for microwave vs conventional-resistance furnace-sintered boron carbide held at 2150°C for 30 min. The maximum density reported, however, is a relatively low 90%. Microwave sintered material was also reported to have a larger average grain size, two to four times larger for the microwave vs hot-pressed and conventional-resistance furnace-sintered material, respectively. This work was performed using an elaborate four-layer coating technique to preclude reaction with the insulation material.

SILICON CARBIDE   Tian et al (53) were able to heat SiC rods to 1700 and 2200°C in argon, pressurized to 0.40 MPa, and nitrogen, pressurized to 0.8 MPa, respectively. These workers used a TE_{111} cylindrical cavity filled with pressurized gas to forestall the formation of a plasma, which ultimately limited the maximum temperature that could be attained. In spite of the high temperatures attained, these workers did not observe any densification.

SILICON NITRIDE   Most nitride ceramics, including silicon nitride, are very low loss and extremely hard to heat using microwaves. Sintering of pure silicon nitride, however, is impractical, and yttrium and aluminum oxides are typically added as sintering aids. The addition of these oxides makes microwave heating achievable, as demonstrated by Tiegs et al (54), who have performed both sintering and annealing studies on silicon nitride. These workers did not report a decrease in the sintering temperature, but did note improved densification of from 5 to 15%, using 2.45 and 28 GHz microwaves. Tiegs and co-workers also report high weight loss for microwave-annealed silicon nitride caused by silicon monoxide vaporization from the grain boundaries. Heating of silicon nitride is believed to occur by coupling of the microwave radiation to the liquid phase located in the grain boundaries, and all processes involving the grain boundaries are reported to be accelerated. The α-to-β silicon nitride transformation and grain growth is reported to occur 300°C lower in microwave-annealed silicon nitride. After minimal analysis, these workers attributed the observed acceleration in reaction kinetics to enhanced diffusion.

TITANIUM DIBORIDE   Katz et al (55) were the first to report heating and limited densification of titanium diboride using microwaves. These workers were not able to stop oxidation of the specimen and abandoned further work. Holcombe et al (56) used a controlled atmosphere cavity and a sintering aid of 3 wt% CrB_2 to achieve densities as high as 98%, which were as much as 8% higher than the density reported for comparative conventional sintering. As was the case with the boron carbide research, this work was also performed using an elaborate four-layer coating technique to preclude reaction with the insulation material.

COMPOSITES

Meek et al (8) reported densities as high as 77% of theoretical for 60 GHz pressureless sintering of alumina-10 vol% silicon carbide whiskers. This finding was significant since it is the highest density reported for the pressureless sintering of alumina at this whisker loading. Katz et al (57) were not able to reproduce these results and only attained 71% of theoretical density. Tian et al (53) were able to microwave sinter Al_2O_3-30 wt% TiC to 95% of theoretical density by heating to 1950°C in a TE_{111} cylindrical cavity pressurized to 1.4 MPa using nitrogen. The nitrogen was pressurized to delay the temperature at which plasma breakdown occurred.

CERAMIC JOINING

A promising area for the application of microwave technology is the joining of ceramics. The first microwave joints were made by Meek & Blake (58), who joined alumina substrates using commercial sealing glass. Fukushima (59) has used a tunable waveguide cavity to butt-join Al_2O_3 and Si_3N_4 rods. Palaith (60) performed similar joining on mullite rods. Al-Assafi et al (61) joined alumina substrates using AlOOH gel. Three point bend tests were performed on the as-joined substrate. The maximum strength achieved was 93% of the original substrate strength. Yu et al (62) performed simultaneous sintering and joining of alumina containing 3 wt% magnesium. These workers were able to obtain good bonds by applying 0.28 MPa to the joint while sintering.

Silberglitt et al (63) have done some preliminary work on the joining of silicon carbide. These workers have successfully used silicon metal to braze silicon carbide and are continuing work to react silicon carbide, using precursors or a combustion synthesis technique in-situ to form a joint. Yiin et al (64) have brazed Si-SiC using aluminum foil in a single mode 2.45 GHz cavity. These workers made three and four point bend tests and examined the fracture path. It is significant to note that fracture did not occur in the joint area. The average strength of the as-joined material is reported to be equal to the unjoined ceramic at about 220 MPa. The utility of joining a high temperature material such as silicon carbide with a low melting metal such as aluminum is, however, questionable.

POWDER SYNTHESIS

Kumar et al (65) have synthesized ultra-fine β-silicon carbide from amorphous silica and carbon in a microwave oven. These workers also performed comparative conventional experiments. They found that microwave synthesized powders were 30–200 nm and conventional powders were 50–450 nm in size. Kozuka & Mackenzie (66) have synthesized SiC, TiC, NbC, and TaC from carbon and their respective oxides using microwave heating. They also observed whisker formation during the synthesis of silicon carbide.

Kiggans et al (67) have nitrided high purity silicon in a microwave oven to produce a reaction-bonded silicon nitride. These workers did not report any differences between microwave and conventional nitridization, except that nitridization begins at a lower temperature, 1200°C for microwave compared to 1250°C for conventional annealing.

SOL-GELS

Roy et al (68, 69), working with Al_2O_3 and SiO_2 gels, were the first to show that microwaves could be used for rapid drying. Komarneni et al (70) crystallized mullite and mixtures of mullite and alumina powders from single-phase and diphasic gels, respectively, in 20 to 25 min using microwave heating. The particle size of powders produced from this procedure was between 10 and 500 nm. Surapanani et al (71) used microwaves to dry monolithic borosilicate gels. These workers did an extensive job of characterizing the gels produced by this technique using FTIR, scanning electron microscopy, and BET surface area analysis. The microwave drying process was found to be much quicker and the microstructure and properties the same as those achieved using conventional techniques. Fang et al (72) sintered hydroxyapatite powders to densities of 86 to 99% by heating to between 1100 and 1300°C. This work was carried out in a commercial microwave oven, using either SiC susceptors or zirconia as a susceptor and insulation.

GLASS

Hassler & Johansen (73) developed a process to draw quartz optical fibers using microwave heating. These workers used a conventional oxyhydrogen torch to heat the quartz to 1000°C at which temperature they could couple microwaves directly to the quartz. Kao & Mackenzie (74) used magnetite additives to increase the microwave loss of soda-lime glass, which was then heated and consolidated using microwaves. Porosity was found to decrease with increases in sintering time, microwave power, and magnetite additions. Sol-gel-derived silica glass has been microwave-sintered to full density by Pope (75). The glass was heated to approximately 1200°C for 10 min using a susceptor.

Several groups of researchers have studied the vitrification of glass containing nuclear waste using 2.45 GHz microwave radiation. Sturcken (76) melted borosilicate glass in an oxide crucible using an SiC susceptor that was heated by microwaves. Jantzen & Cadieux (77) demonstrated microwave vitrification of borosilicate glass by heating to 1150–1200°C for 10–15 min. This is quite an improvement over conventional vitrification, which requires approximately 2 hr at 1150°C. Microwave heating at lower temperatures for longer times resulted in spinel formation. Schulz et al (78) also studied the microwave heating of a borosilicate glass frit containing simulated nuclear waste. A zirconia-lined silicon carbide susceptor was used to heat an alumina crucible containing the glass frit. Vitrification was accomplished by heating the glass to 1150°C for 10 or more min.

SUMMARY

At this time, only one laboratory is claiming enhancements of diffusion, grain growth, and densification kinetics from microwave sintering. Enhancements are claimed when using 2.45 and 28 GHz microwaves to process a variety of ceramics, both with and without the aid of silicon carbide susceptors.

The replacement of fossil fuel fired sintering operations by microwave sintering lacks economic justification on the basis of energy savings. The potential for lower energy usage still exists for ceramics, which are sintered in electric-fired furnaces.

Greater toughness has been demonstrated for microwave-fired ceramics in a few instances. The improved toughness has been attributed to a number of factors including smaller grain size, the increased presence of β-silicon nitride, and localized pore melting.

Microwaves have been used to successfully sinter a variety of oxide and non-oxide ceramics, composites, and glass. In addition, powder synthesis, sol-gel processing, and ceramic joining have been demonstrated using microwaves. In spite of these successes, and with the exception of drying, microwave processing has yet to see wide spread application in the ceramic industry.

Literature Cited

1. Andes J. 1988. Scientists make use of microwaves in ceramics, *The Oak Ridger*, Oak Ridge, Tenn. p. 3
2. Tinga, W. R., Voss, W. A. G. 1968. *Microwave Power Engineering*, ed. E. C. Okress, pp. 189–99. New York: Academic
3. Sutton, W. H., Brooks, M. H., Chabinsky, I. J., eds. 1988. *Microwave Processing of Materials*, *Mater. Res. Soc. Symp. Proc.* 124: 287–95. Pittsburgh: Mater. Res. Soc.
4. Schubring, N. W. 1983. *Microwave Sintering of Alumina Spark Plug Insulators*, *Am. Ceram. Soc., Electron. Div.* Grossinger, NY
5. Haas, P. A. 1979. *Am. Ceram. Soc. Bull.* 58: 873
6. Humphrey, K. D. 1980. Microwave sintering of BaTiO$_3$. MS thesis. Univ. Missouri-Rolla. 77 pp.
7. Krage, M. K. 1981. *Am. Ceram. Soc. Bull.* 60: 1234
8. Meek, T. T., Blake, R. D., Petrovic, J. J. 1987. *Ceram. Eng. Sci. Proc.* 8: 861–71
9. MacDowell, J. F. 1984. *Am. Ceram. Soc. Bull.* 63: 282–86
10. Sutton, W. H. 1989. *Am. Ceram. Soc. Bull.* 68: 376–86
11. Snyder, W. B. Jr., Sutton, W. H., Iskander, M. F., Johnson, D. L., eds. 1991. *Microwave Processing of Materials*, *Mater. Res. Soc. Symp. Proc.* 189. Pittsburgh: Mater. Res. Soc. 531 pp.
12. Clark, D. E., Gac, F. D., Sutton, W. H., eds. 1991. *Microwaves: Theory and Application in Materials Processing*, *Ceram. Trans.* 21. Westerville, Ohio: Am. Ceram. Soc. 698 pp.
13. Deleted in proof
14. Kenkre, V. M., Skala, L., Weiser, M. W., Katz, J. D. 1991. *J. Mater. Sci.* 26: 2483–89
15. Kenkre, V. M. 1991. See Ref. 12, pp. 69–80
16. Stasiw, O., Teltow, J. 1947. *Ann. Phys.* 1: 261–72
17. Iskander, M. F., Andrade, O., Kimrey, H., Smith, R., Lamoreaux, S. 1991. See Ref. 12, pp. 141–72
18. Manring, B., Asmussen, J. Jr. 1991. See Ref. 12, pp. 159–66
19. Gallerneault, C., Lorenson, C. 1991. See Ref. 12, pp. 193–200
20. Meek, T. T., Blake, R. D., Katz, J. D., Bradberry, J. R., Brooks, M. H. 1988. *J. Mater. Sci. Lett.* 7: 928–31
21. Swain, B. 1988. *Adv. Mater. Process.* 2: 76–82
22. Ahmad, I., Clark, D. E. 1991. See Ref. 12, pp. 605–12
23. Fathi, Z., Ahmad, I., Simmons, J. H., Clark, D. E., Lodding, A. R. 1991. See Ref. 12, pp. 623–30
24. Deleted in proof
25. Janney, M. A., Kimrey, H. D. 1991. See Ref. 11, pp. 215–28
26. Katz, J. D., Blake, R. D., Kenkre, V. M. 1991. See Ref. 12, pp. 95–106
27. Das, S., Curlee, T. R. 1987. *Am. Ceram. Soc. Bull.* 66: 1093–94
28. Katz, J. D., Blake, R. D. 1991. *Am. Ceram. Soc. Bull.* 70: 1304–8

29. Patterson, M. C. L., Kimber, R. M., Apte, P. S. 1991. See Ref. 11, pp. 257–72
30. Whittemore, J. O. 1974. *Am. Ceram. Soc. Bull.* 53: 456–57
31. Watters, D. G., Brodwin, M. E., Kriegsman, G. A. 1988. See Ref. 3, pp. 129–34
32. Harmer, M. P., Brook, R. J. 1981. *J. Brit. Ceram. Soc.* 80: 147–48
33. Eastman, J. A., Sickafus, K. E., Katz, J. D., Boekem, S. G., Blake, R. D., et al. 1991. See Ref. 11, pp. 273–78
34. Tiegs, T. N. 1991. *Am. Ceram. Soc. Bull.* 70: 1725
35. Holcombe, C. E., Meek, T. T., Dykes, N. L. 1988. See Ref. 3, pp. 227–34
36. Patil, D. S., Mutsuddy, B. C., Gavulic, J., Dahimene, M. 1991. See Ref. 12, pp. 301–10
37. Katz, J. D., Blake, R. D. 1992. *Am. Ceram. Soc. Bull.* Submitted
38. De, A. S., Ahmad, I., Whitney, E. D., Clark, D. E. 1991. See Ref. 11, pp. 283–88
39. De, A. S., Ahmad, I., Whitney, E. D., Clark, D. E. 1991. See Ref. 12, pp. 319–40
40. Deleted in proof
41. Janney, M. A., Calhoun, C. L., Kimrey, H. D. 1991. See Ref. 12, pp. 311–18
42. Deleted in proof
43. Kimrey, H. D., Kiggans, J. O., Janney, M. A., Beatty, R. L. 1991. See Ref. 11, pp. 243–56
44. Harrison, W. B., Hanson, M. R. B., Koepke, B. G. 1988. See Ref. 3, pp. 279–86
45. Aliouat, M., Mazo, L., Desgardin, G. 1991. See Ref. 11, pp. 229–36
46. McMahon, G., Pant, A., Sood, R., Ahmad, A., Holt, R. T. 1991. See Ref. 11, pp. 237–42
47. Ahmad, I., Chandler, G. T., Clark, D. E. 1988. See Ref. 3, pp. 239–46
48. Cozzi, A. D., Jones, D. K., Fathi, Z., Clark, D. E. 1991. See Ref. 12, pp. 357–64
49. Holcombe, C. E. 1983. *Am. Soc. Bull.* 62: 1388
50. Holcombe, C. E. 1985. *US Patent No. 4,559,429*
51. Katz, J. D., Blake, R. D., Petrovic, J. J., Sheinberg, H. 1988. See Ref. 3, pp. 219–26
52. Holcombe, C. E., Dykes, N. L. 1991. See Ref. 12, pp. 375–86
53. Tian, Y. L., Brodwin, M. E., Dewan, H. S., Johnson, D. L. 1988. See Ref. 3, pp. 213–18
54. Tiegs, T. N., Kiggans, J. O. Jr., Kimrey, H. D. Jr. 1991. See Ref. 11, pp. 267–72
55. Katz, J. D., Blake, R. D., Scherer, C. P. 1989. *Ceram. Eng. Sci. Proc.* 10: 857–67
56. Holcombe, C. E., Dykes, N. L. 1991. *J. Mater. Sci.* 26: 3730–38
57. Katz, J. D., Blake, R. D., Petrovic, J. J. 1988. *Ceram. Eng. Sci. Proc.* 9: 725–34
58. Meek, T. T., Blake, R. D. 1986. *J. Mater. Sci. Lett.* 5: 270–74
59. Fukushima, H., Yamanka, T., Matsui, M. 1988. See Ref. 3, pp. 267–72
60. Palaith, D., Silberglitt, R., Wu, C. C. M., Kleiner, R., Libelo, E. L. 1988. See Ref. 3, pp. 255–66
61. Al-Assafi, S., Ahmad, I., Fathi, Z., Clark, D. E. 1991. See Ref. 12, pp. 515–22
62. Yu, X. D., Varadan, V. V., Varadan, V. K. 1991. See Ref. 12, pp. 497–508
63. Silberglitt, R., Palaith, D., Black, W. M., Sa'adaldin, H. S., Katz, J. D., Blake, R. D. 1991. See Ref. 12, pp. 487–96
64. Yiin, T. Y., Varadan, V. V., Varadan, V. K., Conway, J. C. 1991. See Ref. 12, pp. 507–14
65. Kumar, S. N., Pant, A., Sood, R. R., Ng-Yelim, J., Holt, R. T. 1991. See Ref. 12, pp. 395–402
66. Kozuka, H., Mackenzie, J. D. 1991. See Ref. 12, pp. 387–94
67. Kiggans, J. O., Hubbard, C. R., Steele, R. R., Kimrey, H. D., Holcombe, C. E., Tiegs, T. N. 1991. See Ref. 12, pp. 403–8
68. Roy, R., Komarneni, S., Yang, L. J. 1985. *J. Am. Ceram. Soc.* 68: 392–95
69. Komarneni, S., Roy, R. 1986. *Mater. Lett.* 4: 107–10
70. Komarneni, S., Breval, E., Roy, R. 1988. See Ref. 3, pp. 235–38
71. Surapanani, S., Mullins, M., Cornilsen, B. C. 1991. See Ref. 11, pp. 309–25
72. Fang, Y., Agrawal, D. K., Roy, D. M., Roy, R. 1991. See Ref. 12, pp. 349–56
73. Hassler, Y., Johansen, L. 1988. See Ref. 3, pp. 273–78
74. Mackenzie, J. D., Kao, Y. H. 1991. See Ref. 12, pp. 341–48
75. Pope, E. J. A. 1991. *Am. Ceram. Soc. Bull.* 70: 1777–78
76. Sturcken, E. F. 1991. See Ref. 12, pp. 433–40
77. Jantzen, C. M., Cadieux, J. R. 1991. See Ref. 12, pp. 441–50
78. Schultz, R. L., Fathi, Z., Clark, D. E., Wicks, G. G. 1991. See Ref. 12, pp. 451–59

Z-CONTRAST TRANSMISSION ELECTRON MICROSCOPY: Direct Atomic Imaging of Materials[1]

S. J. Pennycook

Solid State Division, Oak Ridge National Laboratory, Oak Ridge, Tennessee 37831-6030

KEY WORDS: semiconductors, superconductors, epitaxial growth, superlattices, critical currents

INTRODUCTION

Z-contrast electron microscopy provides a new view of materials on the atomic scale, a direct image of atomic structure composition. A direct image can be interpreted without the need for any preconceived model structures. It has the capability of revealing unanticipated atomic arrangements, which provide a new depth of understanding into the origin of materials properties.

Unexpected atomic arrangements generally arise as a result of previously unknown growth mechanisms, and they can often be deduced from the form of the image. Therefore a new level of insight is gained into the atomistic processes of synthesis and growth. Direct imaging represents a powerful new approach to materials science, that is, the ability to connect measured properties to theoretical predictions through the actual atomic structures found at grain boundaries, defects, or interfaces, those critical regions that define a material's bulk properties.

Z-contrast electron microscopy at atomic resolution combines incoherent characteristics with a contrast that is highly sensitive to compo-

---

[1] The US government has the right to retain a nonexclusive, royalty-free license in and to any copyright covering this paper.

sition. In an incoherent image, the object and its image are related in a simple, direct manner, so that given one, it is relatively straightforward to predict the other, at least to first order. An atomically smooth and abrupt interface, for example, will easily be recognized as such, but equally, the formation of new interfacial phases, transition zones, interface defects, or interfacial ordering would also be immediately apparent from the image.

In this review the imaging process is outlined briefly, followed by examples of the insights obtained into the growth mechanisms and properties of semiconducting and superconducting materials. It is seen that the high-T_c materials, with their complex unit cells, have a relatively simple growth behavior, whereas the elemental semiconductors, Ge and Si, show a growth behavior that is remarkably complex. In particular, direct imaging reveals some important mechanisms of semiconductor growth that previously were completely unforeseen.

THE IMAGING PROCESS
An Atomic Resolution Compositional Map

Z-contrast imaging can be visualized very simply, as shown in Figure 1. A scanning transmission electron microscope (STEM) forms a finely focused probe of high energy electrons, which is scanned across a thin sample. Although most of the beam will be diffracted through quite small angles, some will be scattered through much larger angles, and it is this component that is collected by an annular detector and used to form the Z-contrast image. This component represents Rutherford scattering, the intensity of which depends very strongly on composition through the Z^2 dependence of the scattering cross section, where Z represents atomic number. (Hence the term Z-contrast imaging.)

If the material is crystalline, oriented to a major zone-axis, and if the probe size is sufficiently fine compared to the separations of the atomic columns, as the probe is scanned across a thin sample it will directly map out the location of each atomic column. Furthermore, the intensity of each column in the image will directly reflect its composition. Therefore the image can be thought of as a simple map revealing the scattering power of the material at atomic resolution.

Consider, for example, the effect of increasing the crystal thickness. Obviously the intensity scattered by each atomic column will increase as the columns become longer; the scattering is further increased since the beam is attracted by the positive nuclear charge and tends to channel along the atomic columns. Apart from an overall scaling of the image intensity, no other change in the form of the image is expected. This is very close to what is observed in practice. The image does not change in form, but

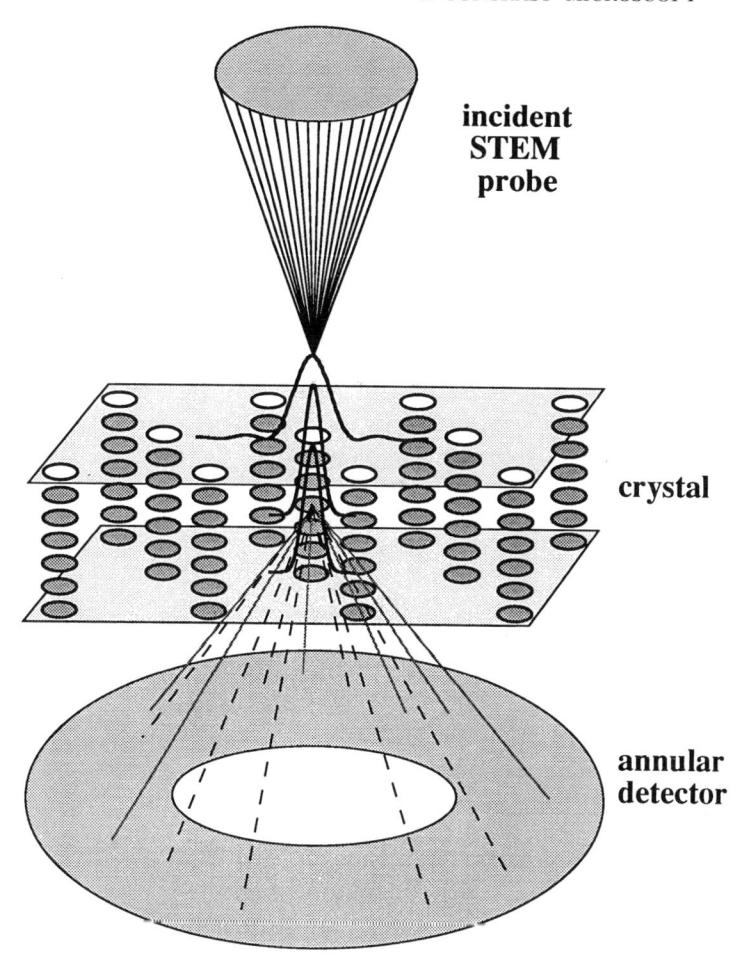

Figure 1 Schematic showing the formation of a Z-contrast image in the STEM. A fine electron probe maps out the location and scattering power of the atomic columns, thus producing a direct image at atomic resolution.

slowly reduces in contrast as the incident beam becomes depleted by the scattering events. Now consider changing the focus of the microscope; the profile of the probe incident on the sample surface will change, altering the relative illumination of neighboring atomic columns. However, since we require the most compact probe to achieve the best resolution, as we move away from this optimum focus condition, the image tends to become more blurred. Again, we generally see a single image form, just as would be expected for an incoherent image seen through a camera, for example.

These characteristics are important since they are responsible for the direct interpretability of Z-contrast images. They are different from the characteristics of a conventional high resolution electron microscope image, which shows many different forms depending on the crystal thickness and microscope focus. In a conventional image, a broad but highly collimated electron beam is used to illuminate the crystal. The transmitted beam, along with a number of diffracted beams generated by the crystal, are collected by an objective aperture and projected onto the microscope screen. The contrast that results from recombining these beams depends on their relative phases after passing through the crystal and the optical system of the microscope. It is because these relative phases are highly sensitive to both the exact crystal thickness and to the exact microscope focus that such phase contrast images show a large variety of different possible forms for the same crystal structure (1). Generally, phase contrast images can only be interpreted by simulating the anticipated image forms for a small set of trial structures, then determining the best fit.

In Z-contrast imaging the relative phases of the incident beam and the scattered beams reaching the annular detector are unimportant. The total intensity reaching the detector depends only on the incident intensity at each atomic site and on the species present, which scatters a small fraction of that intensity to the detector. The inner detector angle must be sufficiently large so that the signal it collects is dominated by thermal diffuse scattering. This is generated very close to the atomic sites, and it is the atomic vibrations themselves that break the coherence of the imaging process. A typical inner detector angle, appropriate for Si at room temperature, for example, would be 75 mrad.

A quantum mechanical description of the imaging process has shown in detail how dynamical diffraction effects are reduced to second order in the Z-contrast image (2). Only one stationary quantum state of the fast electron in the crystal contributes significantly to the image, thus dynamical diffraction effects (that result from the interference of Bloch states of comparable amplitude) are effectively avoided.

In the next two sections variants on this basic imaging mode, more appropriate for certain materials studies, are discussed. This is followed by examples of insights obtained into semiconducting and superconducting materials systems.

Imaging of Small Clusters

The high-angle detector was introduced originally by Howie in order to improve the visibility of small catalyst clusters, particularly when supported on crystalline supports (3). It has proved valuable for such studies, especially for cluster sizes approaching the resolution limit of conventional bright field imaging techniques (4–6).

Figure 2 shows an example of small clusters of Pd, 5–10 Å in diameter, supported on γ-alumina, in which all of the particles present can be seen, a great benefit for the determination of size distributions. Limited information is also available concerning the morphology of individual clusters, even though the atomic columns are not resolved. With improved resolution it seems quite possible that the morphology could be determined with much greater precision and perhaps even linked directly to catalytic activity.

Atomic resolution of small clusters has so far only been achieved by reducing the inner detector angle to ~ 20 mrad, just outside the incident beam semiangle (7). Now a large fraction of the electrons scattered by the sample will reach the detector, which will result in greatly improved signal-to-noise ratio in the images. The vast majority of these electrons have been coherently scattered by the specimen, however, and it has often been considered impossible, in principle, to form an atomic resolution incoherent image in this way (8, 9) despite experimental images that apparently show convincing incoherent characteristics (7).

We now understand that such images show incoherent characteristics only in the transverse plane, that is, each atomic column will appear as a

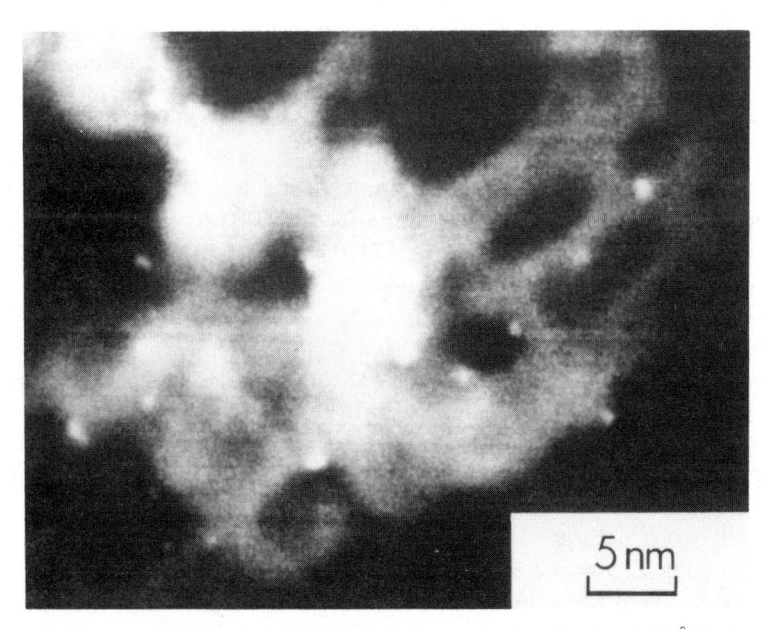

Figure 2 Z-contrast image of a 0.75% Pt catalyst on γ-alumina, showing 5–10 Å Pt clusters (photograph courtesy of S. Bradley).

bright spot in the image, with a size dependent on the probe intensity profile. However, the contributions of atoms along the length of the column must be added coherently, which means that the intensity increases initially as the square of the number of atoms in the column but then, because of destructive interference, becomes oscillatory and never rises above the intensity scattered by a thin raft (10, 10a).

Nevertheless, with awareness of this thickness behavior, this approach represents a useful extension of incoherent imaging to the realm of small clusters and rafts, which otherwise might not survive the higher beam exposure required for adequate signal-to-noise ratios with larger inner detector angles. At the sample thicknesses normally employed in examining bulk materials or thin films, this coherent component suffers strong dynamical effects (11), and incoherent images can only be achieved by increasing the inner detector angle until thermal diffuse scattering again dominates the detected intensity and intuitive interpretability returns.

Imaging of Interfacial Segregation

Segregation of dopants and impurities at internal interfaces in materials is important in understanding an enormous range of bulk phenomena, from the electrical and optical properties of semiconductors to the strength and ductility of high temperature materials. Particularly in the case of heavy impurities in a light matrix, Z-contrast imaging can provide simultaneously high spatial resolution and compositional sensitivity. It is therefore highly complementary to more traditional analytical techniques such as X-ray micro-analysis or energy-loss spectroscopy, for which the scattering cross-sections are orders of magnitude lower, necessitating thicker specimens, which lead to beam broadening. Of course, Z-contrast imaging does not provide a spectroscopic analysis and is therefore most useful when there is a single dominant impurity and the image can then reveal fine scale details in its distribution.

To illustrate this capability, Figure 3 shows cross-section images of Si(100) after ion implantation and annealing (12, 13). For both Sb and Ge implants, recrystallization has occurred part way to the surface, but in the case of the Sb implant, segregation of the dopant ahead of the advancing amorphous/crystalline interface is clearly revealed in the Z-contrast image. In both cases, the diffraction contrast images show break-up of the advancing interface with the formation of extended defects, but the impurity that caused this effect is not visible. Diffraction contrast images are primarily sensitive to structure.

Although Z-contrast images are sensitive to composition, defects can still be visible if they alter the channelling effect of any atomic columns, for example, as a result of their strain field (see Figure 1). To avoid this source

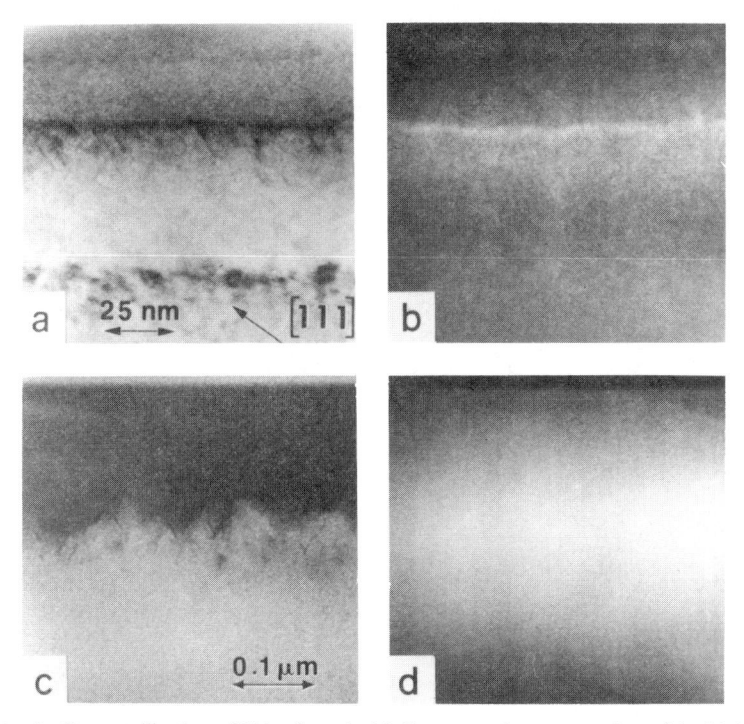

Figure 3 Recrystallization of Si implanted with Sb, at a peak concentration of 6 at% (*a,b*), and Ge, at a peak concentration of 14 at% (*c,d*). The diffraction contrast images (*left*) show the breakdown of epitaxial growth, while the Z-contrast images (*right*) show interfacial segregation in the case of Sb, but not for Ge.

of contrast, the crystals were tilted slightly into a non-channelling orientation, away from any strong low order diffracted beams. The scattering yield is no longer sensitive to the arrangement of the atoms, only to the species present. Amorphous and crystalline material both scatter at the same level, which reveals the profile of the implanted dopant species and any segregation effects that may have occurred at the amorphous/crystalline interface.

Clearly Sb, which has a very low solubility in crystalline Si, shows a substantial tendency to segregate, forming a narrow band of high concentration just ahead of the advancing crystal. From the intensity of this band, the segregation coefficient for non-equilibrium growth can be determined, while the width of the band indicates the diffusion coefficient away from the interface. Interestingly, the values are quite similar to those obtained for nonequilibrium liquid phase recrystallization using pulsed laser annealing (12). Ge, on the other hand, shows no tendency to segregate, as expected since Si and Ge are perfectly miscible.

INSIGHTS INTO ELECTRONIC AND PHOTONIC MATERIALS

III/V Materials

Correlating the electrical and optical properties of semiconducting materials and devices to their atomic scale structure and chemistry would do much to improve our understanding of the key factors limiting performance. The importance of direct imaging is that it can reveal the nature of critical interfaces in quantum wells and superlattices, perhaps even providing insight into the complex processes that occurred during synthesis and processing.

A good example of this capability is seen in Figure 4, which shows GaAs quantum wells, separated by $In_{0.5}Ga_{0.5}P$ barriers, grown by gas source molecular beam epitaxy (14). The GaAs wells show brighter than the barrier layers, so that the interface morphology is visible on the atomic scale. The well seen in Figure 4a is nominally a single unit cell in thickness and reveals a monolayer height interfacial step. This observation strongly suggests that minimal interdiffusion has occurred and that the interfaces are atomically abrupt, but stepped. This correlates well with the observed monolayer splittings of the photoluminescence peaks. From the shape and exact positions of the image features it would be possible, in principle, to

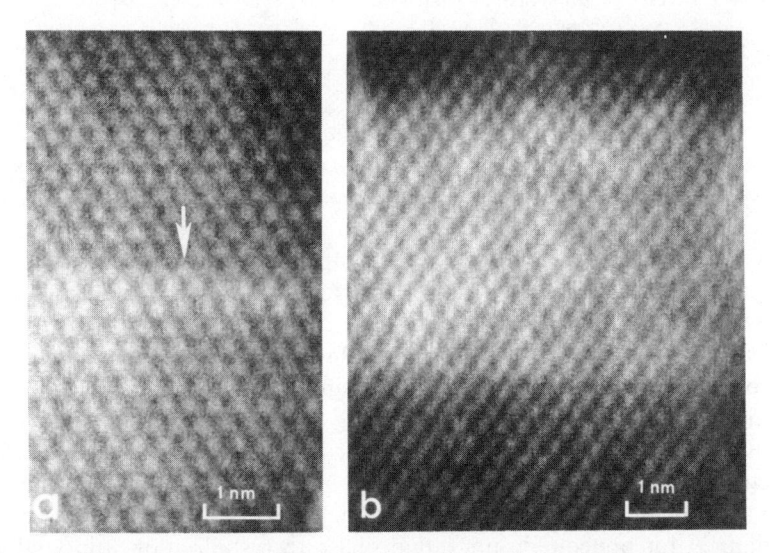

Figure 4 Z-contrast images of GaAs quantum wells in $In_{0.5}Ga_{0.5}P$ barriers: (*a*) shows a single unit cell well with a monolayer height interfacial step (*arrow*); (*b*) shows a thicker well with significant roughening at the upper interface (photograph courtesy of M. F. Chisholm).

determine the detailed interface chemistry, to distinguish between the four distinct abrupt interfaces possible in this system.

In Figure 4b the surface of the GaAs has clearly roughened substantially as its thickness has increased, and possibly has begun to facet, which results in a rather rough upper interface. Since a roughened interface reduces the mobility of the carriers in the wells, this observation immediately suggests that the optical properties would be improved with a growth interruption at the upper interface, which would give time for surface diffusion to minimize the surface energy.

The state of the interface also reflects the growth mode of the layer below. The $In_{0.5}Ga_{0.5}P$ layer clearly grew in a terrace-sweeping mode, maintaining a smooth surface, whereas the surface diffusion length was apparently much smaller for the GaAs, which resulted in island nucleation and the associated roughening of the surface. This problem could perhaps be alleviated by reducing the GaAs supersaturation, decreasing the deposition rate, or increasing the substrate temperature during the GaAs growth cycle.

Direct imaging of interfaces provides insight not only into the electronic properties, but also into the growth mechanisms responsible for the observed structures, which in turn suggests ways to tailor the growth so as to improve those properties.

Si/Ge Materials

Another system that has revealed a whole new level of complexity on the atomic scale is that of Si/Ge interfaces. In recent years there has been a number of diffraction observations indicating the presence of ordering in Si_xGe_{1-x} alloys, and a great deal of effort has been invested in attempting to distinguish between two ordered phases that have been proposed (15–17). However, Z-contrast images of an ultrathin superlattice immediately demonstrate that the situation is far more complicated than had been imagined, as seen in Figure 5. A different ordered arrangement is present at each interface, $2 \times N$ interfacial ordering at the top Si on Ge interface, a $\{111\}$ planar structure in the central Si layer, with Ge threading through to the next Ge layer, while cross-like structures can be seen in the lowest Si layer.

These structures are incompatible with the idea of a stable ordered phase produced by interdiffusion, and again implies that the ordering is an effect resulting from the growth process itself. In particular, the 2×1 periodicity strongly suggests that the 2×1 surface reconstruction is playing a critical role (18). The superlattice was grown at 350°C, where growth proceeds via the successive nucleation of monolayer height islands, which grow and coalesce predominantly through the motion of type S_B steps (19) (Figure 6).

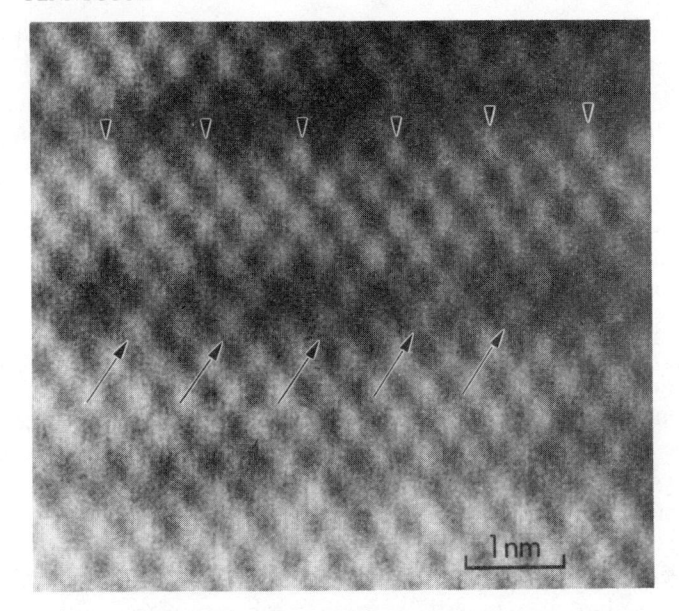

Figure 5 Z-contrast image of an ultrathin Si_mGe_n superlattice showing a different ordered arrangement at each interface (photograph courtesy of D. E. Jesson).

At low temperatures it is reasonable to assume that diffusion of sub-surface atoms is effectively frozen. This is also borne out by the Z-contrast image, which shows a different phase at every interface. Exchange of atoms in different monolayers is likely, therefore, only to take place at the edges of the growing S_B steps themselves. As a step advances over the reconstructed substrate surface, it is forced consecutively through the two different configurations, as shown in Figure 6.

Consider the case of a monolayer of Si growing over a layer of Ge. When the step has the rebonded configuration, interchanging the loosely bound Si and Ge atoms at the step edge will bury the Si dangling bond and replace it with a Ge dangling bond. This results in a considerable saving of energy, approximately 0.1 eV per atom, so that an appreciable driving force exists for this atom pump mechanism (20).

At the other step configuration no dangling bonds can be saved by such an exchange. Thus, we obtain a 2 × 1 compositional ordering along the growth direction of the dimer row. The next monolayer to grow over the first will only pump if Ge appears at the appropriate site at the S_B step edge, which depends on growth direction and the phase of dimerization. The ordering can either propagate, reverse, or terminate with each suc-

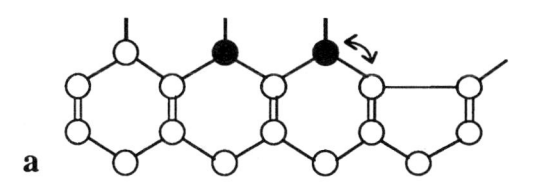

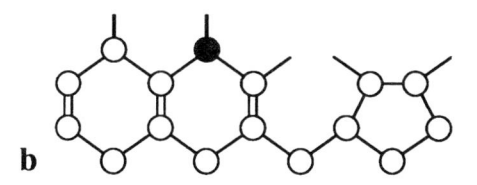

Figure 6 Rebonded (*a*) and nonrebonded (*b*) configurations of the type S_B step.

cessive monolayer, which explains the different ordered structures seen experimentally (20).

In the case of alloys grown by codeposition, the situation is rather different. Ge is now available from the incoming flux, and since both types of step would prefer a Ge atom at the edge position, they compete with each other for the available flux (21). Si tends to be rejected from the advancing steps, in a manner entirely analogous to the segregation of Sb seen in Figure 3.

Ordering results from the nonequilibrium nature of the growth, from the fact that the step velocity is determined by the deposition rate. At low temperatures both steps act as perfect sinks for the arriving species and no ordering results. As the temperature increases, the less stable step can begin to desorb the least bound species. It will reject Si into the reservoir of mobile species on the surface, and the step will become Ge rich. The more stable step will still be acting as a perfect sink and will therefore reflect the composition of the reservoir and become Si rich.

With further increase in temperature, both steps become active in the segregation process, and the ordering will reduce again. The reservoir composition will now become richer in Si to compensate for the preference of both steps for Ge.

Codeposition differs from the case of sequential deposition in that the ordering does not depend on the growth direction of the dimer row. A long range ordered alloy can result, as shown in Figure 7. Note that the unit cell predicted by this ordering mechanism is different from either of the two model structures previously proposed, which highlights the importance of a direct image (21). It would be very difficult to justify this more complex structure on the basis of diffraction evidence alone, since it would be indistinguishable from a mixture of the two simpler phases.

HIGH-T_C SUPERCONDUCTING FILMS AND SUPERLATTICES

Z-contrast imaging is particularly valuable with the high-T_C materials since it directly reveals the location of a defect or interface within the complex unit cell. The strong tendency of these materials to grow cell-by-cell becomes strikingly clear from images of thin films and superlattices and, again, often reveals a microscopic explanation for the measured transport properties.

Cell-by-Cell Growth

Consider the amorphous/crystalline interface shown in Figure 8. Formed by the implantation of oxygen ions at 77 K into a thin film of $YBa_2Cu_3O_{7-x}$

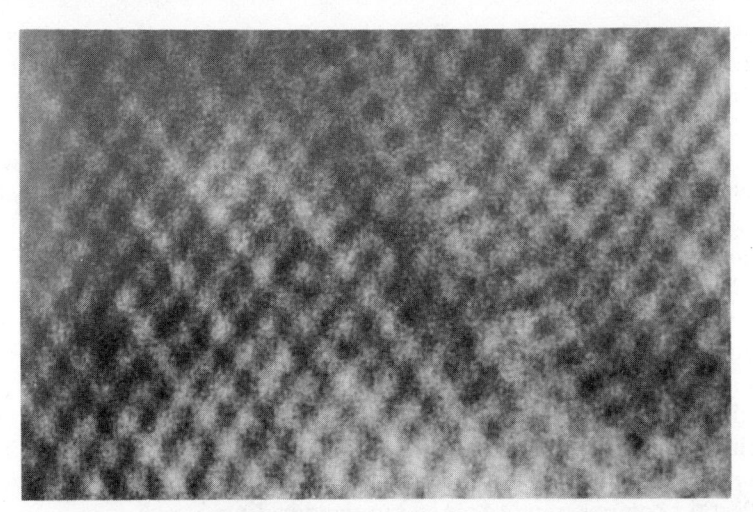

Figure 7 Long range ordering in a $Si_{0.6}Ge_{0.4}$ alloy grown by molecular beam epitaxy. Every alternate {111} plane is Ge rich. A microtwin is also visible (photograph courtesy of D. E. Jesson).

Z-CONTRAST MICROSCOPY 183

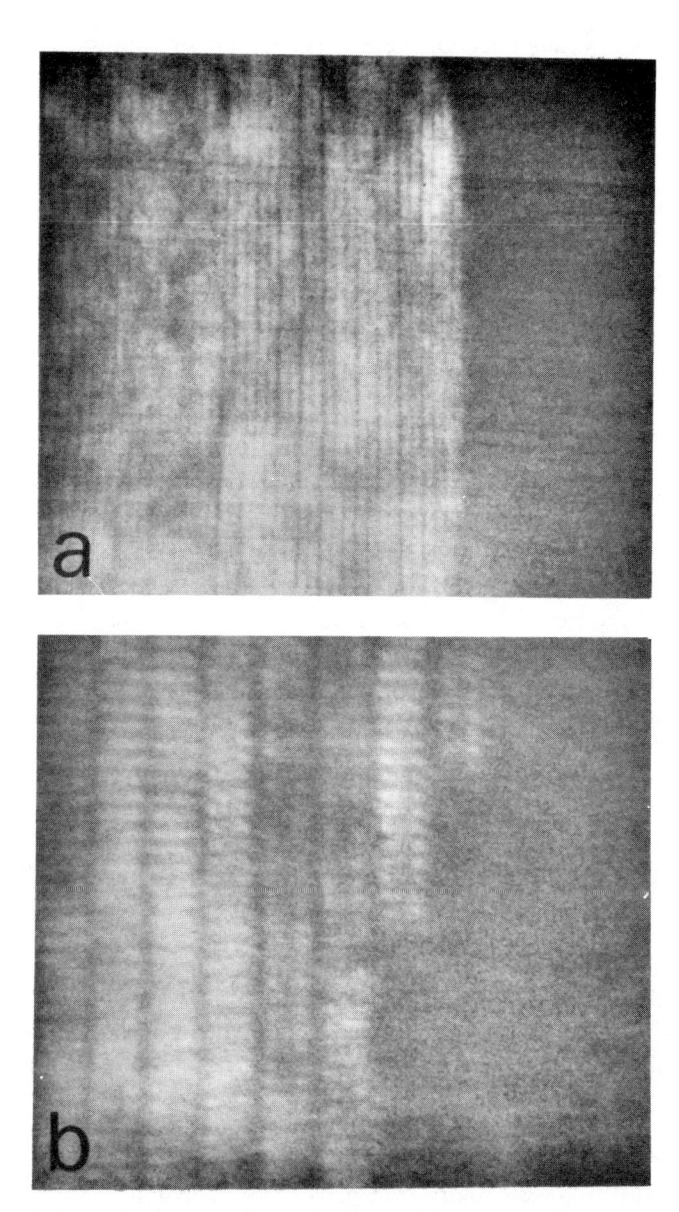

Figure 8 Cell-by-cell amorphization of YBCO via oxygen ion implantation: (*a*) low magnification view of the interface morphology; (*b*) close-up view showing the amorphous/crystal interface located at the Cu-chain plane, but jumping repeatedly by one unit cell.

(YBCO), the interface is remarkably sharp. The macroscopic waviness seen in the low magnification view is seen at higher resolution to comprise discrete interface steps, the height of each step being the full 11.7 Å c-axis lattice parameter.

Although the crystal is quite significantly damaged close to the interface, the Ba atomic columns are still the brightest features in the image and clearly delineate the unit cells. Since the Ba-Ba spacing is greater than the Ba-Y spacing, the Cu-chain plane (between the Ba planes) images darker than the CuO_2 plane. The dark vertical lines locate the Cu-chain planes in the crystal, which are seen to be the preferred planes for terminating the material (22).

This implies that the surface energy of this plane must be substantially lower than the surface energy of the many other possible termination planes. This important property is the origin of the cell-by-cell amorphization. It strongly suggests that growth, the reverse process, will also proceed on a cell-by-cell basis to maintain the same low energy terminating plane. Reflection high-energy electron diffraction (RHEED) oscillations have shown a periodicity of 11.7 Å, although they give no indication of the termination plane (23). The tendency for cell-by-cell growth has also been deduced from TEM studies of ultrathin YBCO films (24). Note that Figure 8 shows only the location of the terminating plane within the unit cell. Current image resolution is not sufficient to determine the number and arrangement of Cu atoms on this plane.

Significant further insight into the growth mechanism can be obtained by imaging superlattices of YBCO and $PrBa_2Cu_3O_{7-x}$ (PBCO). Although isostructural, the heavy Pr columns are easily distinguishable from the lighter Y columns, so that the interface morphology can be imaged directly. Every composition change replicates the surface morphology at that point during the growth cycle. A cross-section image of the superlattice provides a series of snapshots showing the atomic scale evolution of the surface morphology with increasing film thickness.

Figure 9 shows images of a 1 × 8 superlattice grown by laser ablation on a $SrTiO_3$ substrate (25). The YBCO layer is clearly only a single unit cell thick, but frequently jumps by one unit cell in the c-direction. This is exactly as expected for cell-by-cell growth; the single unit cell of YBCO shows a strong tendency to cover the PBCO entirely rather than nucleate a second layer of YBCO (22).

The presence of the jumps, which are uncorrelated in successive layers, and the increasing roughness seen with increasing film thickness, are indicative of an island growth mode, as distinct from a terrace-sweeping growth mode, which would preserve or smooth the surface morphology. Two-dimensional island growth occurs via the sequential nucleation and

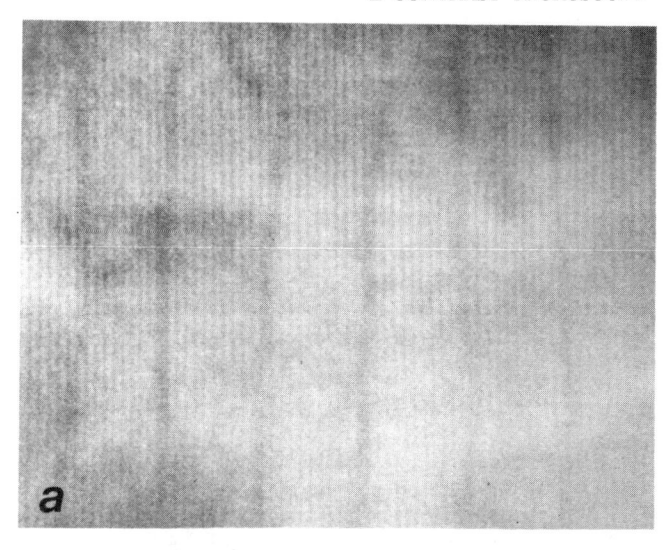

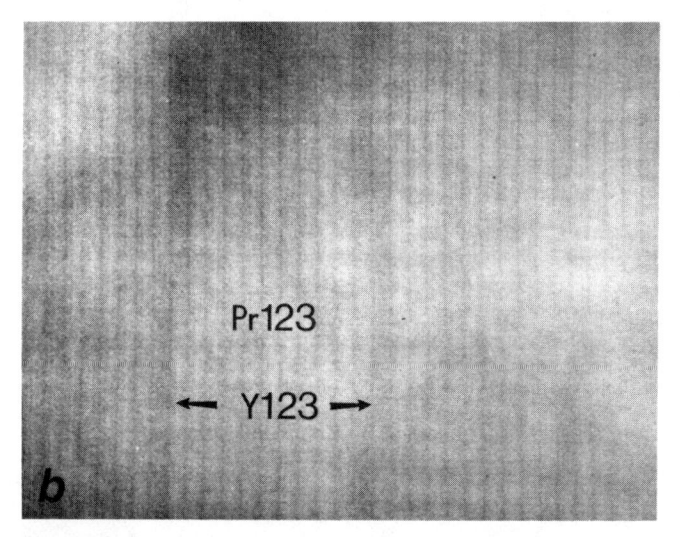

Figure 9 Z-contrast images of a 1 × 8 superlattice: (*a*) low magnification showing YBCO cells as vertical dark lines jumping repeatedly by one unit cell; (*b*) higher magnification showing individual planes.

coalescence of islands, in this case 11.7 Å in height. The statistical nature of this process leads to a gradual roughening of the film surface with increasing thickness, as indicated schematically in Figure 10. The separation of the unit cell jumps is an indication of the intrinsic island size during the nonequilibrium growth, and in Figure 9 is seen to be 20–30 nm.

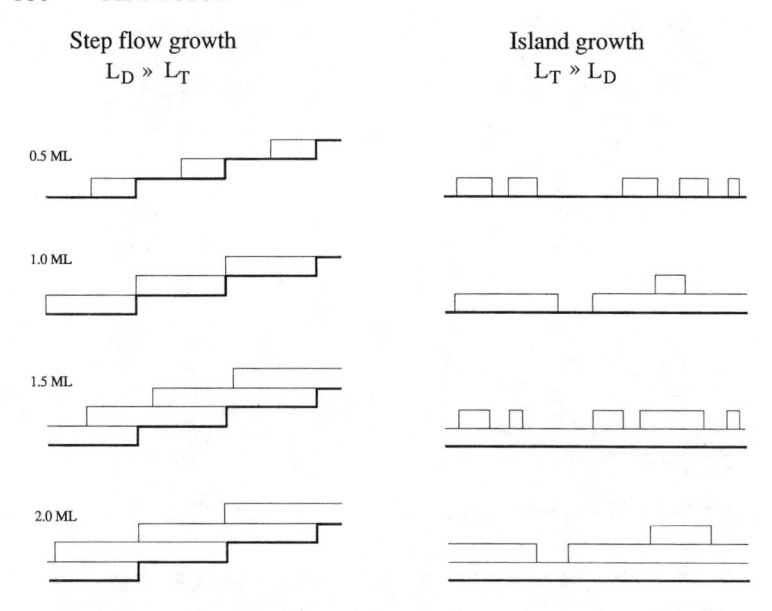

Figure 10 Schematic illustrating the transition from step flow growth to sequential nucleation and coalescence of islands, with associated roughening, as the surface diffusion length L_D becomes less than the terrace width L_T.

Eventually the roughness coarsens into the more macroscopic islands that have been observed by scanning tunneling microscopy (26–29). Now the spacing of the steps on an individual island reflects the surface diffusion length under the depositing flux (assuming that no coarsening occurred after deposition). Whether spiral growth patterns or flat-topped islands are seen is presumably dependent on the details of the strain relaxation mechanisms. Little is known about these processes except that YBCO films on SiTiO$_3$ are strained in the early stages of growth (30), whereas on MgO even a single unit cell layer is relaxed (24).

Transport Properties

Z-contrast images are also valuable in the interpretation of superconducting transport properties. Interdiffusion is a serious concern with the superlattices since the alloys $Y_{1-y}Pr_yBa_2Cu_3O_{7-x}$ show a rapid reduction of T_c with increasing Pr content and become semiconducting at $y \geq 0.5$ (31). While the observation of satellite peaks in X-ray diffraction does indicate the presence of a superlattice modulation, it has not so far been able to determine the composition profile to an accuracy sufficient to exclude interdiffusion effects in the transport properties. Conventional high-resolution TEM has also not been able to distinguish YBCO from PBCO because of their identical crystal structures.

Figure 11 shows a Z-contrast image of an interfacial step in a 3 × 5 superlattice. The Y plane of the YBCO changes into the Pr plane of the PBCO in approximately 4 Å, a distance comparable to the lattice parameter in the a-b plane. The onset of interdiffusion would be seen first at such a step, since diffusion in the a-b plane is significantly higher than along the c-axis. The observation of an abrupt interface step is therefore the best direct evidence that interdiffusion has not affected the transport properties of these structures (22), a common assumption in theoretical interpretations.

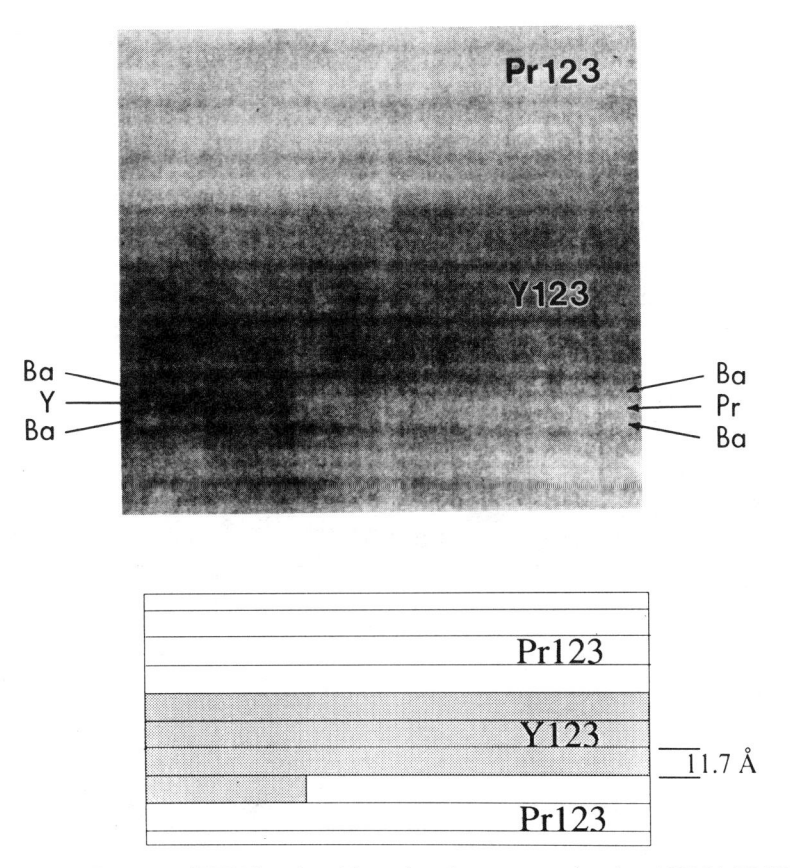

Figure 11 Z-contrast STEM imaging. Abrupt interface step seen in a 3 × 5 YBCO/PBCO superlattice.

The presence of the unit cell-high jumps in the YBCO layers may be an important factor affecting the transport properties. Consider the single unit cell superlattices, as shown in Figure 9. At temperatures well below the mean-field critical temperature, the in-plane coherence length will be much less than the island size, and the resistive behavior is unlikely to be significantly affected. Such a structure is unlikely to carry a substantial critical current, however, since the current would be forced to transfer from island to island at the small regions of overlap around the island perimeters. Very few current paths would exist that did not involve transfer of the current along the c-direction.

The situation with the $2 \times N$ superlattices is substantially different, as indicated schematically in Figure 12. Now the overlap between islands in different layers averages a full island diameter so that transfer from cell to cell in the vertical direction has several hundred Angstroms in which to occur. Excellent film-like critical currents have in fact been measured in 2×4 superlattices (22).

The interface steps are not likely to be the important flux pinning sites, however. Their density is independent of the thickness of the YBCO layer, whereas experimentally it appears that the pinning energy barrier increases linearly with increasing YBCO layer thickness (32). This would suggest that point defects of some kind might be the dominant pinning sites.

Support for this view also comes from images of films grown by co-evaporation and post-annealing in a low oxygen partial pressure (33). A

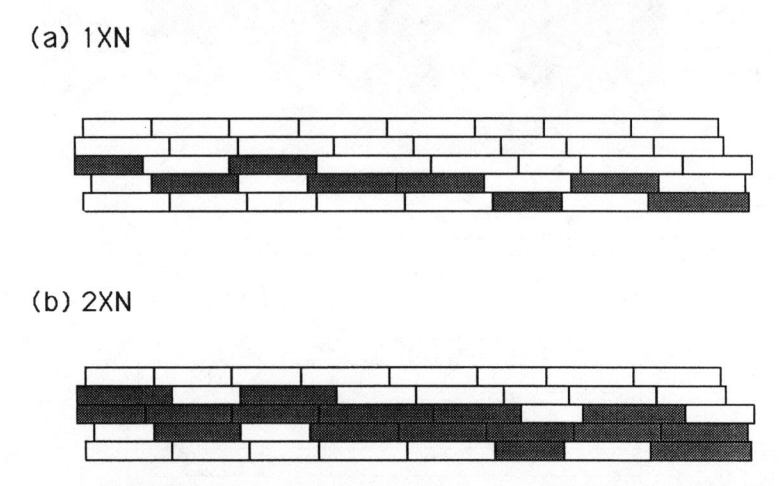

(a) 1XN

(b) 2XN

Figure 12 Schematic indicating the effect of the unit cell high jumps on critical currents. Serious discontinuities occur frequently in $1 \times N$ superlattices, but are effectively removed by growth of one additional unit cell.

high density of planar defects is found, as seen in Figure 13, that take the form of small platelets comprising an extra CuO plane inserted between the Ba planes. Such defects are well-known in these materials and form the basis of the "248" and "247" phases (34), although the small size and high number density seen here are very different from the situation normally found in laser-ablated films. There even seems to be a tendency for the defects to order. However, both microstructures show very high critical currents, thus suggesting that the visible defects are not the dominant pinning sites.

A microstructure that does result in significantly degraded transport properties is shown in Figure 14. This film was grown by laser ablation at a reduced substrate temperature of 620°C (35). The stoichiometry is close to that of the "247" phase, as evidenced by the extra CuO planes inserted between every alternate Ba plane, which would explain the poor super-

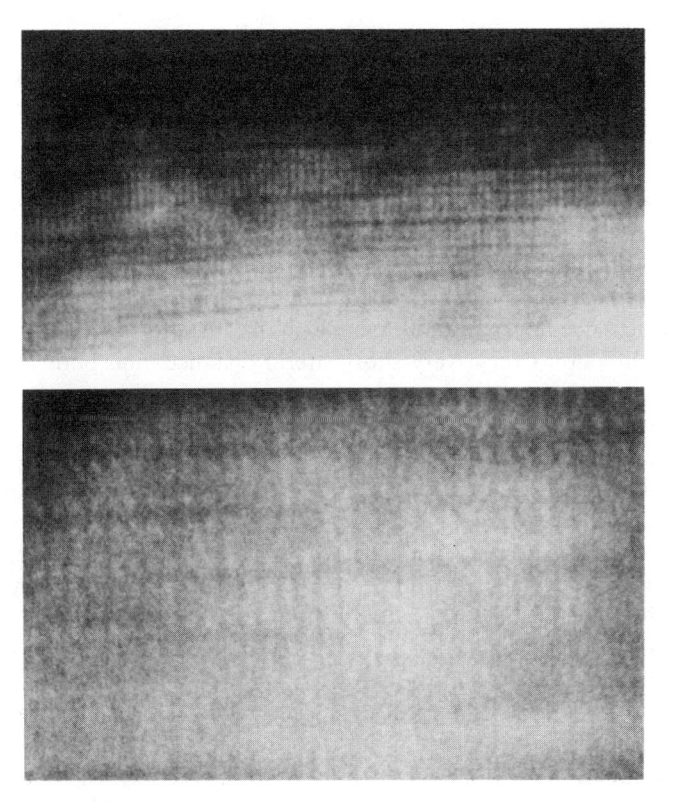

Figure 13 A high density view of planar defects in a film grown by coevaporation and postannealing at low oxygen pressure.

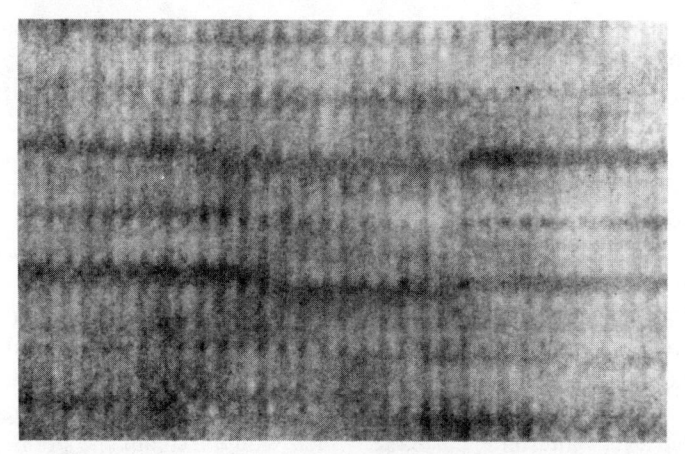

Figure 14 A highly defective c⊥ film resulting from reduced surface mobility during low-temperature growth.

conducting properties. In addition, the central region is shifted by approximately c/3 with respect to the left and right parts of the figure. These defects originate from the substrate as a result of insufficient surface mobility at the low growth temperatures and thread through the entire thickness of the film (36). Defects of this nature will degrade the transport properties, even if the film stoichiometry is correct. If the supercurrent is viewed as being localized primarily on the CuO_2 planes, the c/3 shift will effectively disconnect the superconducting planes. As with the $1 \times N$ superlattices, we have a situation in which the current is forced to transfer in the c-direction over a very short lateral distance, and poor transport properties result.

Much attention has focused on the role of grain boundaries in limiting critical current capacity, in particular, the dramatic effect of a low-angle grain boundary. It has generally been concluded that the dislocation cores comprising the low-angle boundary must have induced impurity segregation or local nonstoichiometry (37–39), hence altering the superconducting properties. Z-contrast images have revealed, however, that compositional variations need not occur at such boundaries (40).

The effect could well be due entirely to the strain field surrounding each dislocation core. Since quite a low strain ($\sim 1\%$) is sufficient to constrain the superconductor in its tetragonal phase, it seems likely that a dislocation will be surrounded by a substantial zone that can never transform to the orthorhombic superconducting phase, whatever the oxygen content may be. This simple model explains the observed critical current behaviour very well (40). The superconducting properties of YBCO appear highly sensitive

Growth at Higher Surface Supersaturation

Although two-dimensional island growth is thermodynamically preferred, since it maintains the lowest energy crystal termination, it is kinetically rather inefficient. A large amount of surface diffusion is required to grow a layer one unit cell in thickness. If the crystal orientation were to change to a⊥, then practically no surface diffusion would be necessary. The incoming flux would automatically find itself at the active crystal growth sites, as indicated schematically in Figure 15. Therefore this growth mode occurs under conditions of high surface supersaturation, for example, low substrate temperatures or high deposition rates (41).

The surface morphology also changes radically with a⊥ growth. Since there is no long-range surface diffusion, the lateral scale of the roughness decreases sharply. Instead of the macroscopic waviness seen with c⊥ growth, a microscopic roughness develops, as preserved by the a⊥ superlattice imaged in Figure 16. This can only be distinguished from interdiffusion in the very thin regions of the sample, so that the image is rather

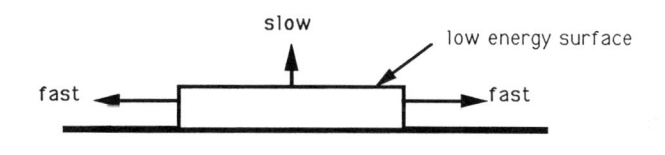

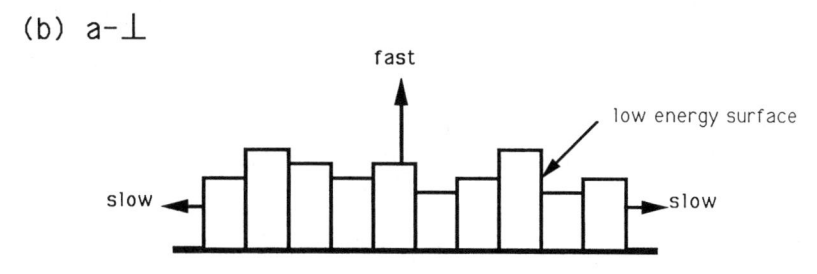

Figure 15 Schematic indicating the transition from c⊥ to a⊥ growth at high surface supersaturation.

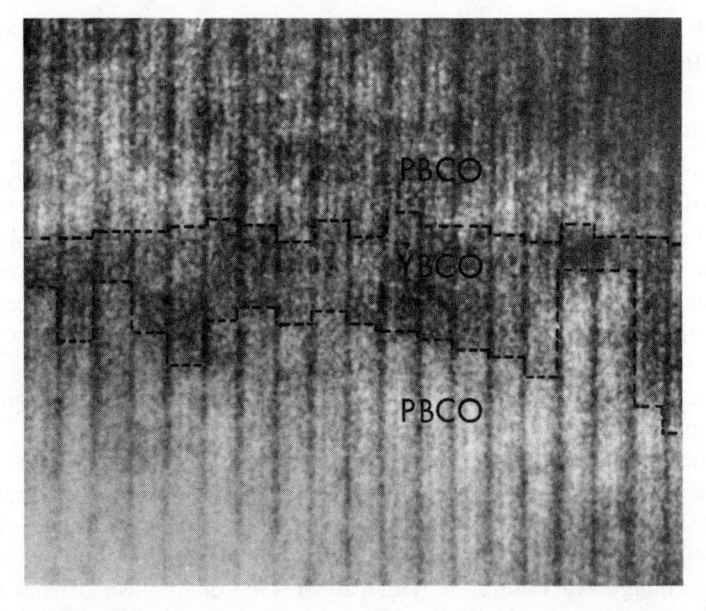

Figure 16 Z-contrast image showing microscopic roughness at the interfaces of an a⊥ 4 × 16 superlattice.

noisy, but the change in composition from PBCO to YBCO, and back, can just be discerned in each individual unit cell column (22).

The anisotropy in surface energies, which leads to the relatively smooth interfaces found for c⊥ growth (and c⊥ amorphization), means that there is rather little driving force for smoothing an a⊥ film; roughness only increases the surface area of the low energy planes (see Figure 15). Obviously, microscopic roughness such as seen in Figure 16 could significantly affect the characteristics of any device structure requiring thin insulating layers, particularly a tunnel junction. A smoother surface could perhaps be obtained using a longer growth interruption between the different materials.

Even higher supersaturations exist during solid-phase epitaxy. For example, annealing the amorphous layer of Figure 1 in 200 mtorr O_2 at 740°C for 30 min leads to the dense cross-hatch microstructure seen in Figure 17. Growth of an a⊥ column from the c⊥ crystal is limited only by the intrinsic crystallization rate, not by any deposition rate. As soon as nucleation occurs at the amorphous/crystal interface, a finger of a⊥ material rapidly grows into the amorphous phase, which creates a fresh amorphous/crystal interface, 90° rotated from the first, at which nucleation can begin again. The amorphous material becomes transformed into a

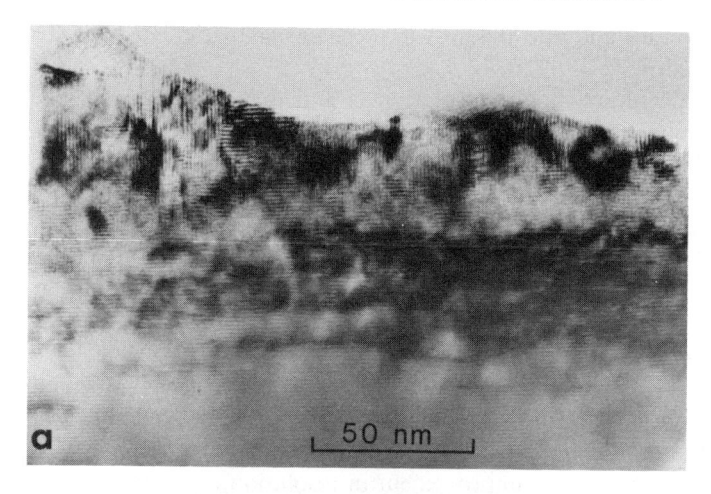

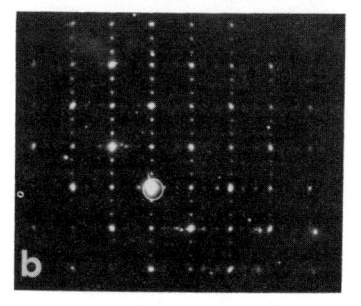

Figure 17 Dense mixture of a⊥ and c⊥ microstructures resulting from an attempt to recrystallize an implanted YBCO film by solid phase epitaxial growth. The conventional TEM image (*a*) shows 11.8 Å lattice fringes in both orientations, corresponding to the diffraction pattern (*b*).

dense array of overlapping crystallites with approximate a⊥ and c⊥ orientations.

CONCLUSIONS

Z-contrast imaging provides the ability to see directly into materials at the atomic scale. Incoherent characteristics and strong compositional sensitivity mean that the images immediately suggest the structure of the material, without the need for any preconceived ideas of likely atomic arrangements. The image represents a direct image of the atomic scale structure of materials. It provides a new level of insight into the properties of materials and into the growth mechanisms that created the structures revealed.

In the future, image resolution is anticipated to improve quite dramatically with the introduction of 300 kV high resolution STEM, which should provide a probe size of 1.3 Å. This would allow all materials to be resolved in the Z-contrast mode, along several different projections, and provide further insight into the three-dimensional nature of defects and interfaces. Spectroscopic techniques will be used to supplement the image, for example, electron energy loss spectroscopy to give information on the local electronic band structure.

Incoherent images are also ideally suited to computer techniques for resolution enhancement since the high spatial frequency information can be reconstructed from the shapes of image features. Preliminary attempts at resolution enhancement through maximum entropy methods suggest that a factor of two improvement in resolution is entirely feasible. Sub-Angstrom electron microscopy might soon be achievable.

Despite the opportunities that direct imaging offers, what is quite certain is that nature will continue to provide surprises in the tremendous diversity of behavior at the atomic scale. Understanding and controlling this behavior now present the greatest challenge to materials science.

ACKNOWLEDGMENTS

Much of the research described here was performed in collaboration with my colleagues at Oak Ridge National Laboratory, D. E. Jesson, M. F. Chisholm, O. W. Holland, D. H. Lowndes, D. P. Norton, and R. Feenstra. Thanks are due also to D. P. Paine, J.-M. Baribeau, D. C. Houghton, S. Bradley, and G. Robinson for provision of samples, and to T. C. Estes, J. T. Luck, and S. L. Carney for technical assistance. This research was sponsored by the Division of Materials Sciences, U.S. Department of Energy, under contract DE-AC05-84OR21400 with Martin Marietta Energy Systems, Inc.

Literature Cited

1. See, for example, Cowley, J. M. 1982. In *High-Resolution Transmission Electron Microscopy*, ed. P. Buseck, J. Cowley, L. Eyring. Oxford: Oxford Univ. Press; for an elementary introduction, Pennycook, S. J. 1988. *Contemp. Phys.* 23: 371
2. Pennycook, S. J., Jesson, D. E. 1990. *Phys. Rev. Lett.* 64: 938; 1991. *Ultramicroscopy* 37: 14
3. Howie, A. 1979. *J. Microsc.* 117: 11
4. Treacy, M. M. J., Rice, S. B. 1989. *J. Microsc.* 156: 211
5. Pan, M., Cowley, J. M., Chan, I. Y. 1990. *Ultramicroscopy* 34: 93
6. Rice, S. B., Koo, J. Y., Disko, M. M., Treacy, M. M. J. 1990. *Ultramicroscopy* 34: 108
7. Isaacson, M. S., Ohtsuki, M., Utlaut, M. 1979. In *Introduction to Analytical Electron Microscopy*, ed. J. J. Hren, J. I.

Goldstein, D. C. Joy, p. 343. New York: Plenum
8. Ade, G. 1977. *Optik* 49: 113
9. Cowley, J. M. 1976. *Ultramicroscopy* 2: 3
10. Jesson, D. E., Pennycook, S. J., Chisholm, M. F. 1990. In *Atomic Scale Structure of Interfaces*, ed. R. D. Brigans, R. M. Feenstra, J. M. Gibson, p. 439. Pittsburgh: Mater. Res. Soc.
10a. Jesson, D. E., Pennycook, S. J. 1992. *Proc. R. Soc. London.* Submitted
11. Kirkland, E. J., Loane, R. F., Silcox, J. 1987. *Ultramicroscopy* 23: 77
12. Pennycook, S. J. 1989. In *Ion Beam Processing of Advanced Electronic Materials*, ed. N. W. Cheung, A. D. Marwick, J. B. Roberto, p. 45. Pittsburgh: Mater. Res. Soc.
13. Paine, D. C., Evans, N. D., Stoffel, N. G. 1991. *J. Appl. Phys.* 70: 4278
14. Hafich, M. J., Quigley, J. H., Owens, R. E., Robinson, G. Y., Li, D., Otsuka, N. 1989. *Appl. Phys. Lett.* 54: 2686
15. Ourmazd, A., Bean, J. C. 1985. *Phys. Rev. Lett.* 55: 765
16. Müller, E., Nissan, H.-U., Ospelt, M., von Kanel, H. 1989. *Phys. Rev. Lett.* 63: 1819
17. LeGoues, F. K., Kesan, V. P., Iyer, S. S. 1990. *Phys. Rev. Lett.* 64: 40
18. LeGoues, F. K., Kesan, V. P., Iyer, S. S., Tersoff, J., Tromp, R. 1990. *Phys. Rev. Lett.* 64: 2038
19. Chadi, D. J. 1981. *Phys. Rev. Lett.* 59: 1691
20. Jesson, D. E., Pennycook, S. J., Baribeau, J.-M. 1991. *Phys. Rev. Lett.* 66: 750
21. Jesson, D. E., Pennycook, S. J., Baribeau, J.-M., Houghton, D. C. 1992. *Phys. Rev. Lett.* Submitted
22. Pennycook, S. J., Chisholm, M. F., Jesson, D. E., Norton, D. P., Lowndes, D. H., et al. 1991. *Phys. Rev. Lett.* 67: 765
23. Terashima, T., Bando, Y., Iijima, K., Yamamoto, K., Hirata, K., et al. 1990. *Phys. Rev. Lett.* 65: 2684
24. Streiffer, S. K., Lairson, B. M., Eom, C. B., Clemens, B. M., Bravman, J. C., Geballe, T. H. 1991. *Phys. Rev. B* 43: 13007
25. Lowndes, D. H., Norton, D. P., Budai, J. D. 1990. *Phys. Rev. Lett.* 65: 1160
26. Gerber, C., Anselmetti, D., Bednorz, J. G., Mannhart, J., Schlom, D. G. 1991. *Nature* 350: 279
27. Hawley, M., Raistrick, I. D., Beery, J. G., Houlton, R. J. 1991. *Science* 251: 1587
28. Schlom, D. G., Anselmetti, D., Bednorz, J. G., Broom, R., Catana, A., et al. 1992. *Z. Phys. B* 86: 163
29. Norton, D. P., Lowndes, D. H., Zheng, X.-Y., Warmack, R. J. 1992. *Phys. Rev. B* 44: 9760
30. Terashima, T., Iijima, K., Yamamoto, K., Hirata, K., Bando, Y., Takada, T. 1989. *Jpn. J. Appl. Phys.* 28: L987
31. Peng, J. L., Klavins, P., Shelton, R. N., Radousky, H. B., Hahn, P. A., Bernardez, L. 1989. *Phys. Rev. B* 40: 4517
32. Brunner, O., Antognazza, L., Triscone, J.-M., Mieville, L., Fischer, Ø. 1991. *Phys. Rev. Lett.* 67: 1354
33. Feenstra, R., Lindemer, T. B., Budai, J. D., Galloway, M. D. 1991. *J. Appl. Phys.* 69: 6569
34. Marshall, A. F., Barton, R. W., Char, K., Kapitulnik, A., Oh, B., et al. 1988. *Phys. Rev. B* 37: 9353
35. Pennycook, S. J., Chisholm, M. F., Jesson, D. E., Norton, D. P., McCamy, J. W., Lowndes, D. H. 1990. In *High Temperature Superconductors: Fundamental Properties and Novel Materials Processing*, ed. D. K. Christen, P. Chu, J. Narayan, L. F. Schneemeyer, p. 765. Pittsburgh: Mater. Res. Soc.
36. Ramesh, R., Inam, A., Hwang, D. M., Ravi, T. S., Sands, T., et al. 1991. *J. Mater. Res.* 6: 2264
37. Babcock, S. E., Larbalestier, D. C. 1989. *Appl. Phys. Lett.* 57: 393
38. Campbell, A. M. 1989. *Supercond. Sci. Technol.* 2: 287
39. Clarke, D. R., Shaw, R. M., Dimos, D. J. 1989. *J. Am. Ceram. Soc.* 72: 1103
40. Chisholm, M. F., Pennycook, S. J. 1991. *Nature* 351: 47
41. Inam, A., Rogers, C. T., Ramesh, R., Remschnig, K., Farrow, L., et al. 1990. *Appl. Phys. Lett.* 57: 2484

OSTWALD RIPENING OF TWO-PHASE MIXTURES

P. W. Voorhees

Department of Materials Science and Engineering, Northwestern University, Evanston, Illinois 60208

KEY WORDS: Ostwald ripening, diffusion, capillarity, elastic stress

INTRODUCTION

Phase-separation processes frequently result in a polydisperse mixture of two phases of nearly equilibrium compositions and volume fractions. Such mixtures can also be created artificially by irradiating materials to create voids or, as is done in liquid phase sintering processes, by mixing together powders of different composition. Despite the nearly equilibrium state of the two-phase system, the mixture is not in its lowest energy state. This is because of the polydisperse nature of the mixture itself and the presence of a nonzero interfacial energy. Thus in the absence of elastic stress, the total interfacial area of the system must decrease with time in order for the system to reach thermodynamic equilibrium. There are many ways the system can reduce this excess interfacial area. The process of interest here is when the interfacial area is reduced via a diffusional mass transfer process from regions of high interfacial curvature to regions of low interfacial curvature. This interfacial area reduction process is commonly called coarsening, or Ostwald ripening, after the physical chemist W. Ostwald, who first described the process (1).

This interfacial energy driven mass transfer process can significantly alter the morphology of the two-phase mixture. In general, the average size-scale of the mixture must increase with time and the number of second-phase domains, or particles, must decrease with time. An example of an Ostwald ripening process is shown in Figure 1. The upper row of micrographs shows the evolution of Sn-rich solid particles in an isothermal Pb-Sn eutectic liquid as a function of time at constant magnification. Evident

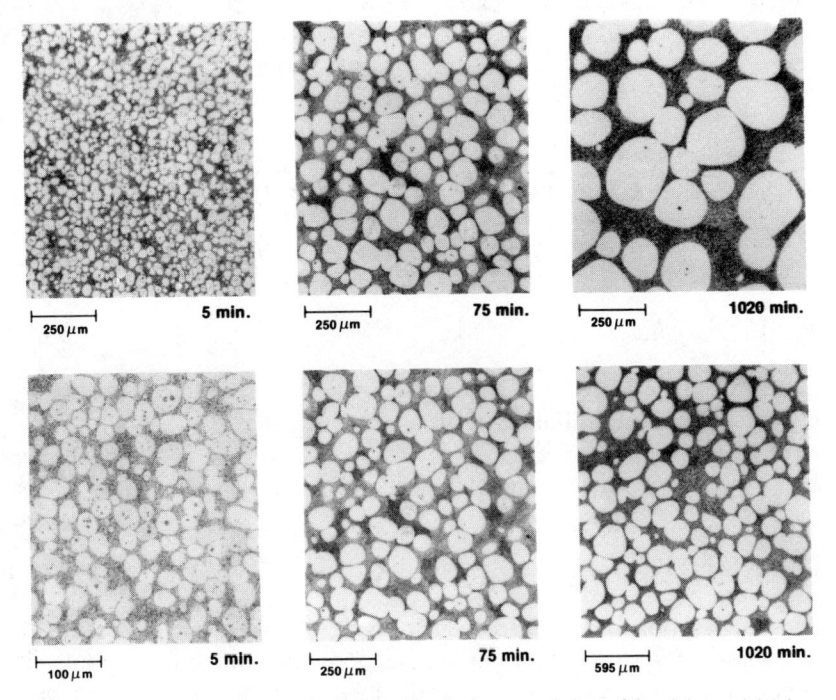

Figure 1 The microstructures of solid-liquid mixtures consisting of Sn-rich particles in a Pb-Sn eutectic liquid. The top row is at constant magnification; the bottom row is at variable magnification (72).

is the decrease in the number of particles per unit cross-sectional area and an increase in the average particle size with time. This change in the morphology occurs as a result of small particles dissolving and transferring their mass to the larger particles. The ripening process shown in these micrographs will continue until the equilibrium state of one large solid particle in a liquid is reached.

While the underlying thermodynamic driving forces for the ripening process are well understood, an understanding of the factors controlling the kinetics of the diffusional coarsening process is not at hand. There has been much theoretical work recently on interfacial energy-driven Ostwald ripening, and we shall attempt to place these more recent works into perspective. Despite the apparent generality of a theory of diffusion-limited ripening, it is possible that other phenomena can play an important role during ripening. As an illustration, the effects of elastic stress on the dynamics of coarsening in coherent solids are discussed. There can be other mediating factors in certain two-phase mixtures. For example, in systems with fluid matrices, particle motion can lead to marked changes in the nature of the ripening process [see the recent review by Ratke (2)].

FUNDAMENTALS

We assume that there is no elastic stress or fluid flow and that the system is isotropic in all respects. The system is composed of two chemical components that have the same intrinsic diffusion coefficients. These assumptions have been relaxed. For example, it is possible to generalize the below analysis to systems with anisotropic interfacial energy (3, 4). In addition, theories have also been developed for ripening in multicomponent alloys (5; C. Kuhemann & P. Voorhees, unpublished research) and for systems in which both heat and mass are transported during ripening (7). There are three equations necessary to describe the ripening kinetics of a two-phase system: a kinetic equation describing the growth or shrinkage rate of a domain of a given size, a continuity equation describing the temporal evolution of a particle size distribution function, and a mass conservation equation.

The Kinetic Equation

Of these three, the kinetic equation is usually the most difficult to determine for it is based upon a solution to a potentially difficult free-boundary problem. It also embodies much of the physics of the ripening process. Specifically, the field equation describing mass flow, which must be solved in both phases, is

$$\nabla^2 C = 0. \qquad 1.$$

The justification for neglecting the time-dependence of the concentration field lies in the small interfacial velocities, which are present during ripening, along with the requirement that an accurate description of the diffusion field is necessary for only small distances away from a particle (8).

One set of boundary conditions is the interfacial concentrations in the matrix and precipitate phases at a curved interface. These boundary conditions, the so-called Gibbs-Thomson equations, reflect the physics behind an interfacial energy-driven ripening process. Using the equilibrium conditions given by Gibbs (9), it is possible to show that the compositions of the α phase, C^α, and β phase, C^β, in an isothermal system at a curved interface are given by (10)

$$C^\alpha = C_e^\alpha + l_c^\alpha \kappa, \qquad 2.$$

$$C^\beta = C_e^\beta + l_c^\beta \kappa, \qquad 3.$$

where l_c is the capillary length in the designated phase,

$$l_c^\alpha = \frac{V_m^\beta \sigma}{(C_e^\beta - C_e^\alpha) G''^\alpha_m}, \qquad 4.$$

$$l_c^\beta = \frac{[V_1^\beta(1 - C_e^\alpha) + V_2^\beta C_e^\alpha]\sigma}{(C_e^\beta - C_e^\alpha) G''^\beta_m}, \qquad 5.$$

V_m^β is the molar volume of β, V_i^β is the partial molar volume of component i in the β phase, C_e denotes the equilibrium mole fraction of component 2 at a planar interface in the noted phase, κ is the sum of the principle curvatures of the interface taken positive for a spherical particle of β, σ is the interfacial energy, and G''_m is the second derivative of the molar free energy of the designated phase with respect to composition. These expressions for the equilibrium interfacial concentrations at a curved interface are valid for a general nonideal-nondilute solution, but are limited by the condition $|C(\kappa) - C_e| \ll 1$ in both phases. In addition, they reduce to the more standard forms for the Gibbs-Thomson equations. For example, in a dilute-ideal solution $l_c^\alpha = V_m^\beta C_e^\alpha / R_g T$, where R_g is the gas constant. These equations show that the concentration at an interface with high curvature will be above that at an interface with low curvature. In systems with nonzero solute diffusivities, this difference will cause mass to flow from an interface with high curvature to an interface with low curvature, thus resulting in the disappearance of regions of high interfacial curvature.

The other boundary condition is that the composition of the matrix is given by a yet to be determined mean-field value of C_∞. Finally, the interfacial velocity is given by the flux conservation condition at the interface,

$$(C^\beta - C^\alpha) V_n = (D^\beta \nabla C^\beta - D^\alpha \nabla C^\alpha) \cdot \mathbf{n}, \qquad 6.$$

where V_n is the local velocity of the interface in the direction of the interface normal, $\mathbf{n}$ is the normal to the interface, which is pointing from α to β, D is the diffusion coefficient in the specified phase, and the concentration gradients are evaluated at the interface in the designated phase.

Unfortunately, the morphology of the second-phase domains is not specified, but must be determined as part of the solution to the problem. As nonlinear multiparticle free-boundary problems are nearly intractable analytically, the morphology of the particles is usually chosen to be spherical and, as discussed below, a theory is developed on this basis.

The Continuity and Mass Conservation Equations

If particles flow through particle size space in a continuous manner, the time rate of change of the number of particles per unit volume of size L to $L+dL$, $f(L,t)$, is given by the following continuity equation

$$\frac{\partial f}{\partial t} + \frac{\partial (f dL/dt)}{\partial L} = 0, \qquad 7.$$

where dL/dt is the growth or shrinkage rate of a particle as given by the kinetic equation, and t is time. The assumption of a continuous flow of particles specifically excludes any process that would give rise to discontinuous jumps in particle size during the coarsening process, such as nucleation or coalescence.

The value of the mean-field concentration in the matrix required in the kinetic equation follows from a constraint that the total amount of solute in the system must be conserved,

$$C_o = (1-\phi)C_\infty + \phi C^\beta, \qquad 8.$$

where ϕ is the mole fraction of β, and C_o is the mole fraction of solute in the alloy. The mass conservation condition must be added explicitly, since the time derivative in the diffusion equation has been neglected. The mass conservation equation implies that if the mean-field concentration is a function of time during ripening, then ϕ must also be a function of time. The mole fraction, or volume fraction in systems with phases of equal molar volume, is related to the particle size distribution function as

$$\phi = G \int_0^\infty L^3 f(L,t) dL, \qquad 9.$$

where G is a geometrical factor that depends on the particle morphology.

THE INFINITELY DILUTE LIMIT

Early theories of ripening either employed somewhat unrealistic assumptions on the diffusion geometry in a coarsening system (11), or only considered the coarsening behavior of a few particles (12). In contrast, the first to attempt a solution to the above stated equations were Lifshitz & Slyozov (13, 14) and Wagner (15). In order to make the free-boundary problem tractable, they chose the β phase to be the precipitate, set $D^\beta = 0$, and considered the limit where ϕ is nearly zero. In this case, there are no interparticle diffusional interactions and this, along with the assumption of isotropy, implies that the particle morphology must be spherical. Thus,

$L = R$, where R is a particle radius, $G = 4\pi/3$, and $V_n = dR/dt$. It is quite straightforward to solve LaPlace's equation in this spherically symmetric system to obtain the kinetic equation. Using the kinetic equation in Equation 7, and using Equations 8 and 9 in the result, yields a nonlinear integro-differential equation that defines the temporal evolution of the particle radius distribution function. This integro-differential equation can be solved numerically for various initial particle radius distribution functions (16–20; M. Chen & P. Voorhees, unpublished research) and, under certain conditions, analytically as well (7, 21a).

In a beautiful piece of asymptotic analysis, Lifshitz & Slyozov were able to find a solution to this integro-differential equation. In particular, they showed that in the limit $t \to \infty$ there is a unique attractor state for this equation in which the behavior of the coarsening system can be characterized by temporal power laws: for example, $\bar{R}^3(t) = Kt$, where $\bar{R}(t)$ is the average particle radius and K is the rate constant. In addition, they showed that a time-independent-scaled particle radius distribution function exists. A manifestation of the existence of a scaled-time-invariant particle radius distribution function is illustrated, for a system with a nonzero ϕ, in the lower row of micrographs shown in Figure 1. These micrographs were constructed by scaling the magnification of each of the photographs in the upper row by a factor related to the average particle radius. Each of the resulting micrographs shown in the lower row appears to be nearly identical, in agreement with the predictions of Lifshitz & Slyozov that the particle radius distribution function should become time-independent under the scaling of the average particle radius. The appearance of such self-similar microstructures implies that the microstructure of a two-phase mixture can be predicted at any time from only a knowledge of the scaled particle radius distribution function and the value of the time-dependent average particle radius. The asymptotic analysis employed by Lifshitz & Slyozov is also discussed in a slightly more tutorial fashion by Lifshitz & Pitaevskii (22). In addition, this attractor state can be found by using other techniques (23–25).

The work of Lifshitz and Slyozov & Wagner represented a significant advance in the understanding of the kinetics of Ostwald ripening phenomena. Nevertheless, there are some shortcomings in their work, the most significant being the assumption of no interparticle diffusional interactions. Since these interactions are neglected, the theory is strictly valid only in the physically unrealistic limit of $\phi = 0$. This is one of the reasons why it is thought that the experimentally measured particle radii distributions rarely agree with the above stated predictions; for a recent review of the experimental literature of coarsening in solids, see Ardell (26). In addition, Lifshitz and Slyozov & Wagner were not able to provide much information on the time required to reach this asymptotic, self-similar, coarsening regime in a realistic two-phase system. It is quite possible that for certain initial particle radii distributions the time required to reach the self-similar coarsening regime is beyond most experimentally accessible times.

NONZERO VOLUME FRACTION SYSTEMS

The restriction of a vanishingly small volume fraction of coarsening phase employed in the theories of Lifshitz and Slyozov & Wagner was recognized immediately as one of its principle shortcomings. As a result, there have been many attempts to construct a theory that describes the ripening process at nonzero volume fractions of coarsening phase. However, only recently have theories been developed that determine the kinetic equation via a solution to the above-stated free-boundary problem. It is helpful to first consider the manner in which interparticle diffusional interactions can affect the dynamics of Ostwald ripening in order to put these more recent theories into perspective.

In nonzero volume fraction systems, the details of the spatial distribution of the particles play a role in determining the ripening behavior of the system. This is seen by considering the difference in interfacial concentration between two particles of different size, as expressed by the Gibbs-Thomson equation, and the distance between these two particles. For example, if a small particle is located near a large particle, it will have a larger dissolution rate than when it is located at the same distance from a particle of nearly the same size, since in the former case the interfacial concentrations of the particles are quite different, whereas in the latter case the interfacial concentrations of both particles are nearly the same. Thus, in contrast to the theories of Lifshitz and Slyozov & Wagner where all particles of a given size will coarsen at the same rate, in a finite volume fraction system, particles of the same size will coarsen at different rates depending on the local distribution of the other particles in the system. Experimental measurements of coarsening rates of individual particles in a finite volume fraction system confirm this fact (27). This example also implies that the spatial distribution of particles during ripening may not be statistically random.

Interparticle diffusional interactions can also influence the morphologies of the coarsening domains. Even under the assumption of isotropic interfacial energy, the particles may not possess the spherically symmetric morphology assumed in the work of Lifshitz and Slyozov & Wagner, but rather a shape that is related to the strength of the local interparticle diffusional interactions. The micrographs of Figure 1 illustrate this phenomenon. Although the solid-liquid interfacial energy in Pb-Sn alloys is nearly isotropic, the particles are not spherical, but have shapes that reflect the shape and spatial distributions of their immediate neighbors. This observation is confirmed by two-dimensional numerical calculations of the morphological development of particles during ripening in high volume fraction systems (28, 29). Here, the entire multiparticle free-boundary problem discussed above was solved when D^β was taken to be zero, and in particular there were no assumptions made on the morphologies of the particles. Shown in Figure 2 are four particles at a dimensionless time, $t = 0$, and the same four particles at $t = 1$. Evident is the distortion in the morphology of the larger growing particles from circles via the development of regions of nearly flat interface in the region between the two particles. The distortion in the particle morphology results from the screening provided by the two particle interfaces; as time progressed very little mass was able to penetrate the region between the particles, and thus the interfacial velocities in this region were nearly zero.

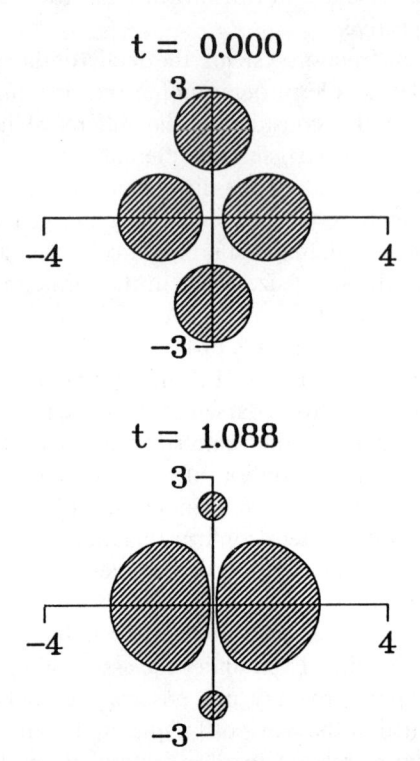

Figure 2 The morphological evolution of four circular particles arranged such that the two growing particles are close together. The smaller particles have a dimensionless radius of 0.9, and the larger particles have a dimensionless radius of 1 (28, 29).

Given the above, not necessarily inclusive, discussion of possible phenomena, it is clear that developing a first-principles theory for the kinetics of Ostwald ripening at an arbitrarily large volume fraction of coarsening phase can be quite difficult. Thus much of the recent work on ripening has centered on systems in which $\phi \ll 1$. In this limit the particles are separated at large but finite distances compared to their radii. By solving Equation 1 in the limit $\phi \ll 1$, with the boundary conditions given by Equation 2 and mass conservation, it can be shown that the particle morphologies remain nearly spherical and that the mass flow from each particle can be represented as a monopole source or sink located at the center of the particle. This approach was first employed in a coarsening problem by Weins & Cahn (30), and the details of the development have been reviewed previously (31). This monopolar solution for the diffusion field in a multiparticle system yields what can be termed a microscopic description of the coarsening rates of each particle since it is necessary to know the detailed spatial distribution of the particles in order to determine a particle's coarsening rate. In many calculations, however, the goal is to determine the statistically averaged growth rate of a particle of a given size since it is the average properties of these coarsening ensembles that are usually of greatest interest. Thus it is necessary to average these microscopic equations.

All of the recent statistical mechanical theories, developed to describe systems in which $\phi \ll 1$, employ the same microscopic equation to describe the coarsening rates of individual particles, but different techniques to perform the statistical averaging. In addition, these theories can be distinguished on yet a finer scale. The theories of Marqusee & Ross (32), Brailsford & Wynblatt (33, 34), and the computer simulations of Voorhees & Glicksman (35) do not account for the effects of interparticle spatial correlations that may develop during ripening, whereas the theories of Marder (36), Tokuyama & Kawasaki (37), and Enomoto et al (38) do consider the influence of these spatial correlations. In addition, simulations by Beenakker (39) and Abinandanan & Johnson (40) also account for the effects of spatial correlations. All of these workers find that the presence of a nonzero volume fraction of coarsening phase does not alter the temporal exponents from those of the theories of Lifshitz and Slyozov & Wagner, but that it does alter the amplitudes of the power laws. As an example, Figure 3 shows the rate constant plotted as a function of the volume fraction of coarsening phase. With the exception of Marder's theory and the simulations of Abinandanan and Johnson, all of the theories predict a rather modest increase in the rate constant for small volume fractions of coarsening phase and are in close agreement. All of the theories predict that the rate constant will vary as $\phi^{1/2}$ in this low volume fraction

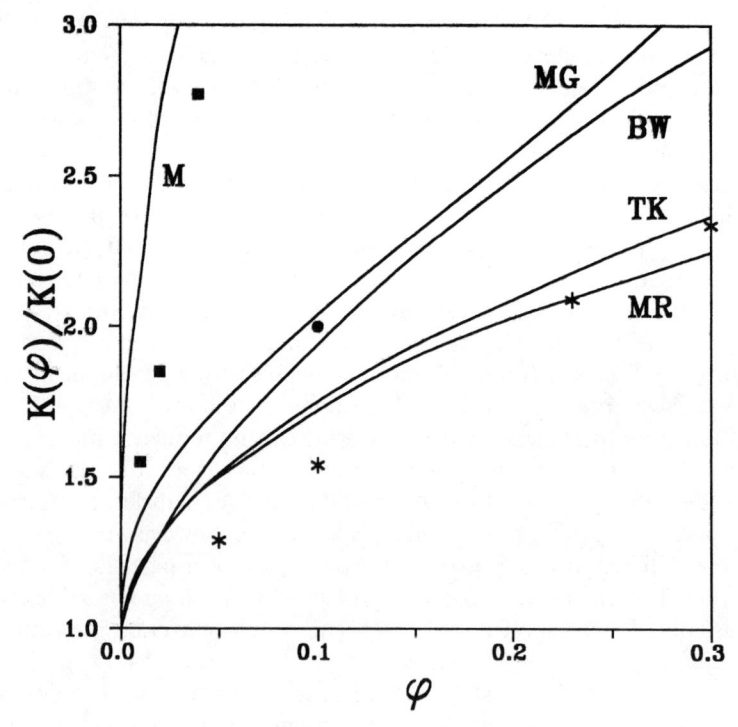

Figure 3 Rate constant, K, as a function of the volume fraction, ϕ, relative to that at zero volume fraction. Shown are predictions of the theories of Marder (M), Brailsford & Wynblatt (BW), Tokuyama & Kawasaki, and Enomoto et al (TK), Marqusee & Ross (MR), and Marsh & Glicksman (MG), along with rate constants from the computer simulations of Abinandanan & Johnson (■), Beenakker (●), and Voorhees & Glicksman (∗).

limit. In addition, all of these theories predict that the scaled time-independent particle radius distributions become broader and more symmetric than those predicted by Lifshitz and Slyozov & Wagner as the volume fraction increases.

It is likely, however, that the particles in these coarsening mixtures are not distributed in a spatially random manner, but become correlated through local interparticle diffusional interactions. Although Enomoto et al show that the spatial distribution of particles is not random because of interparticle diffusional interactions, "soft-collisions" in their terminology, they claim that the presence of these correlations has a rather weak effect on the rate constant. For example, there is a small difference between the rate constants found from the work of Marqusee & Ross, who include interparticle diffusional interactions but assume a random spatial distribution, and those of Tokuyama & Kawasaki and Enomoto et al.

Although Marder also shows that the coarsening mixtures are spatially correlated, he predicts a much larger rate constant at a given volume fraction. While the computer simulation of Beenakker agrees qualitatively with the spatial correlation function predicted by Marder for $\phi = 0.1$, the rate constant found by Beenakker is considerably different from that predicted by Marder. Recent simulations by Abinandanan & Johnson find that the mixtures also become spatially correlated during coarsening. The rate constants are much different from those of Beenakker, Enomoto et al, and Tokuyama & Kawasaki, however, and are much closer to those predicted by Marder.

In addition to theories based upon a solution to the free-boundary problem outlined above, there has been a considerable amount of work aimed at developing mean-field descriptions of the ripening process. Such descriptions have the advantage of being easily extendable to high volume fractions of coarsening phase and are employed in the work of Asimov (11), Ardell (40a), Brailsford & Wynblatt (34), Tsumaraya & Miyata (40b), and more recently in the work of Marsh & Glicksman (40c–e). The difficulty in constructing such mean-field, or effective medium, theories has been to ascribe to the medium properties that will yield a physically realistic description of the statistically averaged interparticle diffusional interactions. For example, Brailsford (33) was able to show that his effective medium theory was consistent with the monopole approximation to the multiparticle diffusion problem. A different approach has been taken by Marsh & Glicksman, who construct the medium in a manner that guarantees that certain of its properties will be invariant under the scaling of the average particle radius. As shown in Figure 3, the rate constants predicted by this theory are close to those of Brailsford & Wynblatt. However, Marsh & Glicksman's theory predicts that $K \sim \phi^{1/3}$ as $\phi \to 0$, in contrast to the statistical mechanical theories discussed previously. Clearly, more work is necessary to resolve the many disagreements between the theories of Ostwald ripening in nonzero volume fraction systems.

An experimental test of the theories describing the effects of a finite volume fraction of coarsening phase on the kinetics of Ostwald ripening is quite difficult to perform. The volume fraction of coarsening phase must be small and, in particular, the system must satisfy all the assumptions of theory. This latter requirement is particularly stringent for there are many other phenomena, one of which is discussed below, that can strongly change the nature of the ripening process from that assumed in these theories. In addition, in order to conclude if the modest factor of 2 or 3, change in the rate constant predicted by some theories exists over this small range in volume fraction, the experimental system must coarsen sufficiently rapidly that a large change in the size-scale of the system is attainable, and thus the data have a small statistical error. The difficulty in producing accurate data is aptly illustrated by the rate constant data in the NiAl system, which have been compiled by Ardell (41). In this case, for many experiments performed using low volume fractions of Ni_3Al particles in a NiAl matrix, the statistical scatter of the data is greater than the factor of 2 or 3 change in the rate constant predicted by some of these theories. Thus it is difficult to conclude on the basis of these data if the volume fraction affects the rate constant in the manner predicted by theory.

It is clear, however, that the volume fraction of coarsening phase does alter the rate constant. Shown in Figure 4 is a compilation of rate constant data for systems consisting of solid particles in a liquid (42). In these systems the ripening rate is quite rapid and a factor of 10 change in the average particle size is easily attainable in most experiments. In addition, in one system, Pb-Sn, the interfacial energy and diffusion coefficient are

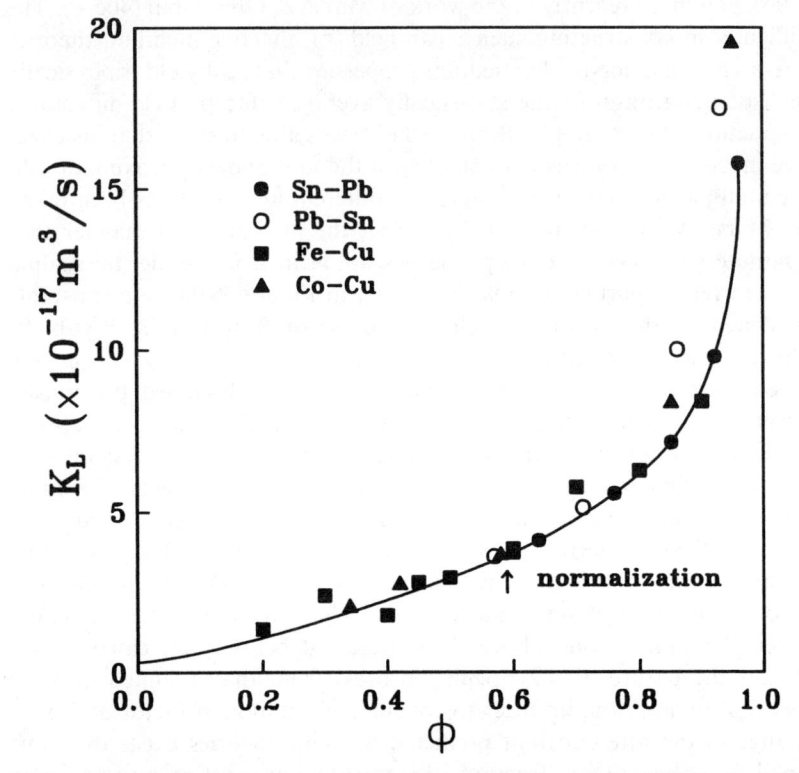

Figure 4 Rate constant, K, as a function of volume fraction, ϕ, for the following solid-liquid systems: Sn-Pb, Pb-Sn (72), Fe-Cu (73, 74), and Co-Cu (74).

known. The data of the other systems, Fe-Ca and Co-Cu, are normalized to the data of Pb-Sn at a volume fraction of 0.6. The line is a cubic spline fit to the data, which employed the rate constant calculated from the Lifshitz and Slyozov & Wagner theory at $\phi = 0$. There is a clear increase in the rate constant with volume fraction and, given the vastly different experimental conditions and alloy systems, it is interesting that all the data appear to fall on one curve. In addition, the relative change in the rate constant with volume fraction shown in Figure 4 agrees quite well with the mean-field predictions of Marsh & Glicksman for $\phi < 0.8$.

OSTWALD RIPENING IN COHERENT SOLIDS

There are many clear examples in the literature of the essential role elastic stress plays in the development of both particle morphology and interparticle spatial correlations during Ostwald ripening in coherent solids. For example, for systems in which stress is generated by a particle-matrix misfit, a nearly random spatial distribution of particles resulting from a nucleation and growth process can evolve into a highly correlated spatial distribution of particles in which the particles are aligned in stringers along elastically soft crystallographic directions of the matrix, or in nearly periodic arrays (43–45). In addition, it has been shown that elastic stress strongly influences the morphology of individual particles as well. In the NiAl system, for example, when the particles are small and the interfacial-to-elastic energy ratio is large, the particles are spherical. As they coarsen, however, the ratio of the interfacial energy to the elastic energy decreases and the morphology of the particles changes from a sphere to a cuboidal shape (43–45). Other illustrations of these shape bifurcations are provided by experiments in which the particle volume is held fixed, but the misfit is increased by the addition of a small quantity of an alloying element. Consistent with the above, when Mo is added to a NiAl alloy, the misfit changes and the particle morphology changes from a sphere to cuboid as the misfit is increased (46, 47). These elastically induced particle shape changes have been discussed extensively in a recent review (48).

A still more striking example of the effects of elastic stress on the morphological development of particles is provided by the work of Miyazaki et al (49) and Kaufman et al (50). Shown in Figures 5 and 6 are micrographs of a Ni-Al alloy taken at different aging times. Figure 5 shows the morphology of the Ni_3Al particles after 1 hr of aging. If a classical coarsening process was operative in this alloy, the average size of the particles would increase with time and the morphology of the particles would not change significantly. This is not the case, however, and Figure 6 shows that after 5 hr of aging the particles spontaneously split, eventually

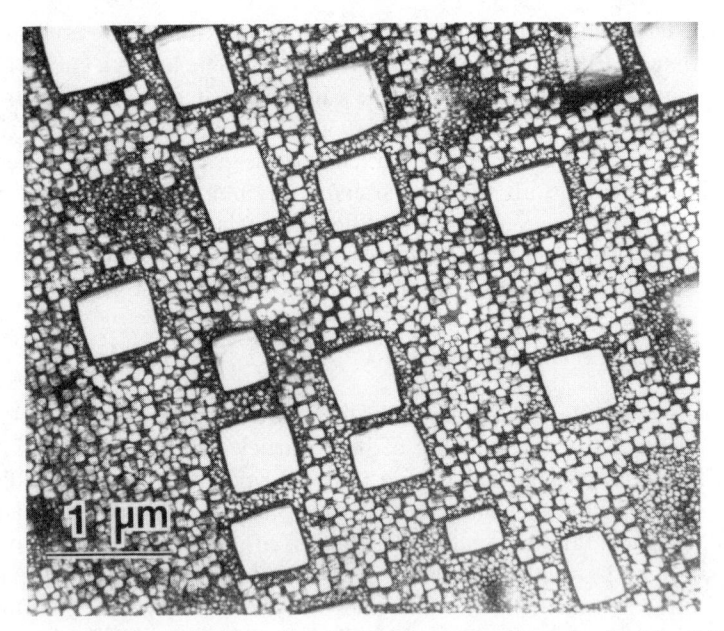

Figure 5 Ni$_3$Al (γ') particles in a disordered Ni-Al matrix after aging for 1 hr at 980°C. Only the large cuboidal γ' particles were present at the aging temperature. The smaller γ' particles formed during the quench from the aging temperature (50).

forming eight smaller particles in place of the one large particle. In this case, the average particle size of the system is decreasing with time, contrary to the classical theories of coarsening. It is interesting to note that the appearance of octets of Ni$_3$Al particles in a Ni-Al matrix has been seen for some time (51), but it was not until the work of Miyazaki et al that particle splitting was identified as the mechanism by which these particle octets formed. In addition, similar observations of this splitting process have been made by Doi et al in multicomponent superalloys (52).

The particle shape changes, particle splitting, and particle alignment processes discussed above are examples of highly nonclassical ripening behavior. The strong spatial correlations seen in these experiments far exceed those predicted by the theories discussed previously for ripening driven solely by interfacial curvature. Attempts at understanding these examples of nonclassical ripening behavior have been largely limited to calculations of the elastic energies of varying particle morphologies and spatial arrangements. These calculations show that elastic stress can be the cause of the observed nonclassical behavior. For example, many workers have shown that the elastic energy of a system of particles in an elastically anisotropic medium decreases upon a decrease in the inter-

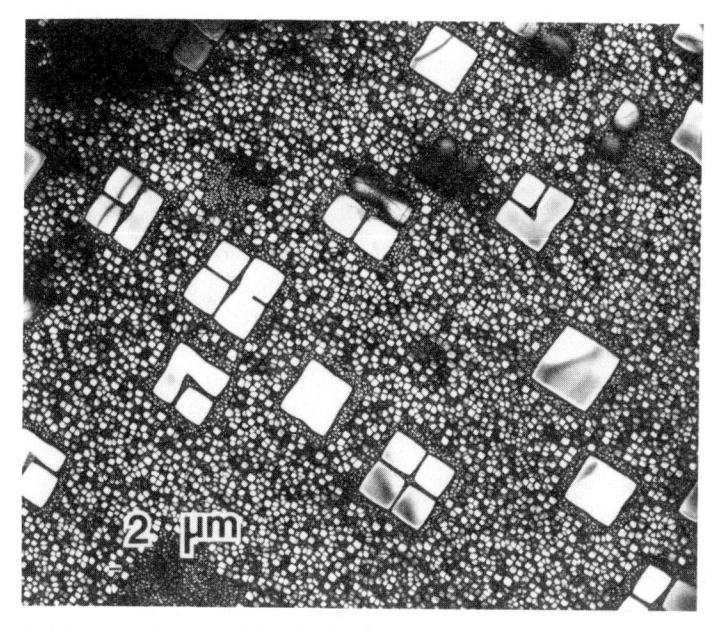

Figure 6 The same alloy after 5 hr of aging. The average size of the γ' particles decreases during this splitting process (50).

particle separation and alignment along the elastically soft crystallographic directions (53–56). In addition, it is clear from other elastic energy calculations that the observed splitting process is probably related to the decrease in elastic energy on forming many small cubes from one large cube (57, 58, 58a). This decrease in elastic energy offsets the increase in interfacial energy that accompanies the splitting process, and thus particle splitting can decrease the total energy of the system.

Recently, there has been much work aimed at investigating the dynamics of the morphological development of misfitting particles where the particle morphology is not restricted to the simple geometric shapes assumed in the energetic calculations. In addition, the dynamics of the overall microstructural development of an elastically stressed multiparticle system has been studied. Two different approaches have been employed. One is quite similar to that described previously for coarsening in the absence of elastic stress, wherein the particles evolve according to solution to a diffusion equation with a Gibbs-Thomson boundary condition, modified for the presence of elastic stress. The other approach has been to use a Cahn-Hilliard equation, which includes the effects of elastic stress. Due to the nonlinearities present in both of these formulations, most of the work has been numerical and, except as noted, in two dimensions.

Both of these techniques have shown that equilibrium particle morphologies in elastically stressed systems may not be simple geometric shapes. For example, in an elastically isotropic system, where the particles are softer than the matrix, Onuki & Nishimori show that the elastic interactions between particles alone can lead to particle morphologies similar to those of the large particles at $t = 1$, as shown in Figure 2 (59). Work by Wang et al has shown that a misfitting particle with isotropic interfacial energy in a cubically anisotropic solid can attain equilibrium shapes that have the fourfold symmetry of the elastic constant tensor, are nonconvex, and can even be multiply connected (60). In this case, if the misfit exceeds a certain value, the equilibrium particle shape consists of a square-like particle with a small region of the matrix phase in the center of the particle. These calculations are also consistent with the recent simulations of McCormack et al (58a). Voorhees et al find that in a system with a cubic elastic anisotropy and isotropic interfacial energy, circular particles are never the equilibrium shapes for any nonzero misfit and that it is possible to have two equilibrium particle morphologies for different particle sizes: a square- or plate-like shape (61). All of these results are in contrast to the classical Wulff construction for the equilibrium shape of a particle in a stress-free system with isotropic interfacial energy: it is a circle, and it is unique.

Two-dimensional simulations of the overall microstructural development in an elastically stressed solid have been performed primarily using the Cahn-Hilliard approach. In an elastically anisotropic system, both Nishimori & Onuki (62) and Wang et al (63) show that elastic stress is the cause of the microstructures commonly observed in Ni-Al alloys; namely, a highly correlated spatial distribution of particles and particles that attain plate-like shapes as their size increases. Nishimori & Onuki's calculations also show that the exponent describing the evolution of the average size of the domains varies with time and is less than the classical value of one third found by Lifshitz & Slyozov. In some cases, they find a strong tendency towards complete stabilization with respect to coarsening, i.e. the exponent tends towards zero with increasing time when the elastic constants of the particle are less than that of the matrix (64). This is qualitatively consistent with experimental observations of a decreasing temporal power law exponent for the average size of misfitting particles with time (65, 66).

Extensions of these calculations to three dimensions in systems with the large numbers of particles necessary to collect accurate statistical data will be quite difficult. Thus the alternate computational scheme has been employed in three dimensions, where a diffusion equation is solved with the modified Gibbs-Thomson boundary condition. Using this approach and assuming statistical self-similarity, Leo et al have shown that in the completely elastic stress-dominated coarsening regime, the temporal exponent for the average domain size should be one half instead of one third (67). This is in agreement with the work of Enomoto & Kawasaki, who employed a different form for the Gibbs-Thomson equation (68). However, both Leo et al and Enomoto & Kawasaki must assume that the two-phase mixtures are either self-similar or randomly distributed, respectively. In addition, Leo et al indicate that the particle sizes for which the $t^{1/2}$ law can be observed may be beyond the sizes at which the particles can be coherent with the matrix. Enomoto & Kawasaki assume that the particles are always spherical. This is a serious shortcoming of their work for, as pointed out by Onuki & Nishimori (59), once the particles change their shape in response to the interparticle elastic fields, the elastic interaction energy is quite different from that calculated for elastically interacting spheres. Since, as described above, misfitting particles are frequently not randomly distributed in alloys undergoing coarsening, Abinandanan & Johnson solved the multiparticle elastic and diffusion problem for an array of spherical particles in an isotropic system where the elastic constants of the particles are less than the matrix, without making any assumptions on particle location (40). They find, in agreement with earlier work on a simpler two-particle problem (69–71), that the elastic interactions between particles can lead to movement of the particles through the matrix via a dissolution-reprecipitation process and a spatial correlation function that is both non-uniform and time-dependent. It also appears that when the elastic constants of the particles are less than those of the matrix, the exponent of the temporal power law for the average particle radius is less than one third.

CONCLUSIONS

Despite the large amount of theoretical work aimed at understanding the effects of interparticle diffusional interactions, much remains to be done both theoretically and experimentally. For example, it is not clear why certain theories, which begin from an identical description of the diffusion field in a coarsening mixture, arrive at contradictory results for the average behavior of the system. These theories also need to be extended to the higher volume fractions of coarsening phase, which are easily accessible experimentally and will permit a more critical comparison to existing mean-field theories. However, the existing finite volume fraction theories for Ostwald ripening remain untested. Thus there is a great need to locate a system that satisfies all the assumptions employed in the theory and is suitable for use in an experiment to measure the kinetics of Ostwald ripening.

With the advent of understanding the manner in which elastic stress influences the thermodynamics of crystalline solids, it has become possible to predict some of the qualitative features of systems that coarsen in the presence of elastic stress generated by a particle-matrix misfit. It remains for future work to improve the computational techniques to the point that quantitative tests of these calculations are possible. In particular, the calculations must be performed in three dimensions and with a theory that depends on parameters that can be determined a priori.

ACKNOWLEDGMENTS

We thank N. Akaiwa and M. Chen for a review of an early version of this manuscript and M. E. Glicksman, S. P. Marsh, and A. Onuki for many helpful comments. The financial support of the National Science Foundation, grant number DMR-8957219 and the Microgravity Sciences and Applications Division of NASA is gratefully acknowledged.

Literature Cited

1. Ostwald, W. 1900. *Z. Phys. Chem.* 34: 495
2. Ratke, L. 1990. *Low-Gravity Fluid Dynamics and Transport Phenomena*, pp. 661. Washington, DC: Am. Inst. Aeronaut. Astronaut.
3. Mullins, W. W. 1986. *J. Appl. Phys.* 59: 1341
4. Zwillinger, D. 1989. *J. Cryst. Growth* 94: 159
5. Björklund, S., Donaghey, L. F., Hillert, M. 1972. *Acta Metall.* 20: 867
6. Deleted in proof
7. Voorhees, P. W. 1990. *Metall. Trans. A* 21: 27
8. Coriell, S., Parker, R. L. 1965. *J. Appl. Phys.* 36: 632
9. Gibbs, J. W. 1906. *The Collected Works of J. Willard Gibbs*. Longmans, Green & Co. Reprinted by Dover, New York
10. Hillert, M. 1975. *Lectures on the Theory of Phase Transformations*. New York: Am. Inst. Min., Pet. Metall. Eng.
11. Asimov, R. 1963. *Acta Metall.* 11: 72
12. Greenwood, G. W. 1956. *Acta Metall.* 4: 243
13. Lifshitz, I. M., Slyozov, V. V. 1960. *Sov. Phys.: Solid State* 1: 1285
14. Lifshitz, I. M., Slyozov, V. V. 1961. *J. Phys. Chem. Solids* 19: 35
15. Wagner, C. 1961. *Z. Elektrochem.* 65: 35
16. Venzl, G. 1983. *Ber. Busenges. Phys. Chem.* 87: 318
17. Enomoto, Y., Kawasaki, K., Tokuyama, M. 1987. *Acta Metall.* 35: 915
18. Tsang, T. H., Brock, J. R. 1984. *Aerosol. Sci. Technol.* 3: 283
19. Atwater, H. A., Yang, C. M. 1990. *J. Appl. Phys.* 67: 6202
20. Ratke, L., Host, M. 1992. *J. Physicochem. Hydrodynam.* In press
21. Rios, P. R. 1992. *J. Mater. Sci. Lett.* In press
21a. Markworth, A. J. 1971. *Ber. Bunsenges. Phys. Chem.* 75: 533
22. Lifshitz, E. M., Pitaevskii, L. P. 1981. *Physical Kinetics*, p. 432. London: Pergamon
23. Marqusee, J. A., Ross, J. 1983. *J. Chem. Phys.* 79: 373
24. Coutsias, E. A., Neu, J. A. 1984. *Physica D* 12: 295
25. Mullins, W. W. 1991. *Acta Metall.* 39: 2081
26. Ardell, A. J. 1988. *Phase Transformations '87*, p. 485. London: Inst. Metals
27. Voorhees, P. W., Schaefer, R. J. 1987. *Acta Metall.* 35: 327
28. McFadden, G. B., Voorhees, P. W.,

Boisvert, R. F., Meiron, D. I. 1986. *J. Sci. Comp.* 1: 117
29. Voorhees, P. W., McFadden, G. B., Boisvert, R. F., Meiron, D. I. 1988. *Acta Metall.* 36: 207
30. Weins, J. J., Cahn, J. W. 1973. In *Sintering and Related Phenomena*, ed. G. C. Kuczynski, p. 151. London: Plenum
31. Voorhees, P. W. 1985. *J. Stat. Phys.* 38: 231
32. Marqusee, J. A., Ross, J. 1984. *J. Chem. Phys.* 80: 536
33. Brailsford, A. D. 1976. *J. Nucl. Mater.* 60: 257
34. Brailsford, A. D., Wynblatt, P. 1979. *Acta Metall.* 27: 489
35. Voorhees, P. W., Glicksman, M. E. 1984. *Acta Metall.* 32: 2013
36. Marder, M. 1987. *Phys. Rev. A* 36: 858
37. Tokuyama, M., Kawasaki, K. 1984. *Physica A* 123: 386
38. Enomoto, Y., Tokuyama, M., Kawasaki, K. 1987. *Acta Metall.* 35: 907
39. Beenakker, C. W. J. 1986. *Phys. Rev. A* 33: 4482
40. Abinandanan, T. A. 1991. PhD thesis. *Coarsening of elastically interacting coherent particles*. Carnegie Mellon Univ.
40a. Ardell, A. J. 1972. *Acta Metall.* 20: 61
40b. Tsumaraya, K., Miyata, Y. 1983. *Acta Metall.* 31: 437
40c. Marsh, S. P. 1991. PhD thesis. *Kinetics of diffusionally limited microstructural coarsening*. Rensselaer Polytech. Inst.
40d. Marsh, S. P. 1990. In *Simulation and Theory of Evolving Microstructures*, ed. M. P. Anderson, A. D. Rollet, p. 167. Warrandale, Penn.: Min. Met. Mater. Soc.
40e. Glicksman, M. E., Smith, R. N., Marsh, S. P., Kuklinski, R. 1992. *Metall. Trans.* In press
41. Ardell, A. J. 1990. *Scr. Metall.* 24: 343
42. Hardy, S. C., Akaiwa, N., Voorhees, P. W. 1991. *Acta Metall. Mater.* 39: 2931
43. Ardell, A. J., Nicholson, R. B., Eschelby, J. D. 1966. *Acta Metall.* 14: 1295
44. Taypkin, Y. D., Travina, N. T., Kozlov, V. P., Ugarova, Y. V. 1976. *Fiz. Met. Metalloved.* 42: 1294
45. Miyazaki, T., Seki, K., Doi, M., Kozakai, T. 1988. *Mater. Sci. Eng.* 77: 125
46. Biss, V., Sponseller, D. L. 1973. *Metall. Trans.* 4: 1953
47. Conley, J., Fine, M. E., Weertman, J. R. 1989. *Acta Metall.* 37: 1251
48. Johnson, W. C., Voorhees, P. W. 1992. *Non-Linear Phenomena in Materials Science*, Vol. 2. Submitted

49. Miyazaki, T., Imamura, H., Kozakai, T. 1982. *Mater Sci. Eng.* 54: 9
50. Kaufman, M. J., Voorhees, P. W., Johnson, W. C., Biancaniello, F. S. 1989. *Metal. Trans. A* 20: 2171
51. Westbrook, J. H. 1958. *Z. Kristallogr.* 110: 21
52. Doi, M., Miyazaki, T., Wakatsuki, T. 1985. *Mater. Sci. Eng.* 74: 139
53. Khachaturyan, A. G. 1969. *Phys. Status Solidi* 35: 119
54. Khachaturyan, A. G. 1983. *Theory of Structural Transformations in Solids*, Ch. 10. New York: Wiley
55. Johnson, W. C., Lee, J. K. 1979. *Metall. Trans.* 10: 1141
56. Seitz, E., de Fontaine, D. 1978. *Acta Metall.* 26: 1671
57. Johnson, W. C., Voorhees, P. W. 1987. *J. Appl. Phys.* 61: 1619
58. Khachaturyan, A. G., Semenovskaya, S. V., Morris, J. W. 1988. *Acta Metall.* 36: 1563
58a. McCormack, M., Khachaturyan, A. G., Morris, J. W. 1992. *Acta Metall.* 40: 325
59. Onuki, A., Nishimori, H. 1991. *J. Phys. Soc. Jpn.* 60: 1
60. Wang, Y., Chen, L. Q., Khachaturyan, A. G. 1991. *Scr. Metall. Mater.* 25: 1387
61. Voorhees, P. W., McFadden, G. B., Johnson, W. C. 1992. *Acta Metall.* In press
62. Nishimori, H., Onuki, A. 1990. *Phys. Rev. B* 42: 980
63. Wang, Y., Chen, L. Q., Khachaturyan, A. G. 1991. *Scr. Metall. Mater.* 25: 1969
64. Onuki, A., Nishimori, H. 1991. *J. Phys. Soc. Jpn.* 60: 1208
65. Miyazaki, T., Doi, M. 1989. *Mater. Sci. Eng. A* 110: 175
66. Miyazaki, T., Doi, M., Kozakai, T. 1988. *Solid State Phenom.* 38\1: 227
67. Leo, P. H., Mullins, W. W., Sekerka, R. F., Viñals, J. 1990. *Acta Metall. Mater.* 38: 1573
68. Enomoto, Y., Kawasaki, K. 1989. *Acta Metall. Mater.* 37: 1399
69. Voorhees, P. W., Johnson, W. C. 1988. *Phys. Rev. Lett.* 61: 2225
70. Johnson, W. C., Voorhees, P. W., Zupon, D. E. 1989. *Metall. Trans. A* 20: 1175
71. Johnson, W. C., Abinandanan, T. A., Voorhees, P. W. 1990. *Acta Metall. Mater.* 38: 1349
72. Hardy, S. C., Voorhees, P. W. 1988. *Metall. Trans. A* 19: 2713
73. Niemi, A. N., Courtney, T. H. 1981. *Acta Metall.* 3: 1393
74. Kang, C. H., Yoon, D. N. 1981. *Metall. Trans. A* 12: 65

RMC: MODELING DISORDERED STRUCTURES

R. L. McGreevy and M. A. Howe

Department of Physics, Oxford University, Clarendon Laboratory, Parks Road, Oxford OX1 3PU, England

KEY WORDS: Monte Carlo, liquid, glass, fast ion conductor, diffuse scattering

INTRODUCTION

When a new crystalline material that shows interesting properties is discovered, for example the high T_c superconductors found a few years ago (1), the first step in understanding these properties is almost always a determination of the structure. X-ray or neutron diffraction experiments are performed and the data are used, possibly in conjunction with some prior knowledge about similar materials, to determine a structure that is consistent with the available data (the structure cannot actually be determined directly). In the case of high T_c materials, it was eventually found that they all have in common a structure of Cu-O planes within which superconductivity occurs. Clearly this knowledge is crucial in determining the coupling mechanism and hence the origin of superconductivity at such high temperatures.

If, however, a new disordered material is discovered, for example a liquid, glass, or polymer, many experiments are often performed without any clear knowledge of the structure. In fact data that have no direct relation to the structure are used to predict structural features. This approach can often be misleading, so why is it used? The problem is that there are numerous general methods available for predicting crystal structures from diffraction data, but there have been no such methods available for disordered structures. Since many materials of technological interest are both complex and disordered, this has been a considerable disadvantage. Herein we describe a recently developed technique, reverse Monte Carlo (RMC) modeling (2), that overcomes such problems. We review the present applications and possible future developments. Because the technique is new much of the work has only been published recently or, particularly for new developments, is still unpublished.

RMC METHOD
Monte Carlo Methods

RMC is a variation of the standard Metropolis Monte Carlo (MMC) method (3). For those unfamiliar with such Monte Carlo methods a brief introduction is useful. The principle of MMC is that we wish to produce a statistical ensemble of atoms (configuration) with a Boltzmann distribution of energies. Rather than simply generating and sampling configurations completely at random, which would be highly inefficient, we make use of a weighted sampling procedure known as a Markov chain. This improves the sampling efficiency considerably while still allowing access to all possible states of the system, i.e. no configurations are excluded as possibilities. For an ensemble in which the number of particles, volume, and temperature are fixed (NVT), this is achieved by the following algorithm.

1. N atoms are placed in a cell with periodic boundary conditions, i.e. the cell is surrounded by images of itself. Normally a cubic cell is used, but other geometries may also be chosen. For a cube of side L, the density N/L^3 must equal the required density of the system. The probability of this particular configuration in a Boltzmann distribution (old = o) is given by

$$P_o \propto e^{-U_o/kT}, \qquad\qquad 1.$$

where U_o is the total potential energy, which may be calculated on the basis of a specified form of the interatomic potential, and T is the specified temperature. The initial configuration may be chosen in any manner; for instance it may be generated randomly, a crystal structure or a configuration from a previous simulation.

2. One atom is moved at random. The probability of the new (n) configuration is

$$P_n = e^{-U_n/kT}, \qquad\qquad 2.$$

and hence

$$P_n/P_o = e^{-(U_n-U_o)/kT} = e^{-\Delta U/kT}. \qquad\qquad 3.$$

3. If $\Delta U < 0$, the new configuration is accepted and becomes the next

starting point. If $\Delta U > 0$, then it is accepted with probability P_n/P_o. Otherwise it is rejected and we return to the previous configuration.
4. Repeat from step 2.

As atoms are moved, U will decrease until it reaches an equilibrium value about which it will fluctuate. In equilibrium, configurations are considered to be statistically independent if separated by at least N accepted moves, and they may be saved. In this way an appropriate ensemble is generated that may then be used to calculate various structural or thermodynamic properties.

RMC Algorithm

In RMC we wish to generate an ensemble of atoms that corresponds to a structure factor (set of experimental data) within its errors, which are assumed to be purely statistical and to have a normal distribution. The algorithm is as follows:

1. Start with an initial configuration as for MMC. Calculate the radial distribution function

$$g_o^C(r) = \frac{n_o^C(r)}{4\pi r^2 dr \rho}, \qquad 4.$$

where ρ is the atom number density and $n^C(r)$ is the number of atoms at a distance between r and $r+dr$ from a central atom, averaged over all atoms as centers. Transform to the structure factor

$$A_o^C(Q) - 1 = \rho \int_0^\infty 4\pi r^2 [g_o^C(r) - 1] \frac{\sin Qr}{Qr} dr, \qquad 5.$$

where Q is the momentum transfer. Calculate the difference between the measured structure factor, $A^F(Q)$, and that determined from the configuration, $A^C(Q)$,

$$\chi_o^2 = \sum_{i=1}^m [A_o^C(Q_i) - A^E(Q_i)]^2 / \sigma^2(Q_i), \qquad 6.$$

where the sum is over the m experimental points and σ is the experimental error (in practice an estimate).

2. Move one atom at random. Calculate the new $g_n^C(r)$ and $A_n^C(Q)$ and

$$\chi_n^2 = \sum_{i=1}^m [A_n^C(Q_i) - A^E(Q_i)]^2 / \sigma^2(Q_i). \qquad 7.$$

3. If $\chi_n^2 < \chi_o^2$, then the move is accepted. If $\chi_n^2 > \chi_o^2$, the move is accepted with probability $\exp[-(\chi_n^2 - \chi_o^2)/2]$. Otherwise it is rejected.
4. Repeat from step 2.

As this procedure is iterated, χ^2 will initially decrease until it reaches an equilibrium value about which it will oscillate. The resulting configuration should be a three-dimensional structure that is consistent with the experimental structure factor within experimental error. Configurations may then be collected under the same conditions as for MMC.

It can immediately be seen that χ^2 in RMC replaces U/kT in MMC, but otherwise the two algorithms are identical. It is particularly important that RMC uses a proper Markov chain so that the final structure is independent of the initial configuration. The use of the RMC algorithm in practice has been discussed in detail by McGreevy et al (4).

In Figure 1 we show a simple two-dimensional example of the course of an RMC run. In this case we start with a configuration that is a square lattice, although any other configuration could have been chosen. The data come from an MMC simulation using a Lennard-Jones potential. As the RMC modeling progresses, the $g(r)$ of the configuration gradually gets closer to that of the data until, in equilibrium, a good fit is obtained. The deviations are purely statistical errors due to the small simulation size in this example. For a real RMC model we typically use approximately 4000 atoms, and configurations as large as 30,000 atoms have been used.

Variations on the Basic Method

The algorithm described above is specifically for modeling of a set of diffraction data that could be obtained using either X-rays, neutrons, or possibly electrons. In such a case the fit may be either to the structure factor or to the radial distribution function. However, the RMC method is more general in that any set(s) of data that can be directly related to the structure can be modeled. Application to isotopic substitution in neutron diffraction (or equivalently to anomalous scattering in X-ray diffraction), to EXAFS, and possibly to NMR data are described later. Other information that cannot be used directly can be made use of in the form of constraints; this may include NMR, EPR, Raman scattering, and chemical knowledge. All data and constraints can be used simultaneously to obtain a single structural model simply by adding the appropriate χ^2 values.

Uniqueness

One question that is continually raised with regard to the RMC method concerns the uniqueness of the structures generated. The structures are not unique, they are simply possible structures that are consistent with the data. In this it is no different from any other method of modeling or theory. The non-uniqueness of RMC is not necessarily a disadvantage, it can be an advantage. For instance, the constraints described above can be used in an empirical fashion to map out the range of possible structures that

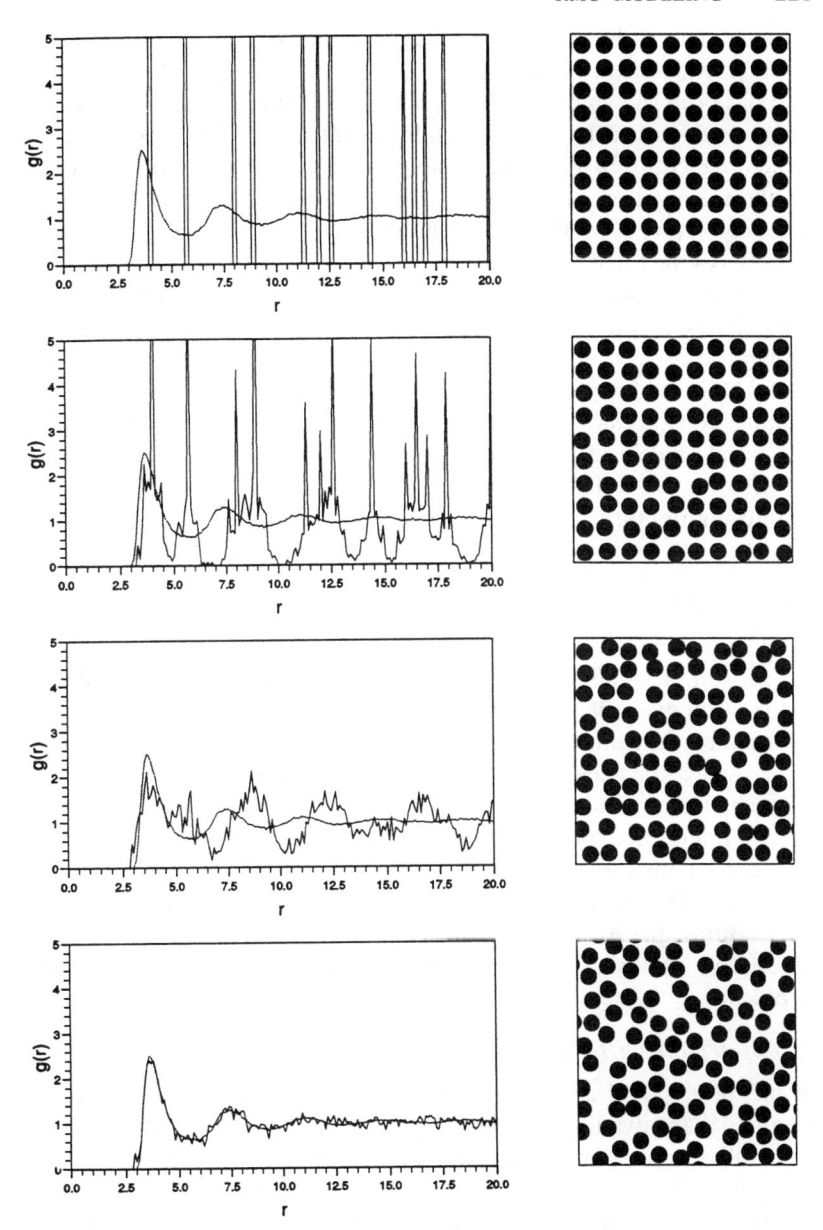

Figure 1 Example of RMC modeling of a simple test system. On the left is the configuration (two dimensional), and on the right is the $g(r)$ calculated from the configuration (*solid curve*) compared to the data (*broken curve*). The starting configuration (a square lattice) is at the top, and an equilibrium configuration is at the bottom, with intermediate configurations inbetween.

are consistent with the data, and these can then be considered in the light of other available information.

It has been shown by Evans (5), however, that in the particular case of a system described by pairwise additive potentials, the one-dimensional structure factor of a macroscopically isotropic system does in fact contain all the information necessary to define the three-dimensional structure uniquely. Howe & McGreevy (6) have shown that for such systems RMC reproduces the three body correlations correctly, i.e. the algorithm works. They have also shown that in systems where the potentials are definitely not pairwise additive it may nevertheless be possible to use constraints to obtain the correct answers. However, this should not be taken to mean that RMC can always produce the correct answer; often we may not know suitable constraints, or they may not even be definable. In the end the structural models obtained must be judged in the light of all the available information, although it should be noted that what is sometimes viewed as information turns out to be more in the nature of historically accepted opinion, or even personal bias, and must be treated with due caution.

Advantages of RMC

RMC modeling has many advantages that are illustrated by examples in the section on application. These are summarized below: (*a*) Full use is made of all available data, not just particular features, in a quantitative rather than a qualitative manner. Many other models are based on particular structural features such as peak positions and average coordination numbers from radial distribution functions, which can be misleading (7). (*b*) No interatomic potential is required. (*c*) The model is self-consistent and corresponds to a possible physical structure. This is not true of all models, particularly those which only concern short range structural correlations. (*d*) Many different types of data can be combined simultaneously. (*e*) RMC is general and easily adapted to different physical problems.

Comparison with Simulation Methods Using Potentials

It has been stated above that the fact that no interatomic potential is required is an advantage of RMC. The determination of suitable potentials that will accurately reproduce the experimental diffraction data on even comparatively simple systems is a complex task. If done, however, then the structural models produced this way are clearly equally as possible as those produced by RMC. The availability of potentials is then advantageous because (*a*) all thermodynamic states of the same system can be investigated, whereas RMC would require separate data sets, and (*b*) some parts of potentials are transferable between systems. Molecular dynamics also enables the calculation of dynamical correlations, whereas Monte Carlo procedures only provide structural information (however, some dynamical information can be inferred, see below).

APPLICATION TO MATERIALS
Liquids

RMC modeling has been applied more to the study of liquid structures than in any other area. A considerable body of this work is summarized below. It is worth noting that for hardly any of the systems described, in this and the following subsections, have models that agree comparably well with the data been produced by any other method.

Howe et al (8) recently completed a major survey of the structures of 51 elemental liquids, using the available neutron and X-ray diffraction data, and systematically studied changes in structure in relation to changes in bonding across the periodic table. They found that the large majority of elemental liquids can be classified as simple, in that the structure factors and radial distribution functions can be approximately scaled to one another. In these cases the systems are almost certainly well described by pairwise additive potentials. As many body potentials begin to become important, e.g. for elements such as Ga, then RMC will not necessarily be expected to produce the right answer (see above), but nevertheless will provide useful information on trends in bond angle distributions, which later may be useful for assessing angle-dependent potentials. [Where systems are not truly molecular, i.e. the idea of a bond is not well defined, we normally consider two atoms to be bonded if separated by a distance less than that of the first minimum in $g(r)$.] For elements such as S, Se, and Te, where the structures are complex and considered to be either weakly molecular or polymeric, the use of constraints in the RMC model can determine the range of structural models that are consistent with the data. For molecular liquids such as the halogens, molecules can be included explicitly; the intermolecular forces are then largely pairwise additive, so RMC can be used to study orientational correlations between molecules (9).

There have been two slightly more specialized investigations of elemental liquids. Nield et al (10) studied the structure of expanded caesium at the approach to the critical point in relation to the metal-non-metal transition. By using a simple model in which bonds less than a certain length are considered to be metallic, they have created bond networks, which are then treated as resistor (conductor) networks. In this way the conductivity can be calculated, and is found to be in surprisingly good agreement with the experimental data. Bond networks at two temperatures are shown in

Figure 2. As the critical point is approached, it is found that the bond network approaches a percolation transition that is the metal-non-metal transition. Pusztai & McGreevy (11) looked at the possibility of structural change in liquid ^{4}He at the superfluid transition. They concluded that the errors in the available data are as large as any possible changes. However, they note the interesting result that whereas the angular correlations in all simple liquids, including the condensed inert gases such as neon and argon, indicate that the structures are essentially (hexagonal) close-packed, in helium there is evidence of icosahedral packing.

There have been three studies of different types of binary liquids. McGreevy & Putsztai (12) modeled the structures of twelve molten salts using sets of three structure factors obtained by neutron diffraction with isotopic substitution. They conclude that the local liquid structures tend to be disordered versions of the crystalline structures below the melting point. In the cases of some alkaline earth and nickel halides, this conclusion contradicts what has been suggested by other authors. Howe & McGreevy

Figure 2 Networks of metallic bonds in Cs at (*a*) 323 K and (*b*) 1673 K (9).

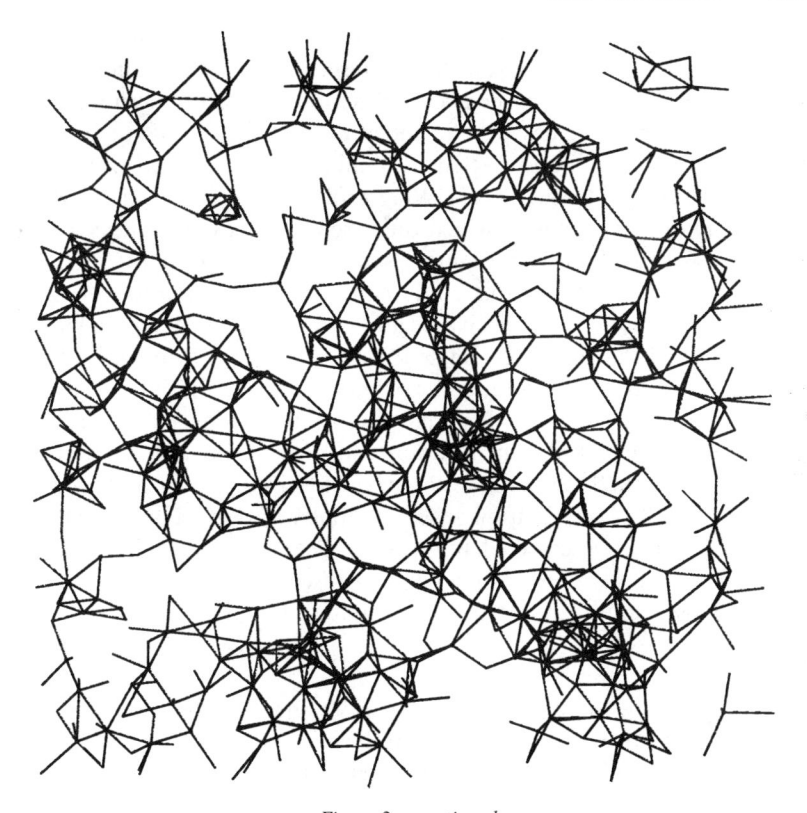

Figure 2—continued.

(13) modeled the structures of some molten alkali-Pb alloys and related these to changes in the electronic conductivity with composition. In this case they obtain the remarkable result, from modeling only single structure factors that contain contributions from both Pb and alkali correlations, that the existence of a first sharp diffraction peak at $Q \approx 1$ Å^{-1} is the result of Pb-Pb clustering in the form of "Zintl" ions. By using constraints they have been able to assess the degree of Zintl ion formation and come to the conclusion that the unconstrained RMC model has features which, as a function of sample composition, correlate most strongly with the experimentally determined resistivity. Howe (14) studied possible complex formation in the liquid semiconductor CuSe (isotopic substitution data). His results confirm the suggestion that the data are consistent with the existence of some Se_2 pairs.

Howe (15) modeled the structure of $NiCl_2$ in aqueous solution. We discuss this application of RMC in more detail because it represents the most complex liquid that has yet been studied and because solutions are clearly of considerable importance in materials science. Five different data sets were used, obtained by neutron diffraction with isotopic substitution (16), and water molecules were included explicitly in the configurations with parameters (i.e. bond angle and bond lengths) obtained from other experimental diffraction data (17). When molecules are included, the RMC algorithm contains a step in which they are rotated randomly, as well as having their centers moved. Figure 3 shows the data and RMC fits. The most interesting result of this study concerns the orientation of water molecules with respect to the solvated anions and cations. It had previously been suggested that the mean angle of tilt between the symmetry axis of the molecule and the line joining the oxygen atom and the cation was 42°; however, Howe found that this angle had no particular significance and was merely the mean of a distribution that peaked (i.e. had its mode) at 0°, that is with the lone pairs on the oxygen pointing towards the cation (see Figure 4). The distribution around anions was peaked in the opposite direction, as might be expected, but was significantly broader. These results clearly illustrate the danger of predicting three-dimensional correlations from average features of the radial distribution functions.

Glasses

The numerous possible applications of RMC in modeling the structures of glasses are obvious. However, there has as yet been less work in this area than for liquids. The reason for this is that most of the liquids studied were likely to be well-described by pairwise additive potentials, and hence the application of RMC was straightforward. Many glasses, however, particularly network glasses with strong covalent bonding, definitely do not have pair potentials and thus constraints must be used. While this has been successful in some cases, the technique for applying constraints is not as obvious in others.

$Ni_{1-x}B_x$ metallic glasses have been studied by three different groups (L. Pusztai & R. McGreevy, private communication; B. Thijsse & J. Sietsma, private communication; J.-M. Dubois, private communication). These samples are chosen because the neutron diffraction technique of isotopic substitution is optimized by the large differences in scattering lengths for Ni isotopes, and some excellent data are available. The number density of metallic glasses is high and therefore motions of the atoms are restricted. Thus it is unlikely that any configurations are truly in equilibrium and hence the results will be dependent on the choice of initial configuration. Different research groups have made different choices, with interesting results. When corresponding crystalline structures have been chosen, the data can be fitted well while retaining local order similar to that in the

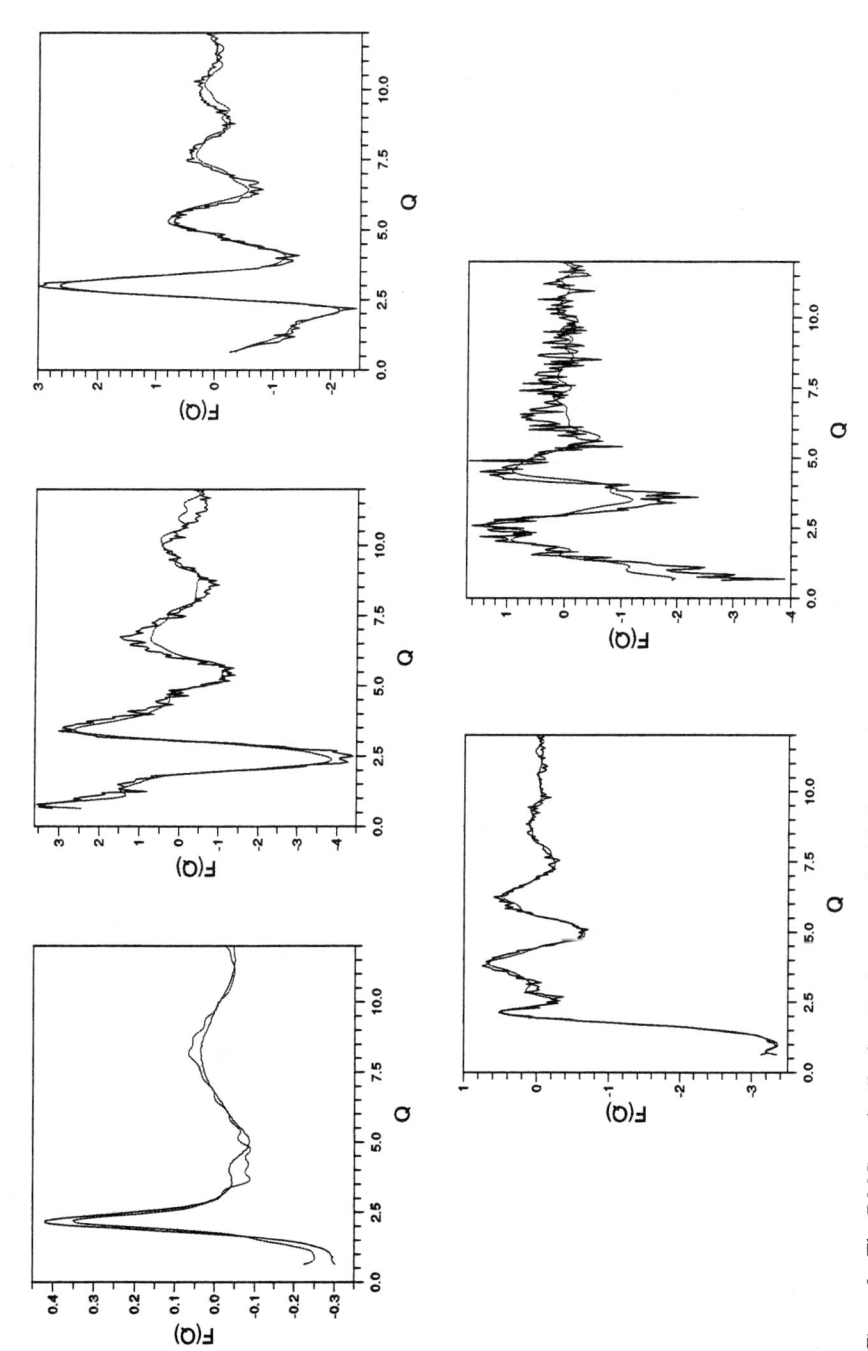

Figure 3 The RMC results (*broken lines*) compared with the experimental data (*solid lines*) for a total structure factor for $NiCl_2$ in heavy water (*top left*), the Ni-O partial structure factor (*top center*), the Ni-H partial structure factor (*top right*), the difference function ($\times 100$) for NaCl in heavy water (*bottom left*) and the difference function ($\times 100$) for $NiCl_2$ in light water (*bottom right*) (15).

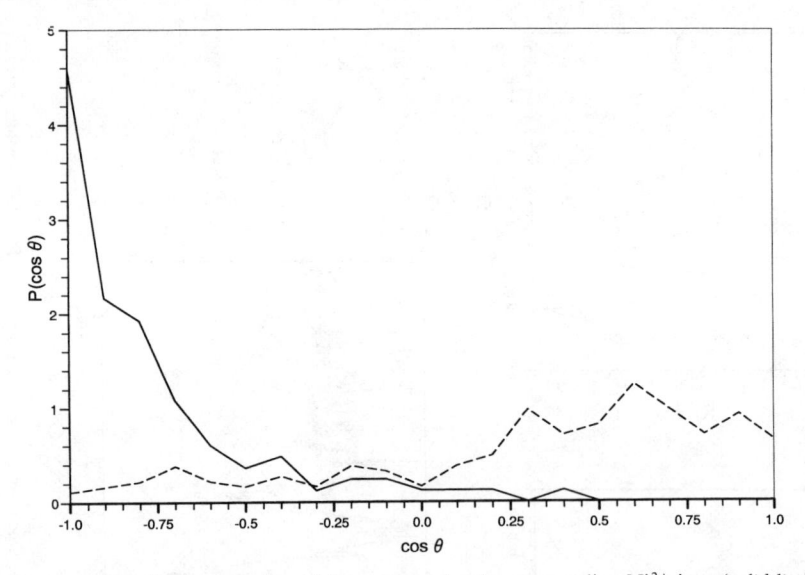

Figure 4 The distribution of $\cos\theta$ for water molecules surrounding Ni^{2+} ions (*solid line*) and Cl^- ions (*broken line*) (15).

crystal. When random starting configurations are used, the local order around B atoms is found to be slightly icosahedral, which is of particular interest in relation to the formation of quasi-crystals. Without additional information, however (see below), it is not possible to decide between the different models. Figure 5 shows fits to three structure factors for $Ni_{0.65}B_{0.35}$.

Keen & McGreevy (21) used the combination of neutron and X-ray diffraction data to model the structure of the archetypal network glass SiO_2. The structure factors and fits are shown in Figure 6. It is interesting to note that, even in the absence of any constraints to specify the type of covalent bonding expected, the structure obtained is a continuous random network consisting (mostly) of corner-sharing SiO_4 tetrahedra, as was proposed many years ago by Zachariasen (22). However, the model contains too many coordination defects (in the sense that not all Si are coordinated to 4 O) in comparison to results from NMR and EPR. The method of coordination constraints is being used to remove these. Here we can define a required coordination, C_r, of e.g. O atoms within a specified interatomic distance of a Si atom, and then an additional term,

$$\chi^2 = (N_{Si}(C_r) - \alpha N_{Si})^2/\sigma_c^2, \qquad 8.$$

can be added to χ^2 in the RMC algorithm. $N_{Si}(C_r)$ is the number of Si

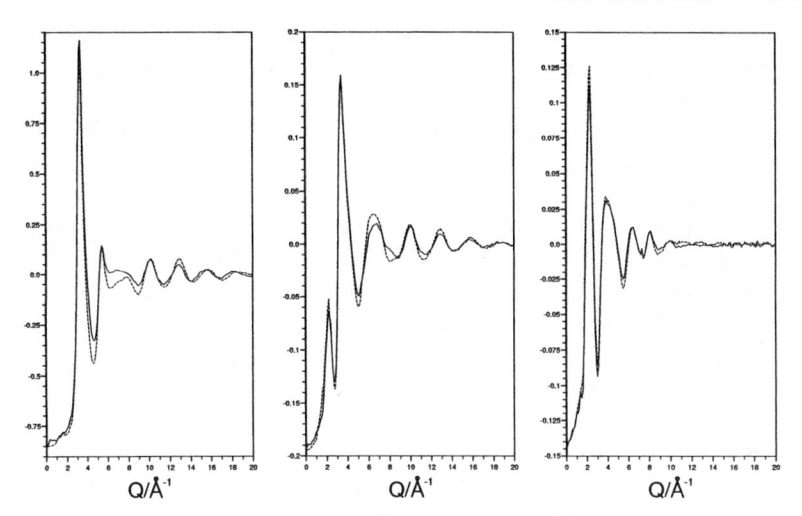

Figure 5 RMC fits to total structure factors measured for $Ni_{0.65}B_{0.35}$ using neutron diffraction with isotopic substitution; ^{58}Ni (*left*), ^{60}Ni (*center*), and $^{(0)}Ni$ (*right*). $^{(0)}Ni$ refers to the isotopic composition with zero neutron scattering length.

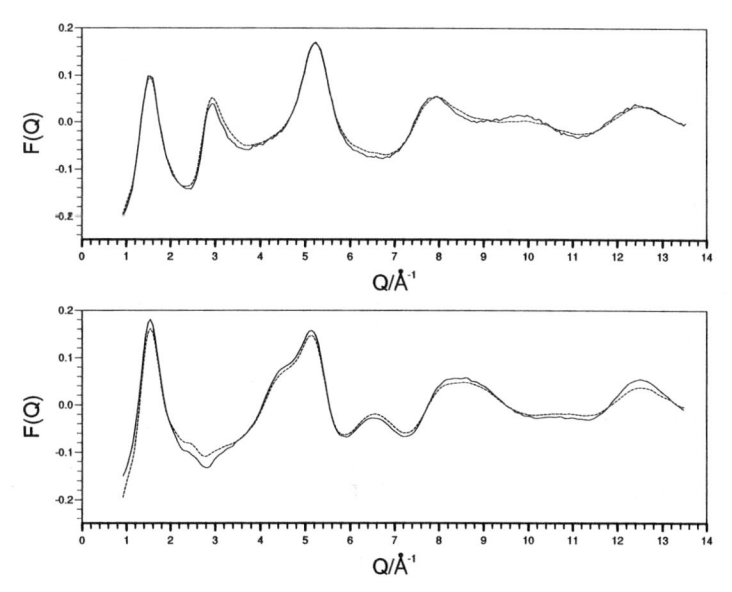

Figure 6 Structure factors $F(Q)$ for vitreous silica: neutron diffraction (*top*) and X-ray diffraction (*bottom*). The dashed line is experimental and the solid line is the RMC fit (21).

atoms in the configuration with the required coordination, N_{Si} is the total number of Si atoms, and α is the proportion of these required to have that coordination (this is not necessarily 1). σ_c is essentially a weighting factor that determines the severity of the constraint. Other constraints can obviously be included in a similar fashion.

There have been several proposals that magic angle spinning nuclear magnetic resonance (MAS NMR) data can be used to derive information on bond angle distributions in glasses (e.g. 23, 24). To test the possibility of using such data directly as input to RMC modeling, we calculated the spectra for ^{29}Si using the SiO_2 model of Keen & McGreevy (21). This is compared with an experimental spectrum (23) in Figure 7. It can be seen that the linear model is in quite good agreement with the data, but this is only one of four possible models that have been suggested to relate the NMR signal to the bond angle distribution (24); the secant model is in considerably worse agreement. Unfortunately until there is some definitive way of relating the data to the structure, such data cannot be used by RMC.

When alkali oxides are added to SiO_2, some of the oxygens are no longer shared between two tetrahedra. The proportions of tetrahedra with one or more non-bridging oxygen can be determined from MAS NMR (25). (They can also be determined, although less accurately, from Raman scattering spectra.) In contrast to the determination of the bond angle

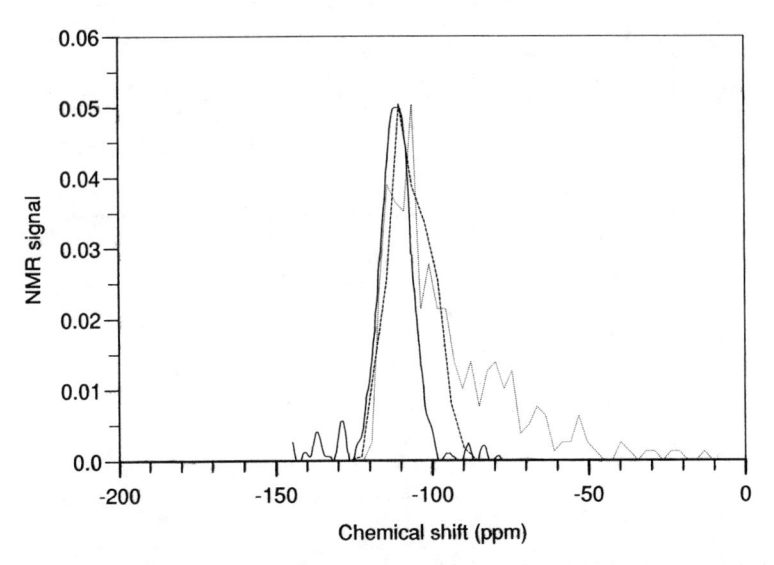

Figure 7 ^{18}Si NMR spectrum for vitreous silica: experiment (23) (*solid line*) and calculation from RMC model using linear theory (*broken line*) or secant theory (points).

distribution, which requires accurate fitting of the shape of a single peak, this method requires the determination of the relative areas of different overlapping peaks; inaccurate knowledge of the peak shape introduces some errors, but does not invalidate the whole procedure. The proportion of numbers of non-bridging oxygens (together with some realistic estimate of the errors) can be used as constraints in RMC modeling of neutron and/or X-ray diffraction and EXAFS data. J. Wicks et al (private communication) are using this method to model the structures of potassium, rubidium, and caesium silicates, sometimes with five different constraints at once.

Chemical knowledge can also be used as a constraint in RMC modeling. The fast ion-conducting glasses based on AgI doped into phosphate or borate glass host materials are of considerable technological interest. However, they are complex materials, with at least four components. It would not be expected, therefore, that RMC modeling of only a single structure factor measured by neutron diffraction would be able to provide much useful information. One of the most interesting features of the structure factor for these materials is the growth of a sharp diffraction peak at $Q \approx 0.7 \text{ Å}^{-1}$ upon doping, whose intensity is approximately proportional to the AgI content (27). It was suggested that this peak was associated with the structure of the salt component. An initial configuration was constructed with the chemical requirement that each P be tetrahedrally coordinated to 4 O with two corners being shared, thus forming a polymeric chain network as in the crystal structure of $AgPO_3$. RMC modeling, using coordination constraints to allow the phosphate network to deform, but not to break any bonds, then showed that the sharp low Q peak was in fact produced by local density fluctuations in the phosphate network, not in the salt. These fluctuations are caused by the requirement to maintain connectivity (bonding) and yet decrease average density with increasing salt content, and are not the direct result of scattering from the salt (28, 29). This was then confirmed by X-ray diffraction. These materials are now being studied by combination of neutron and X-ray diffraction and (multiple edge) EXAFS data to determine the mechanism of Ag^+ ion conduction through the expanded network. It is interesting to note that the structural model produced for the doped glass showed some evidence of a weakly fractal nature (28, 29).

The technique of EXAFS (extended X-ray absorption fine structure) has been used to a wide extent to study the local structure in disordered materials (see e.g. 30). The strength of this technique is that it is atom-specific, but the weaknesses are that only short range information can be obtained and the relationship of the data to the structure is complex. In studies of glasses another problem has been that the common analysis method of assuming Gaussian coordination shells is often invalid and can lead to errors and misleading results. Gurman & McGreevy (31), in a study of amorphous silicon, have shown that RMC modeling can be applied to EXAFS data. In Figure 8 we show a fit to the experimental data; the level of agreement is comparable with that obtained by standard analysis methods. Figure 9 shows the bond angle distribution for amorphous silicon in comparison with that for crystalline silicon (also modeled by RMC from EXAFS data); it has not previously been possible to derive bond angle distributions from EXAFS data.

Disorder in Crystals

The RMC method can also be applied to the study of disorder in crystals. The diffraction pattern of a disordered crystal consists of sharp Bragg peaks and a broad diffuse background. The Bragg peaks, the result of elastic scattering, provide information on the time average structure of the material, i.e. on the crystal structure, with the Debye-Waller factors being related to the average deviations from this structure. Bragg scattering cannot provide information on local structural correlations such as those due to defects. This information is contained in the diffuse scattering. If one measures the total structure factor, that is both Bragg and diffuse

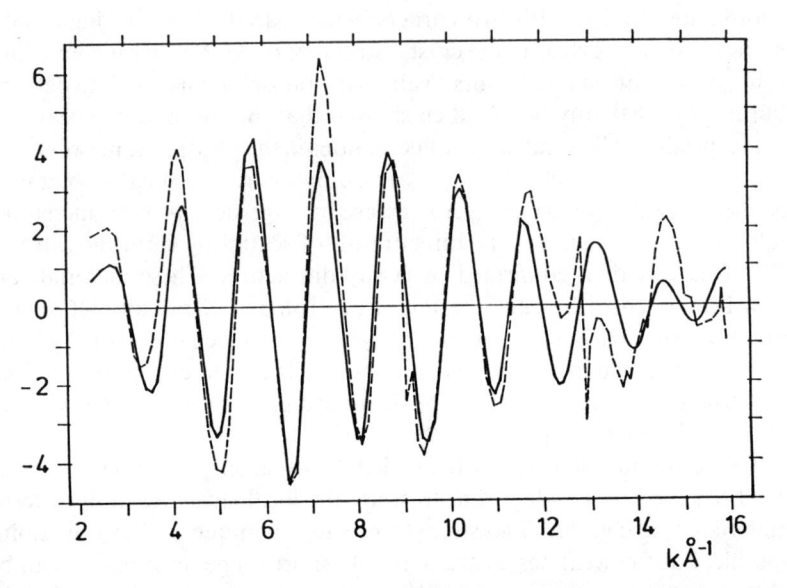

Figure 8 Experimental (*broken curve*) and RMC fitted (*solid curve*) EXAFS spectrum (weighted by k^3) for amorphous silicon (31).

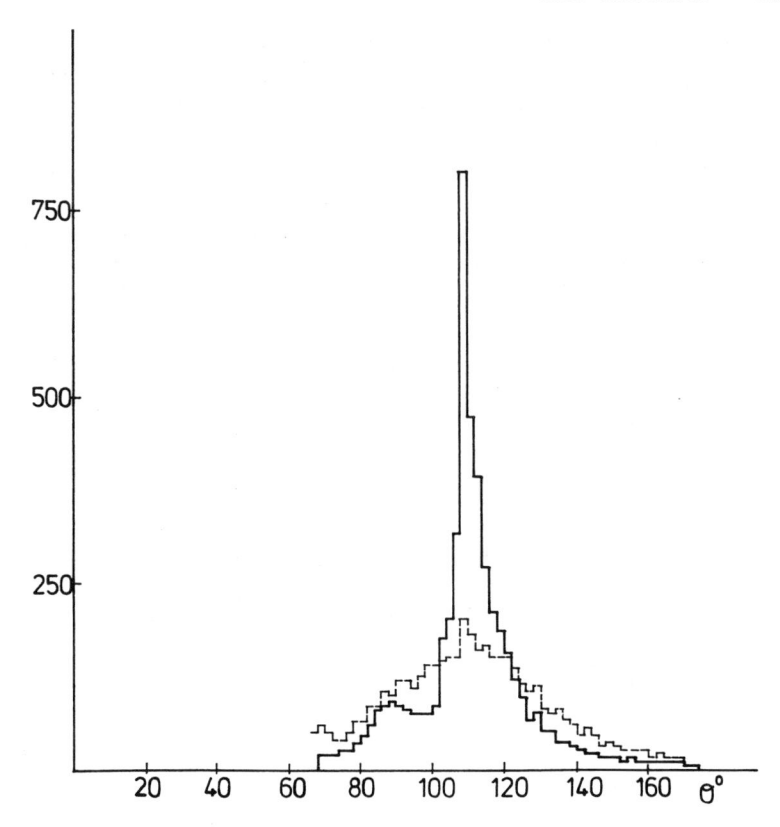

Figure 9 Bond angle distributions in crystalline (*solid line*) and amorphous (*broken line*) silicon obtained from RMC modeling (31).

scattering integrated over all scattered energies, then this is related to a time zero (snap-shot) picture of the structure and contains information on local correlations. It is also worth noting that as the disorder increases, the Bragg intensities decrease, but the diffuse scattering increases; for highly disordered crystals only a few Bragg peaks may be measured. Therefore in these cases it makes more sense to analyze the largest signal, i.e. diffuse scattering, rather than the smallest, i.e. Bragg scattering.

One class of crystals that shows considerable structural disorder and is also of considerable technological interest is fast ion (or superionic) conductors. Keen et al (32, 33) have made a detailed study of disorder in AgBr close to the melting point, T_m. AgBr, which has a rocksalt structure, is not normally classified as a fast ion conductor, but its conductivity starts to increase anomalously about 100 K below T_m, and just below T_m is only a factor of three smaller than that of the well known fast ion conductor α-AgI. RMC modeling of the total structure factor, measured by powder diffraction data, produces a configuration consisting of many (e.g. $8 \times 8 \times 8$) unit cells. This can be used to produce a space average distribution of ions within the unit cell. By examining the distribution in detail, one can obtain considerable information on the nature and extent of disorder in the material. Figure 10 shows a cut through the Ag$^+$ ion distribution in the unit cell in the $\langle 111 \rangle$ direction at three temperatures, the highest being only 2 K below T_m. As T increases through the region with anomalous conductivity, a small tail on the distribution is seen to grow into a distinct peak. Such a tail/peak is not found in directions other than $\langle 111 \rangle$ and is caused by anharmonic vibrations that eventually result in occupation of the $(\frac{1}{4}, \frac{1}{4}, \frac{1}{4})$ interstitial site. By integrating the peak area, the number of interstitials can be determined. The interstitial concentration is small, only $\approx 3\%$ at T_m, thus showing the remarkable sensitivity of this type of modeling. Nield et al (34) concluded that the disorder in AgBr is not actually indicative of pre-melting behavior, but rather of an incipient second order structural phase transition to a true fast ion phase, which happens to be incomplete when melting occurs, in agreement with the earlier proposal of Andreoni & Tosi (35).

It is widely believed that Monte Carlo methods only provide information on the static structure and not on dynamics, but this is not entirely true.

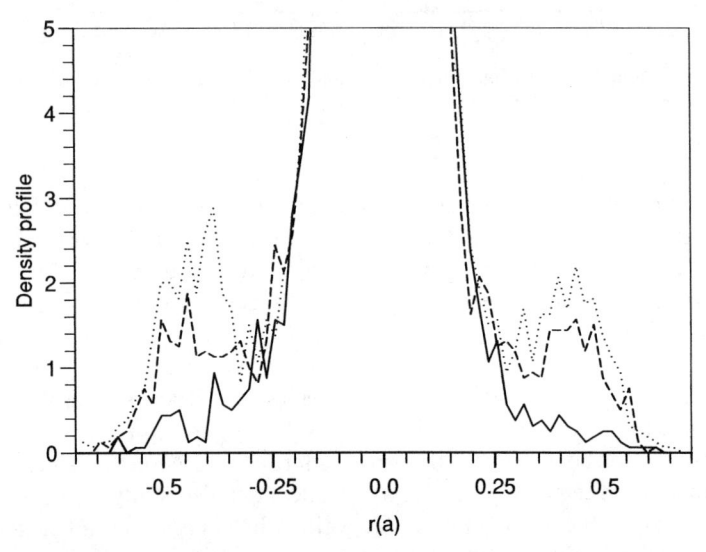

Figure 10 The density profile across Ag$^+$ lattice sites in $\langle 111 \rangle$ directions in AgBr at $T = 669$ (*solid line*), 689 (*broken line*), and 699 K (*points*) (32).

The density distribution can also be used to provide information on the mechanism of ionic conduction. If ions conduct by moving along particular pathways in the crystal, then a snap-shot of the structure will catch some ions while moving, and thus there will be finite density along these pathways even if they do not correspond to equilibrium sites in the crystal. Pathways that are not active will have zero (or very low) density. The fast ion conductor α-AgI has a bcc I$^-$ structure with Ag$^+$ ions hopping rapidly between a variety of alternative sites. Crystallographic studies have concluded that the most likely site for Ag$^+$ ions is tetrahedrally coordinated to I$^-$, although trigonal and octahedral sites are also possible. RMC modeling (V. Nield et al, private communication) shows unambiguously that it is indeed the tetrahedral sites that are occupied. It is also able to show that conduction occurs via the trigonal sites, and not via octahedral sites, as illustrated in Figure 11. The same method, when used for AgBr, shows that the colinear interstitialcy jump is the dominant conduction mechanism, although other mechanisms do contribute slightly close to T_m (32).

For crystalline materials, modeling of single crystal data might be preferable to powder diffraction data. In fact, for the two materials discussed above it is difficult to produce high quality single crystals large enough for diffuse scattering measurements. Even in cases where it is possible, single

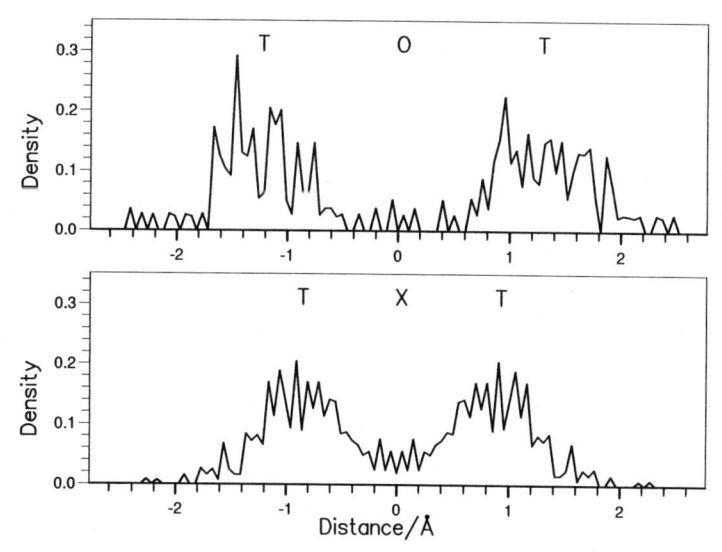

Figure 11 The density profile of Ag$^+$ ions in α-AgI at 420 K (36). (*Top*) across octahedral (O) sites in $\langle 100 \rangle$ directions, through tetrahedral (T) sites. (*Bottom*) across trigonal (X) sites in $\langle 110 \rangle$ directions, through tetrahedral (T) sites.

crystal measurements are not always better in practice. One advantage of powder diffraction is that data can be measured accurately over a wide Q range; both range and accuracy are important for modeling. On the other hand, single crystal data are normally only measured over a restricted range and with less accuracy (except at Bragg peaks) simply because there are far more points to measure in three dimensions than in one. This can introduce serious problems for modeling. There are also problems introduced by the considerably greater ratio of Bragg peak intensity to diffuse scattering. This matter is being considered by V. Nield et al (private communication), who are modeling test data on single crystal Pb. While the diffuse scattering can be modeled accurately (Figure 12), there are difficulties in doing this consistently with Bragg scattering. It may prove advantageous to model single crystal and powder data simultaneously, thus taking advantage of the strengths of both techniques.

Magnetic Disorder

The magnetic structure of materials that exhibit either long range order or complete disorder are well understood for systems where the spins lie on regular lattice sites. For systems where these extremes of order/disorder are not displayed, or where the spins lie on an irregular atomic arrangement, the magnetic structures are still not well characterized. Spin glasses, frustrated magnets such as certain dilute magnetic semiconductors, and magnetic amorphous materials all fall into this category. Because of the wide interest in these materials, both academic and applied, it would be of considerable value to have a method for determining their magnetic structures. However, one of the difficulties has been that such materials tend to show diffuse scattering from both structural and magnetic disorder, and these must first be separated. This can be done by polarization analysis in neutron diffraction, but such experiments are difficult and relatively inaccurate. Isotopic substitution can also be used, but this is limited to a very few cases.

Keen & McGreevy (38) have shown that RMC modeling can be applied to diffuse magnetic scattering. In a study of amorphous Dy_7Ni_3, they used a unique set of data obtained by Hannon et al (39), who used neutron diffraction with double null isotopic substitution. This enables the separation of magnetic and nuclear scattering directly from the data. The nuclear scattering has been used to produce a structural model, as discussed above, and the magnetic structure has then been modeled on the basis of the positions of magnetic (Dy) ions in this model. Figure 13 shows fits to the magnetic scattering at temperatures above and below the ordering temperature ($T_c \approx 40$ K). The measured macroscopic magnetization has been used as a constraint in the modeling. This is particularly important

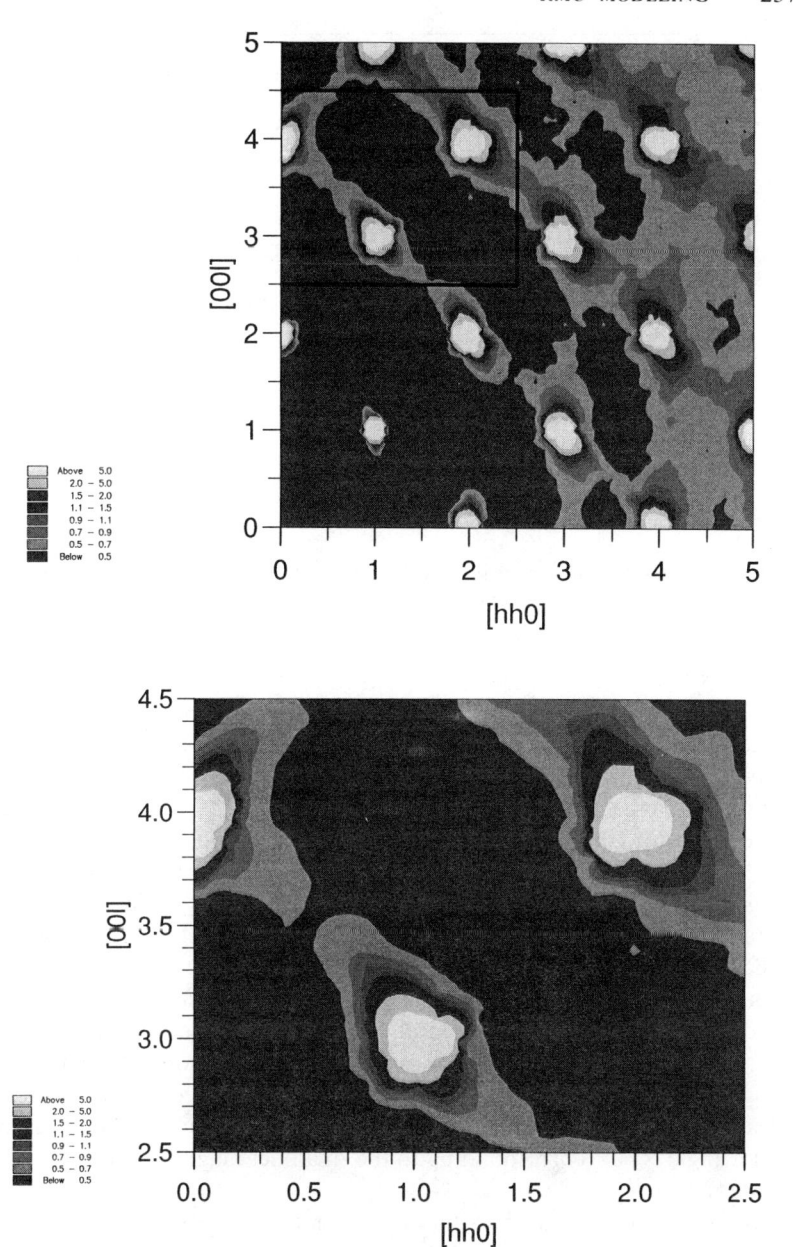

Figure 12 Contour plot of the structure factor of single crystal Pb in the [hh0] plane at 300 K. (*Top*) Experimental data; lighter shadings indicate higher scattering intensity. (*Bottom*) RMC fit to the area indicated at the top by the black line.

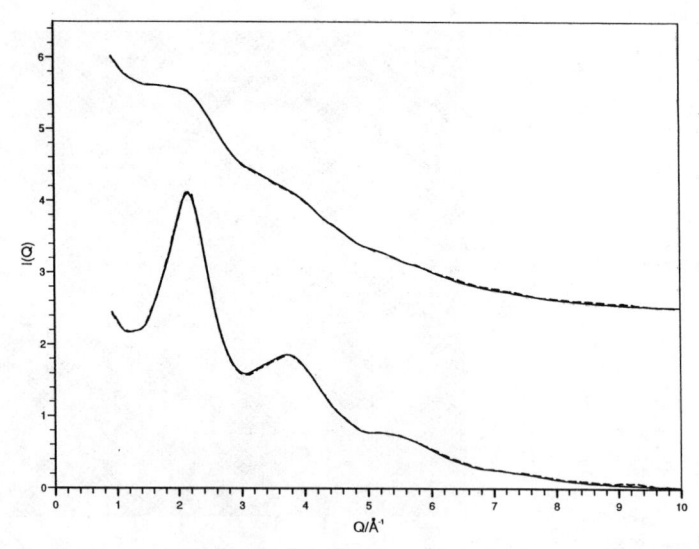

Figure 13 Total structure factors for $^0\text{Dy}_7{}^0\text{Ni}_3$ at $T = 7$ and 96 K (shifted by 2.5). (*Broken line*) experimental data, (*solid line*) RMC fit (38).

since the diffraction data only extend down to $Q \approx 0.5$ Å^{-1} and hence only contain information on correlations over ≈ 12 Å, while the correlation length for magnetic order becomes infinite at T_c. Even if lower Q data were available, the limiting correlation length would be determined by the simulation size. Figure 14 shows the angular correlations between neighboring spins above and below T_c; at 96 K this is flat, indicating paramagnetism, while at 7 K it peaks at $\cos\theta = 1$, indicating ferromagnetic order. The spin distribution in a section of a configuration is shown in Figure 15. It appears almost totally disordered. The magnetization of the model agrees with the experimental value since it is used as a constraint, but it should be noted that this is only $\approx 20\%$ of the moment per Dy ion. This picture shows that the magnetic structure is not consistent with the idea that the spins are canted with respect to the magnetization axis; the mean angle is 80° (corresponding to the small net magnetization), but the mode is 0°.

While a set of data such as that of Hannon et al (39) is rare, the RMC technique can be applied much more generally to the study of magnetic disorder. For instance, the combination of X-ray diffraction and EXAFS could be used to determine the atomic structure, and neutron diffraction to determine the magnetic structure. If necessary both atomic and magnetic

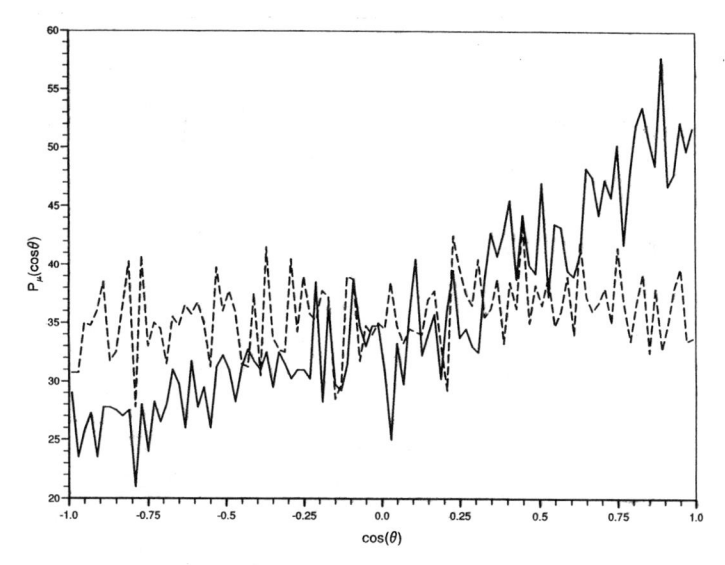

Figure 14 Angular correlation $P_\mu(\cos\theta)$ between pairs of spins within the first Dy-Dy coordination shell for Dy_7Ni_3 at $T = 7$ K (*solid line*) and 96 K (*broken line*) (38).

structures can be modeled simultaneously. J. Wicks et al (private communication) are currently collecting X-ray and neutron diffraction and EXAFS data on the magnetic glass $Fe_{0.91}Zr_{0.09}$ to test such a procedure.

FUTURE DEVELOPMENTS AND PROSPECTS

As is illustrated in the previous section, RMC modeling has been applied to a very wide range of materials and experimental techniques within a short time. While only simple materials were studied at first, the applications are not becoming increasingly complex because of the ability of RMC to combine data from different experimental techniques quantitatively. Two current applications that exemplify this are polymers (P. Jedlovsky et al, personal communication) and high T_c superconductors (R. Hadfield et al, private communication), both real (rather than model) materials with important applications. In the latter case the amount of disorder is small, but it is clearly crucial to the superconducting behavior and represents a considerable challenge for both experiment and modeling.

The use of constraints in RMC modeling has been found to be valuable in certain cases. In the most general sense one can view all sets of data as constraints. Coordination constraints, as used in modeling of some network glasses, are simply a crude type of interatomic potential. It is possible to directly combine RMC and MMC, i.e. to minimize $\chi^2 + \alpha U/kT$, and

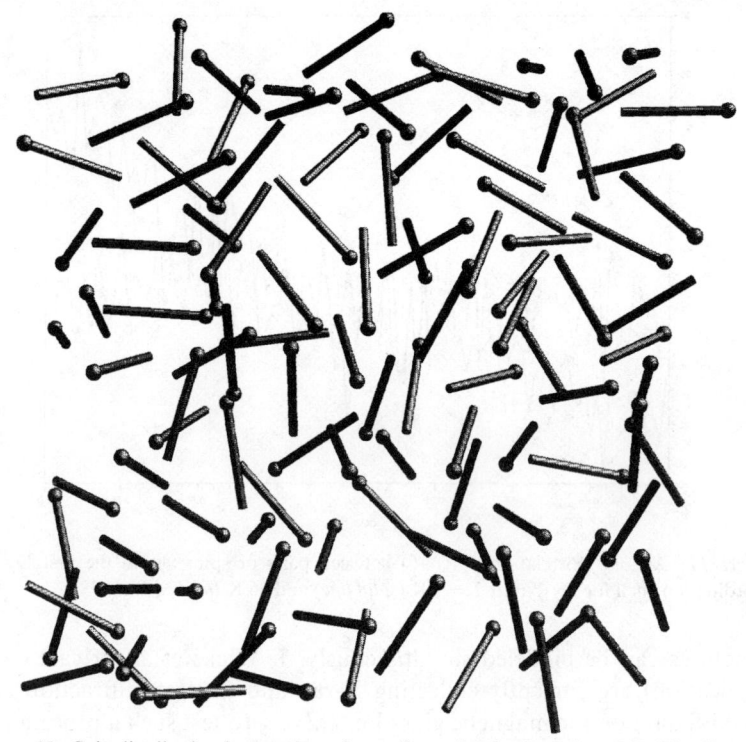

Figure 15 Spin distribution in a section of a configuration for Dy_7Ni_3 at 7 K. The direction of the macroscopic magnetization is almost exactly to the right and in the plane of the page (38).

hence to use a more refined constraint; however, if the potential available is a good one, then clearly this is unnecessary (molecular dynamics should be used), and if it is not, then care should be taken not to weight it too strongly (i.e. small α) with respect to the data.

It is possible that RMC modeling could be applied to the study of ordered as well as disordered materials, i.e. to crystallography. While many refinement techniques already exist, the ab initio determination of crystal structures is still not an automatic process and thus there is scope for new ideas. The strength of RMC in this area would be the ease of applying constraints, or indeed the possibility of directly combining crystallographic data (i.e. neutron and/or X-ray diffraction) with other data, as described above. Perhaps the most interesting application would be the combination of RMC and simulated annealing techniques for predicting crystal structure from crystal chemistry rules [e.g. see Pannetier et al (43)]. The weakness of RMC would be the amount of computing power/time required.

In a more general sense the RMC method can be applied to any physical problem where the relationship between the data measured and the model required is complex. There are many example of Monte Carlo inversion methods in the literature, ranging from oil exploration to order in metallic alloys. However, the RMC method described here is the only one, as far as we are aware, that uses a physical model (in the sense that for this application we have a collection of atoms that cannot do clearly unphysical things such as be in the same position, at the correct density) and a proper Markov chain. The first characteristic is important in the sense that the simple constraints inherent in having a physical model are very powerful, and the second is important because the result is independent of starting position, i.e. the process is not purely a refinement, although it may be used as such if required. With the rapidly decreasing price/performance ratio of computers, and the correspondingly increasing expense of collecting experimental data, it is clear that sophisticated modeling methods such as RMC will become a feature of most research programs in materials science.

Acknowledgments

RLM wishes to thank the Royal Society (UK) for continued support of his work, and the Science and Engineering Research Council (UK) for supporting the development of RMC.

Literature Cited

1. Bednorz, J. G., Muller, K. A. 1986. *Z. Phys.* B64: 189
2. McGreevy, R. L., Pusztai, L. 1988. *Mol. Sim.* 1: 359–67
3. Metropolis, N., Rosenbluth, A. W., Rosenbluth, M. N., Teller, A. H., Teller, E. 1953. *J. Phys. Chem.* 21: 1087
4. McGreevy, R. L., Howe, M. A., Keen, D. A., Clausen, K. N. 1990. *IOP Conf. Ser.* 107: 165–84
5. Evans, R. A. 1990. *Mol. Sim.* 4: 409–11
6. Howe, M. A., McGreevy, R. L. 1991. *Phys. Chem. Liq.* 24: 1–12
7. McGreevy, R. L. 1991. *J. Phys: Condens. Matter* 3: F9–21
8. Howe, M. A., McGreevy, R. L., Pusztai, L., Borzsak, I. 1992. *Phys. Chem. Liq.* In press
9. Howe, M. A. 1990. *Mol. Phys.* 69: 161–74
10. Nield, V. M., Howe, M. A., McGreevy, R. L. 1991. *J. Phys: Condens. Matter* 3: 7519–25
11. Pusztai, L., McGreevy, R. L. 1991. *Phys. Chem. Liq.* 24: 119–25
12. McGreevy, R. L., Pusztai, L. 1990. *Proc. R. Soc. London Ser. A* 430: 241–61
13. Howe, M. A., McGreevy, R. L. 1991. *J. Phys: Condens. Matter* 3: 577–91
14. Howe, M. A. 1989. *Physica B* 160: 170–76
15. Howe, M. A. 1990. *J. Phys: Condens. Matter* 2: 741–48
16. Powell, D. H., Neilson, G. W., Enderby, J. E. 1989. *J. Phys: Condens. Matter* 1: 8721–33
17. Soper, A. K., Phillips, M. G. 1986. *Chem. Phys.* 107: 47
18. Deleted in proof
19. Deleted in proof
20. Deleted in proof
21. Keen, D. A., McGreevy, R. L. 1990. *Nature* 344: 423–24
22. Zachariasen, W. H. 1932. *J. Am. Chem. Soc.* 54: 3841

23. Gladden, L. F., Carpenter, T. A., Elliot, S. R. 1986. *Philos. Mag.* B53: L81
24. Pettifer, R. F., Dupree, R., Farnan, I., Sternberg, U. 1988. *J. Non-Cryst. Solids* 106: 408–12
25. Murdoch, J. B., Stebbins, J. F., Carmichael, I. S. E. 1985. *Am. Miner.* 70: 332–41
26. Deleted in proof
27. Borjesson, L., Torell, L. M., Dahlborg, U., Howells, W. S. 1989. *Phys. Rev.* B39: 3404
28. Borjesson, L., McGreevy, R. L., Howells, W. S. 1992. *Philos. Mag.* In press
29. Borjesson, L., McGreevy, R. L. 1992. *Phys. Rev. Lett.* Submitted
30. Hayes, T. M., Boyce, J. B. 1982. *Solid State Phys.* 37: 173–351
31. Gurman, S. J., McGreevy, R. L. 1990. *J. Phys: Condens. Matter* 2: 9463–74
32. Keen, D. A., McGreevy, R. L., Hayes, W. 1990. *J. Phys: Condens. Matter* 2: 2773–86
33. Keen, D. A., Hayes, W., McGreevy, R. L., Clausen, K. N. 1990. *Philos. Mag. Lett.* 61: 349–57
34. Nield, V. M., Keen, D. A., Hayes, W., McGreevy, R. L. 1992. *J. Phys: Condens. Matter.* Submitted
35. Andreoni, W., Tosi, M. P. 1983. *Solid State Ionics* 11: 49
36. Deleted in proof
37. Deleted in proof
38. Keen, D. A., McGreevy, R. L. 1991. *J. Phys: Condens. Matter.* 3: 7383–94
39. Hannon, A. C., Wright, A. C., Sinclair, R. N. 1991. *Mater. Sci. Eng.* A134: 883
40. Deleted in proof
41. Deleted in proof
42. Deleted in proof
43. Pannetier, J., Bassas-Alsina, J., Rodriguez-Carvajal, J., Caignaert, V. 1990. *Nature* 346: 343–45

PHASE TRANSITIONS OF GELS

Yong Li[1] and Toyoichi Tanaka

Department of Physics and Center for Materials Science and Engineering, Massachusetts Institute of Technology, Cambridge, Massachusetts 02139

KEY WORDS: gel, phase transition, critical phenomena, swelling volume transition

INTRODUCTION

The study of polymer network systems can be traced back to the series of experiments on rubbers by Gouth in 1805 (1). These experiments demonstrated that a stretched rubber shrinks upon heating. This is opposite of the way in which most other materials behave. In 1920, Staudinger correctly pointed out that polymers are flexible chains of covalently connected molecules (2). Based on Staudinger's finding, the entropic nature of the rubber elasticity was revealed by Meyer et al (3).

The quantitative treatment of the rubber network was started in 1934 by Guth & Mark (4) and Kuhn (5, 6), and was extended later by James, Flory, and others (7–13). Their efforts resulted in the Phantom Network Theory (14) and the Affine Network Theory (7, 8). Both theories give similar results on the rubber elasticity. For detailed discussion on these theories and rubber elasticity, see References 15, 16.

A gel is a three-dimensional polymer network swollen with a solvent (Figure 1). Its mechanical properties are similar to those of natural rubbers. It has high deformability and nearly complete recoverability. Depending upon the solvent, temperature, and other environmental conditions, the polymer chains can either repel each other and be swollen, or attract each other and be very compact. The first observation of a large gel volume change brought about by a solvent change was reported by Kuhn and his colleagues in 1950 (17). In their study, polyacrylic acid and polymethylacrylate acid gels were used. However, the gel phase transition was not observed in these experiments.

[1] Current address: Kimberley-Clark Corp., 2100 Winchester Road, Neenah, Wisconsin 54957

collapsed **swollen**

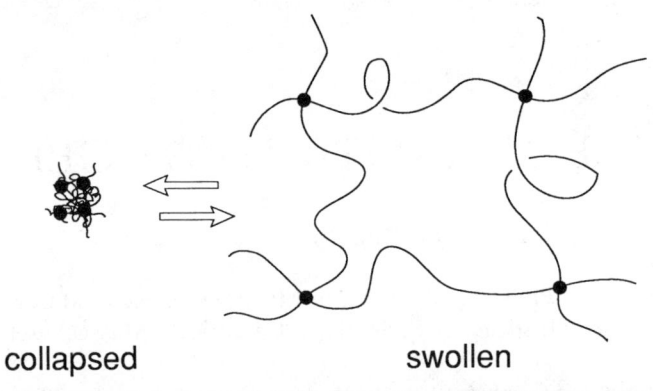

Figure 1 Gel is defined as a cross-linked polymer network swollen with a liquid. Gels undergo reversible volume transitions in response to changes in external conditions.

The first mean-field treatment of gel network systems was given independently by Flory (18) and Huggins (19). In 1968, based on the Flory-Huggins theory, Dusek & Patterson predicted a volume phase transition of the network system between a dense phase and a dilute phase (20). (A volume transition should not be confused with the sol-gel transition or gelation process. It is rather analogous to the coil-globule transition of polymer chains.) In 1973 Tanaka et al were able to observe the density fluctuations of a polymer network using dynamic light scattering spectroscopy (21). They and de Gennes (22) theoretically established that the mode that scattered light was a collective diffusion of the polymer network. In 1977 Tanaka et al (23) observed the critical behavior of polyacrylamide gel; the amplitude and relaxation time of the mode diverge as a certain temperature is approached. This suggested the possible density phase transition in a cross-linked polymer network. Shortly after this, Tanaka observed the volume phase transition of slightly ionized acrylamide gels in an acetone-water mixture (24). These observations marked the beginning of the intensive study of the gel phase transition and related critical phenomena.

The volume phase transition of gels is universal, i.e. the phenomenon has been observed in all kinds of gels with different chemical compositions (25). The gel phase transition can be induced by many factors, including temperature (26), solvent composition (27–29), pH (30), electric field (31), external pressure (32), ultraviolet (33), and light (34).

In this article, we consider only the chemically cross-linked gels. The networks formed solely by physical contacts are not discussed. Readers who are interested in the subject of physically cross-linked gels should see References 35–38 for details. The purpose of this paper is to provide a general review of the gel phase transition research in both scientific understanding and technological applications.

CONFORMATIONS AND STRUCTURE OF POLYMER NETWORK

Networks in Good Solvent

The conformation of a polymer chain is determined by the competition of the effective chain-solvent interaction, whose free energy is expressed in the reduced form $\chi = \Delta F/2kT$, where ΔF is the free energy of the contact of polymer segments and the entropic free energy of the chain. Good solvents ($\chi < 1/2$) of certain polymers or their networks refers to the types of solvents in which the polymers or networks are swollen. In a good solvent, the thermal energy kT is larger than the energy ΔF associated with polymer-solvent interaction.

Some of the most successful theories on gel systems are those related to the gels in a good solvent (22, 39). We do not have an ideal theory yet to describe gel phase properties in all the solvent conditions, which include poor, theta, and good solvents, at all polymer network densities from zero to one.

The viscoelastic properties of a polymer network strongly depend on the conformation of the constituent chains, which is represented, for the first approximation by the average end-to-end distance, R_F (F denotes Flory radius), or the radius of gyration, R_G of the chain. The R_F of a single polymer in a good solvent can be most easily obtained using the Flory method (1). The total free energy of the system consists of the elastic energy ($\sim R^2/Na^2$) and the excluded volume interaction ($\sim V\phi^2 \sim R^3(N/R^3)^2$). The minimization of this free energy with respect to R yields the equilibrium end-to-end distance R_F.

$$R_F \approx (\tfrac{1}{2}-\chi)^{1/5} N^{3/5} a, \qquad\qquad 1.$$

where χ is the effective polymer-solvent interaction parameter, a is the persistent length, N is the length of the polymer in the unit of a. The renormalization group theory by de Gennes gives the exponent value 0.592 (22). As the concentration of the solution is increased to the value at which polymers start to overlap ($\phi^* \sim Na^2/R_F^2 \sim N^{-4/5}$), the interaction among different polymers becomes important. Beyond this point, the polymers begin to inter-penetrate each other. In the semi-dilute concentration regime, i.e. $\phi^* \ll \phi \ll 1$, the relevant parameter is the screening length ξ, which is related to the density as $\xi \sim \phi^{-3/4}$ (40–42). This length scale measures the average distance between the nearest chain contacts, and beyond this length scale, the system is homogeneous. Within the time scale much shorter than the relaxation time for an entanglement to be unfolded (43), the semi-dilute solution can also be visualized as a polymer network with a mesh size equal to ξ. The osmotic pressure of the system is $\pi \sim \phi^{9/4}$ (42, 44, 45).

It has been shown that the swelling equilibrium of the networks in a good solvent is similar to that of semi-dilute polymers in good solvent (22, 40). The equilibrium concentration ϕ_e of the network is proportional to the overlap concentration of the corresponding polymer solution (ϕ^* theorem),

$$\phi_e \propto \phi^*. \qquad 2.$$

Here we use the subscript e to emphasize the difference between the concentration of the network at preparation and the concentration of the final equilibrium state. The degree of swelling, relative to the concentration at preparation, depends on the solvent condition and the network polymerization process (46). The structure factors (pair density correlation function) and other properties measured from the network system are almost identical to the corresponding polymer solution (39, 47). The pair correlation length ξ, which represents the effective mesh size, is similar to that of semi-dilute polymer solution (48–50).

$$\xi_e \propto \phi_e^{-3/4}. \qquad 3.$$

This relation is shown in Figure 2a (from Reference 48). The average end-to-end distance of a network chain and the other length quantities are scaled the same way as that of a free polymer chain,

$$R_F \propto N^{3/5}. \qquad 4.$$

The elasticity of gels in a good solvent is

$$\mu \approx K \approx \phi^{9.4}, \qquad 5.$$

where μ and K are the shear and bulk moduli, respectively. These power laws have been confirmed experimentally (47, 51–54). The proportionality between μ and K were shown recently by Geissler and co-workers (54, 55). Figure 2b is the shear modulus of acrylamide gel as a function of concentration and deswelling ratio, measured by the mechanical method (taken from 54).

The Affine Network Theory has been shown to be valid on a macroscopic scale. It has been demonstrated that the average distance h between two nearest cross-links varies with the network concentration as $h \sim \phi^{-1/3}$ (56) (also see Figure 3). The theory also predicts the same ϕ dependence of the radius of gyration of the network chains, i.e. $R_G \sim \phi^{-1/3}$. However, the recent small angle neutron scattering (SANS) results showed that R_G

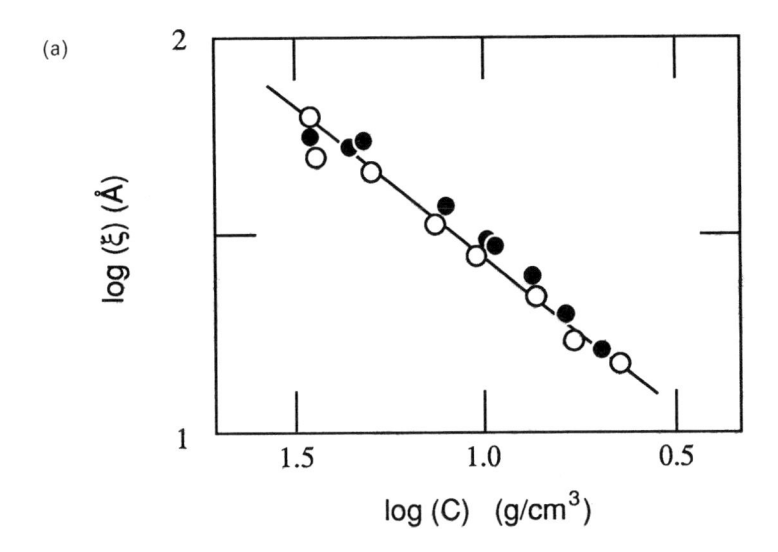

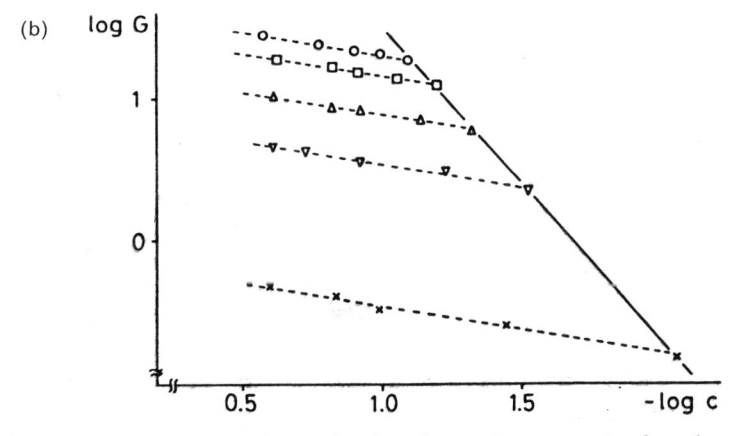

Figure 2 (*a*) Screening length ξ as a function of network concentration for polyacrylamide gels measured by the small angle neutron scattering technique. The cross-link density is 1/37.5. The open circles denote the unswollen gels in the prepared state. The solid circles denote the data after they were swollen in water. The straight line is from the light scattering experiment (see Figure 2, Reference 48). (*b*) Shear modulus G (kPa) vs concentration c as a function of the degree of deswelling. The mean slope of the dotted lines describing the deswelling of each sample is 0.334. The line through the final equilibrium swelling point is 2.19 (data taken from Figure 2 of Reference 54).

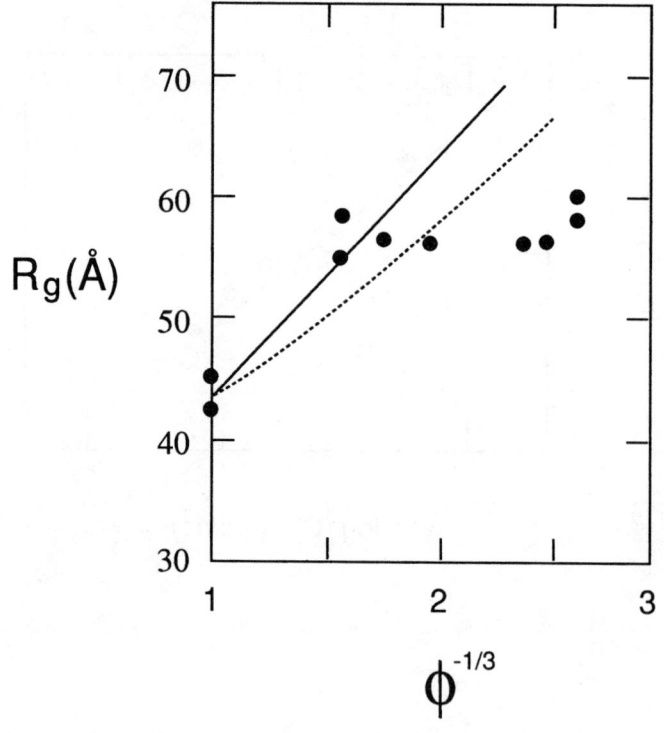

Figure 3 Dependence of radius of gyration R_G on network density ϕ for a polystyrene network in an osmotic deswelling experiment. The full and dotted lines refer to the affine behavior and the phantom network model, respectively (Figure 2 of Reference 39).

is independent of ϕ for a wide range of ϕ (55, 57), as shown in Figure 3. This result indicates that when the gel shrinks, the chains fold and overlap each other and maintain their size R_G (55). This is an important result to remember when constructing the equation of state of gels.

Defects in Gels

The structure of an ideal network should be topologically similar to an ideal lattice. A network with the functionality four corresponds to a simple cubic lattice. The lattice sites correspond to the network cross-links, and the bonds between adjacent sites correspond to flexible polymer chains of equal length. The ideal network can be treated theoretically with relative ease. In reality, however, there are many defects in a real polymer network, including dangling chains, loops, entanglements, and large polymers trapped inside the gel (58). Also, since the gelation process quenches the randomness of the structure, there are permanent structural fluctuations in the network. Figure 1 schematically shows some of the defects in a real network. Many properties of gels, including permeability (59), elasticity (60), swelling ratio (61), solvent molecules diffusion rate (62), etc. are directly dependent on the defects of the network. Below are some of the commonly encountered defects.

ERGODICITY Since each polymer chain is permanently connected to the network, and because of the topological constraint (63), each individual chain in the network cannot sample all of the possible states a free polymer can. Thus the system is not exactly ergodic. Pusey & van Megen (64) showed that the time-averaged scattered light intensity correlation function of the gel system is different from the ensemble-averaged correlation function. They point out that for some gels the collective diffusion coefficient determined from the time-averaged correlation function of scattered light could be larger than the real value of the diffusion coefficient.

DISTRIBUTION OF CROSS-LINK POINTS Solvent-phobic cross-link molecules tend to aggregate to form clusters upon a free radical copolymerization process. This effect has been reported in the case of poly(acrylamide) and poly(acrylamide)-derivitive gels with cross-linking molecules N'N'-methylene-bis-acrylamide (BIS) (61, 62, 65) and poly(dimethyl siloxane) gels, with ethyl triacetoxy silane (ETAS) as cross-link molecules (66). This aggregation causes the formation of structural inhomogeneities and makes the gel more permeable (67–69). Gels prepared with relatively high BIS or ETAS concentrations become opaque because of the frozen density fluctuations. It is interesting to point out that although the numerical value of the elastic modulus depends on the cross-link concentration, the power law relation between the elastic modulus and the network concentration is preserved (70, 71).

ENTANGLEMENTS The networks made by cross-linking polymers, rather than starting with monomers, have many entanglements. The entanglements of the network chains effectively increase the number of cross-links (22, 43, 72). The high frequency response of a polymer melt of entangled long chains is similar to that of a gel system (73).

PENDENT CHAINS The pendent (side) chains exist practically in all the gels made by different methods. Although the equilibrium swelling ratio in a good solvent is relatively insensitive to the concentration of the pendent chains, the elasticity and other properties (51, 60, 74) crucially depend on the pendent chain concentration. The loops in a network can be effectively treated as pendent chains.

SWELLING INHOMOGENEITIES   A uniform distribution of the cross-link molecules in a network at preparation can become non-uniform and inhomogeneous upon swelling in good solvent (75, 76). This phenomenon was experimentally observed by Bastide et al, using SANS technique (77).

PRE-GEL SOLUTION FLUCTUATIONS   Free radical copolymerization is a quenching process. The fluctuations are permanently frozen and memorized in the network structure (78, 79). The effect of the thermal fluctuations of the pre-gel solution on the final network structure and the dynamic properties of the network have been studied using a light scattering technique (S.-T. Sun et al, in preparation).

IONIZATION EFFECTS   Candau and co-workers recently reported another type of inhomogeneity in the partially ionized poly(acrylic acid) gel (57) probed by SANS. Because of the hydrophobicity of the cross-linking agent (BIS) and the repulsion between the ionized groups, the network will periodically form a dense region rich in BIS and a dilute region rich in ionization. This periodic modulation of the structure causes the peak in the structure factor in neutron scattering experiments.

VOLUME PHASE TRANSITION OF GELS

Similar to the liquid-gas transition of a fluid system, the volume (or its inverse density) of a gel can change either continuously or discontinuously upon a change of the environmental conditions such as temperature and solvent composition. The volume transition is directly related to the coil-globule transition of a dilute polymer system (81, 82). It has been shown by Amiya & Tanaka that the volume phase transition is quite general in all kinds of polymer networks (25). For practical reasons, thus far the transition has been only observed in water or a mixture of water and other solvents such as alcohol and acetone.

The phase transition is a result of a competitive balance between a repulsive force that acts to expand the polymer network and an attractive force that acts to shrink the network. The most effective repulsive force is the electrostatic interaction between the polymer charges of the same kind. It can be imposed upon a gel by introducing ionization into the network. The osmotic pressure by counter ions adds to the expanding pressure. The attractive forces can be van der Waals, hydrophobic interaction, ion-ion interactions between opposite kinds of charges, and hydrogen bonding. The phase transition was discovered in gels induced by each of the fundamental forces.

Flory-Huggins Theory

The gel volume phase transition was first predicted by Dusek & Patterson, using the Flory-Huggins (FH) theory for polymer solution (20). The Flory-Huggins theory is a mean-field theory, which qualitatively describes the phase transition, but it is not quite satisfactory quantitatively (18, 19). There have been improved theories based on the FH model (83–86), especially in the case of polyelectrolyte gels (87, 88). These improved theories usually are molecular thermodynamic theories that involve details in the chemistry of the model gel. The polyelectrolyte effect is discussed in the next section.

The Flory-Huggins free energy of a gel is written as

$$\frac{F}{vkT} = N\frac{1-\phi}{\phi}[\ln(1-\phi)+\chi\phi] + \frac{1}{2}[\alpha_x^2 + \alpha_y^2 + \alpha_z^2 - \ln(\alpha_x\alpha_y\alpha_z)] + f\ln\phi, \quad 6.$$

where v is the total number of chains in the network, N is the total number of persistent (monomer if freely jointed) units, χ is the reduced chain-solvent interaction energy $\alpha_i = r_i/r_i^0$ is the linear swelling ratio along the i-th direction, and f is the number of ionized groups per chain.

The three terms on the right hand side of Equation 6 corresponds to the free energies of mixing, rubber elasticity, and ionization, respectively. The ionization term represents only the ideal solution type contribution of counterions.

Gel experiments are often carried out under the condition of zero osmotic pressure, $\pi = \phi^2(\partial F/\partial\phi) = 0$. Figure 4 shows the swelling curves related to a volume phase transition, using the FH theory. By expanding the logarithmic term of ϕ to the order of ϕ^3, we obtain the isobar in a simple reduced form (27),

$$t = S(\rho^{-5/3} - \rho^{-1}/2) - \rho/3, \quad 7.$$

where

$$t \equiv \frac{(1-2\chi)\cdot(2f+1)^{3/2}}{2\phi_0}, \quad 8.$$

$$\rho \equiv \left(\frac{\phi}{\phi_0}\right)\cdot(2f+1)^{3/2}, \quad 9.$$

and

$$S \equiv \left(\frac{vv}{N\phi_0^3}\right)\cdot(2f+1)^4. \quad 10.$$

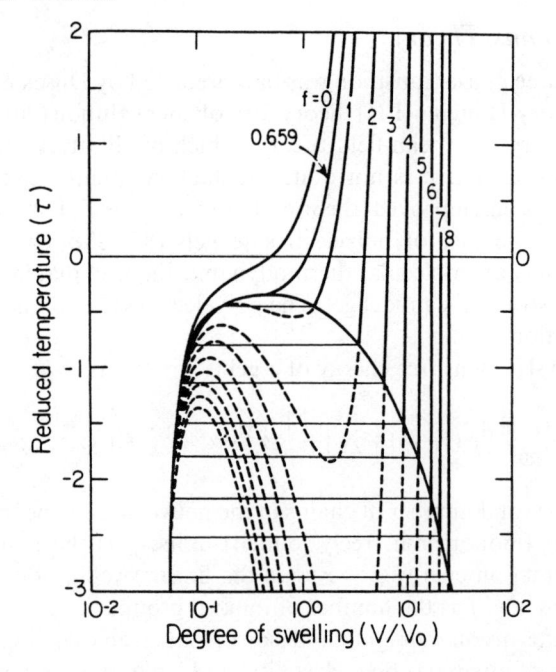

Figure 4 Theoretical swelling curves (equilibrium volume as a function of reduced temperature) for various degrees of ionization based on the Flory-Huggins formula. Here f is the number of ions per chain, τ is the reduced temperature, $\tau = (1-2\chi)$ (from Reference 50).

Notice that there is only one adjustable parameter, S, in the equation. Whether the transition is continuous or discontinuous is, therefore, solely controlled by S. For the S value larger than a certain value S_c, the isobar has the Maxwell's loop, and the transition becomes discontinuous. For smaller S, the transition is continuous. The critical point of this equation is defined by $\partial t/\partial \rho = \partial^2 t/\partial \rho^2 = 0$, which gives $S_c = 234.1$. Equation 10 clearly demonstrates the strong ionization dependence of the phase transition. Using sensible known parameters, Equation 7 predicts that the neutral polyacrylamide and N-isopropylacrylamide (NIPA) gels (25) undergo a continuous volume phase transition. Careful experimental studies, however, have shown that the neutral NIPA gel undergoes a discontinuous volume phase transition depending on the cross-linking density (89).

The Flory-Huggins theory predicts the volume phase transition qualitatively (20). Probably because of oversimplification of this theory, experimental results cannot be quantitatively analyzed (26). We briefly examine how each term of the equation of state in the mean-field approximation should be improved. The elasticity term assumes that each chain behaves as a Gaussian chain (random walk chain) and that the volume change of the network is accompanied by the proportional size change of the chains. However, recent works by Candau and co-workers (57) suggest that in a swelling process, the radius of gyration of the most chains in the network does not change (see above). They suggest that the network deformation is absorbed by the large deformation of a small amount of chains. These chains have a small elastic moduli and are associated with some of gel defects as described in the previous section. A typical example of these types of defects would be weakly cross-linked volume elements. These elements can take a large deformation and keep the size of the neighboring chains relatively unchanged. Under these circumstances, the elasticity of the network should be more weakly dependent on the polymer density than the term in Equation 6.

To take this into account, we introduce a parameter σ to represent the fraction of chains that actually deform when the gel size changes. This parameter should appear in Equation 6 as a pre-factor of the rubber elasticity. With this factor, Equation 7 could be rewritten as

$$\bar{t} = \bar{S} \cdot (\bar{\rho}^{-5/3} - \bar{\rho}^{-1}/2) - \bar{\rho}/3, \qquad 11.$$

where

$$\bar{t} \equiv \sigma^{-3/2} \cdot \frac{(1-2\chi) \cdot (2f+1)^{3/2}}{2\phi_0} = \sigma^{-3/2} \cdot t, \qquad 12.$$

$$\bar{\rho} \equiv \sigma^{-3/2} \cdot \left(\frac{\phi}{\phi_0}\right) \cdot (2f+1)^{3/2} = \sigma^{-3/2} \cdot \rho, \qquad 13.$$

and

$$\bar{S} \equiv \sigma^{-3} \cdot \left(\frac{vv}{N\phi_0^3}\right) \cdot (2f+1)^4 = \sigma^{-3} \cdot S. \qquad 14.$$

Note that the controlling factor, S, strongly depends on σ. Even for a neutral gel, $f = 0$. As long as σ is small enough, one can still have a large enough S and observe a discontinuous phase transition.

The cross-link concentration is directly related to the value of σ. The higher the concentration, the larger the value of σ, thus the more continuous the phase transition should be. This prediction is supported by the experiment presented in Figure 5, wherein the cross-linking (BIS) density of NIPA gels determines whether the transition becomes continuous or discontinuous. A critical value of the BIS concentration exists above which the transition becomes continuous. This behavior is explained by Equation 14.

One of the most obvious failures of the FH theory is its inability to predict the scaling behavior of networks in a good solvent (see above). Studies of gels in good solvents suggest that the interaction has a stronger concentration dependence ($\phi^{9/4}$) than that assumed in the FH theory (ϕ^2).

Ionic Interactions

A frequently used means of varying the degree of the discontinuity of the gel phase transition is the ionization of the network. The greater the ionization, the larger the volume change at a discontinuous transition. Ionizable polymer networks can be obtained by several ways; copolymerizing ionizable molecules into the network (26), hydrolysis (27), and

Polymer Concentration									
1.0	D		D		D		M	M	C
1.2	D		D		D		M	M	C
1.4	D		D		D		M	M	C
1.6	D		D		D		M D D M	M	C
1.8	D	D	D	D	D	D	D D M D D M	*	C C
2.0	D	D	D	D	D	D	D D M D M	M	C C
2.2	D	D	D	D	D	D	D	M	M C C
2.4	D	D	D	D	D	D	D	M	M C C
	1.0		1.4		1.8		2.2	2.6	3.0

Cross-link Concentration

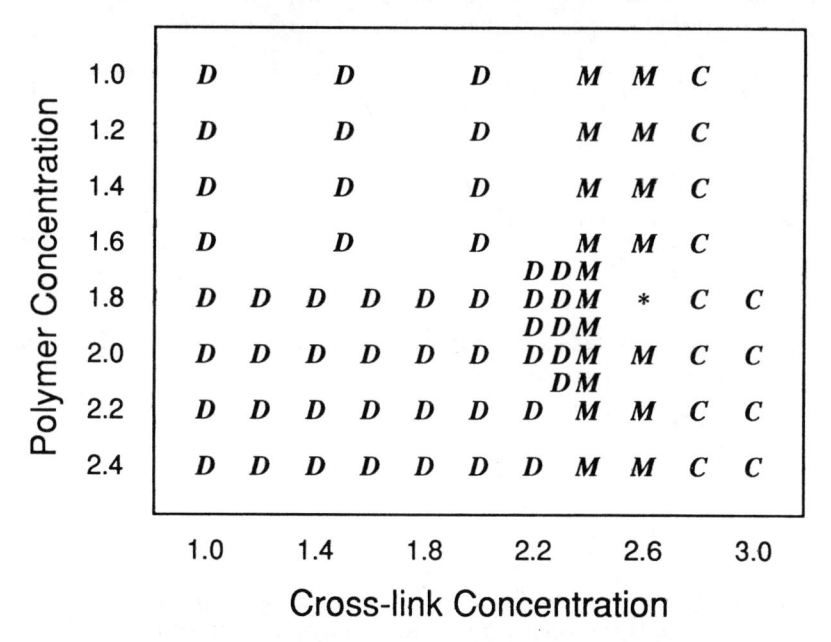

Figure 5 Monomer and cross-link concentration-dependence of the N-isopropylacrylamide gel phase transition. C, D, and M denote the continuous, discontinuous, and marginal (difficult to determine) transitions. Whether a transition becomes continuous, C, discontinuous, D, or marginal, M, depends on the composition of the N-isopropylacrylamide cross-linker, bisacrylamide (191).

light illumination (33). Ionized networks are sensitive to pH (90, 91), salt (91, 92), electric field (31, 93, 94), and light (95). Figure 6 shows the effect of network ionization on the swelling behavior of poly(acrylamide) gels (91).

Once the ionized gels are prepared, the degree of ionization can be controlled in several ways, including introducing deassociable chemicals into the network (97; E. Sato-Matsuo et al, in preparation), and varying pH (98, 99) and salt concentration (100, 101).

Ionization contributes to the free energy of the network in two ways (87, 88). In the FH theory, the mobile counterions are confined inside the gel by the Donnan potential barrier (100). These mobile counterions exert a gas-like pressure on the wall of Donnan charge bilayer. Screened electrostatic interaction among charges account for another means of enhancing free energy in the polymer network. There have been many theoretical works on the ionized gels. Here we mainly discuss the results of Konak & Bansil (KB) (87).

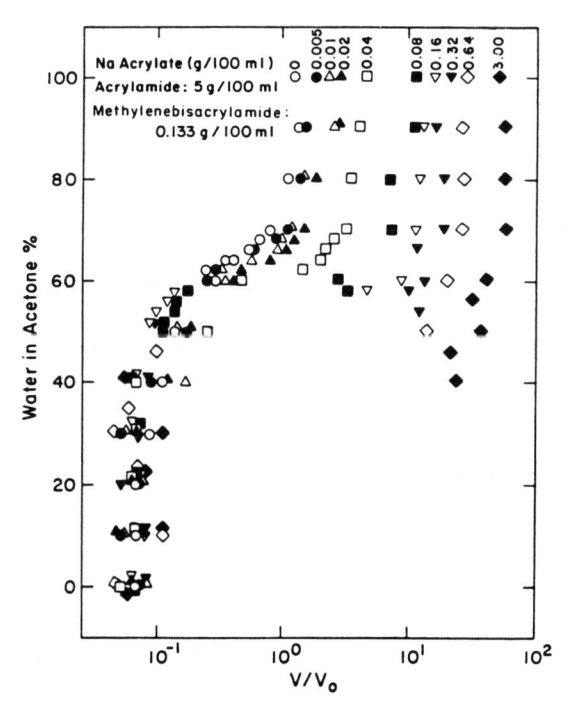

Figure 6 Effect of ionizable groups (sodium acrylate) incorporated in the network on the swelling of acrylamide gels (from Reference 91).

The ionization effect on the equation of state is oversimplified in the FH treatment. It is known from the study of polyelectrolytes that charged polymer chains behave quite differently from neutral ones. The elastic energy of a charged chain scales as $\sim kT (R/R_F)^{5/2}$ (102). Ionization makes the chains more rigid, the persistent length longer (103), and the excluded volume effect more significant. Furthermore, the screening effect makes the effective number of charges smaller. As has been pointed out by Oosawa (104) and Mannings (105), when the ionization exceeds a certain degree, counterions condensate on the polymer backbone and the effective number of free charges that contribute to the osmotic pressure will saturate. The counterion condensation has been observed in the polymer network system (E. Sato-Matsuo et al, in preparation).

In an ionized gel network system, the average interaction between charges is given by Debye-Hückel theory with screening length k^{-1} (106),

$$\frac{q^2}{\varepsilon r} \exp(-kr). \qquad 15.$$

Two energies are involved when a pair of charges of opposite signs approach each other. One is the Coulomb energy $q^2/\varepsilon_0 r$, and the other is the thermal energy $k_B T$. The balance of these two energies defines the Bjerrum length, $l_B = q^2/\varepsilon_0 k_B T$ (107). The pair of charges will be in the bound state if their separation is less than l_B. The Debye-Hückel radius k^{-1} is related to the concentration of the charges c through l_B, $k^2 = 4\pi l_B c$. As we attempt to increase the ionization more and more, the average distance A between two adjacent charges on a chain decreases. According to Mannings (105) and Oosawa (104), when $A < l_B$ (strong coupling limit), a significant number of counterions will be in the bound state with the network chain, which leaves the total number of free counterions constant. The coupling constant l_B/A remains unity (E. Sato-Matsuo et al, in preparation; Y. Hirose, in preparation).

The total free energy of ionization is then given by

$$F_{\text{ion}} = \left[1 - \frac{l_B k e^{-kA}}{1 - e^{-kA}}\right] v f k_B T \cdot \ln V. \qquad 16.$$

The volume V is related with the network volume fraction ϕ by $V/V_0 = \phi_0/\phi$.

The osmotic pressure associated with this free energy given by the KB theory is

$$\pi_{\text{ion}} = \left[1 - \frac{l_B k \, e^{-kA}}{1 - e^{-kA}}\right] \frac{vf k_B T}{V}. \qquad 17.$$

The second term of this result is different from that of Hasa et al and Ilavsky in the high ionization region (88, 97).

For weakly ionized networks, $A \gg k^{-1}$, l_B, Equations 16 and 17 reduce to the Flory form. In the original Flory-Huggins theory, only the first term for counterions was included. Due to the condensation effect, the total number of counterions has an upper limit, $V/A^3 \geq vf$. This limit plays an important role in the shrunken state of a gel.

The degree of ionization is often combined with other interactions, such as hydrophobic, hydrophilic, or hydrogen bonding, which affect the degree of the discontinuity of the phase transition.

Hydrophobic Interactions

The hydrophobic interaction arises between non-polar molecules in water, which is a polar solvent. A hydrophobic group is shielded by water molecules, which form a cage around the group (109–111). The water molecules in the cage are arranged in a certain order and may be considered frozen. When the temperature is increased, the cage of frozen water molecules is partially melted and the protection of the hydrophobic groups becomes weakened. This is the reason why the hydrophobic interaction increases as the temperature is increased. The entropy decrease due to the network collapse is compensated by the larger entropy decrease associated with the melting of the cage. Therefore, the entropy of the collapsed gel increases upon heating.

A thermally induced gel phase transition in pure water has been observed in several systems, including poly(N,N-diethylacrylamide) (112) and poly(N-isopropylacrylamide) (29), and other systems (113). Figure 7 shows the phase transition of N-isopropylacrylamide gels in water.

The hydrophobic interaction can also be weakened by introducing small additives such as alcohol, DMSO, and DMF. In water these molecules form hydrates and effectively reduce the available water molecules for the cage. This effect has been observed by several groups (114–116).

The hydrophobic interaction can also be modified using surfactants (117, 118). A drastic change in the phase transition of hydrophobic gels when an anionic surfactant is added has recently been reported (119).

Van der Waals and Hydrogen-Bonding Interactions

Phase transition of hydrophilic networks in water has also been observed. The phase transition is induced by overcoming the hydrophilic interaction

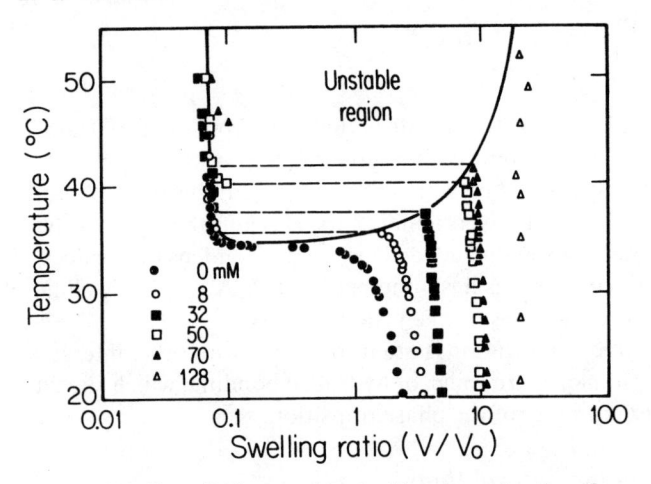

Figure 7 Degree of swelling of N-isopropylacrylamide gels as a function of temperature for various degree of ionization (Reference 29).

between polymers and solvent by the chain-chain attraction. This is achieved by mixing water with other solvents such as alcohol, acetone, DMSO, etc. to enhance the Van der Waals interaction (see Figure 6).

Another important attractive interaction is hydrogen bonding. It is known that polymer complexes are formed between poly(acrylic acid) and some polybases such as poly(oxyethlene) (POE), poly(vinylmethylether) (PVME), and poly(vinylpyrrolidone) (PVP) (120). Based on this principle, Okano and his colleagues developed an interpenetrating polymer network of poly(acrylic acid) and poly(acrylamide) and observed a sharp swelling of the gel upon increasing the temperature (26). The phase transition of the gel was recently observed by Ilmain et al (121), who slightly ionized the gel developed by Okano's group (122). The mechanisms of the polymer complexation and gel phase transition are believed to be the same, where the zipping effect seems to play an important role in enhancing the attractive interaction.

KINETICS OF GELS

The kinetics of gel swelling and shrinking is directly related to the viscoelastic properties of the system. In this section, we discuss the kinetic processes of gels when they undergo a small volume change or a discontinuous phase transition.

Kinetics of the Gel Swelling/Shrinking Process

The kinetics of the gel swelling/shrinking process is directly related to the gel elasticity and the friction between the network and the solvent (123–126). It has been shown that the relaxation time of swelling and shrinking is proportional to the square of a linear size of the gel (123). Which linear size to be used is an important question and is discussed below. The square law relation has been confirmed experimentally (123–125). (A smaller power than two was also reported (126, 127) and a possible explanation is discussed below.) One of the most important features of the gel swelling process is that it is isotropic; i.e. for a long cylindrical gel the swelling ratios are the same in both the radial and axial directions. When the radius increases 10%, the axial length increases 10%. The absolute increment of the length is much larger than that for the radius.

The high friction coefficient f between the network and the solvent overdamps the motion of the network, thus resulting in the diffusion-like relaxation. The macroscopic motion of a network element during the swelling and shrinking can be described by (50)

$$\frac{\partial \mathbf{u}(\mathbf{r},t)}{\partial t} = D_0 \cdot \nabla^2 \mathbf{u}(\mathbf{r},t). \qquad 18.$$

Here $\mathbf{u}(\mathbf{r},t)$ is the displacement vector measured from the final equilibrium state, $d_0 = (K+4\mu/3)/f$ is the collective diffusion coefficient. This equation has been used to analyze the swelling/shrinking of gels, with some success (123, 124), as well as to analyze the dynamic fluctuations of network density, as measured by dynamic light scattering spectroscopy. However, these analyses did not properly treat the shear deformation that occurs within a gel during the swelling kinetics, and hence cannot explain, for example, the isotropic swelling of a cylinder gel (128, 129).

For gels with symmetric geometry such as a sphere, a long cylinder, or a flat disk, we can define a dimension d_d as the number of the orthogonal directions along which the gel has the smallest length scale. For disk, cylindrical, and spherical gels, d_d is equal to 1, 2, and 3, respectively. Since the relaxation time of a diffusion process is proportional to the square of the length scale, the swelling should be the fastest along these directions if the process is purely diffusion-like, which involves the relative motions between the network and the fluid. Under these circumstances the gel swells anisotropically. Since the shear modulus of a gel network is finite, such an anisotropic deformation would cost shear energy. Therefore a correction is needed to minimize the overall shear energy at any point in the kinetic process. Such a relaxation of shear energy can be done without involving relative motion between the network and water, since shear relaxation is done at constant volume.

Introducing operators D and S to represent the diffusion process and the shear energy relaxation process, an infinitesimal swelling process can be described by the product of S and D. The displacement vector can then be written as (129)

$$\mathbf{u}(\mathbf{r}, t) = \left[\prod_i S(\mathbf{r}, i) \cdot D(\mathbf{r}, i)\right] \mathbf{u}(\mathbf{r}, 0), \qquad 19.$$

where the index i represents time, $t = i\Delta t$ (i = 1, 2, ..., $t\Delta t$; $\Delta t \to \infty$). Notice that the two operators S and D do not commute, which can be seen from Figure 8. Following this idea, a new theory of swelling/shrinking kinetics was developed by Li & Tanaka. This theory explains the isotropic swelling of a gel in all directions and shows that the effective collective diffusion constant $D_e(\mathbf{r}, t)$ depends not only on the time and position within the gel, but also on the shape of the gel. It is interesting that on the surface of the gel, D_e depends on the shape of the gel only,

$$D_e = D_d \cdot \frac{d_d}{d}, \qquad 20.$$

where $d = 3$ denotes d_d for a spherical gel.

This result can also be derived qualitatively by a simple dimension-counting argument. The diffusion in which a network undergoes a relative motion with respect to a fluid occurs in all three dimensions for spheres ($d_d = 3$), two dimensions (radial) for cylinders ($d_d = 2$), and only one dimension for discs (axial) ($d_d = 1$). Because of the existence of the shear modulus, the volume change caused by the diffusion occurring only in the d_d directions is shared by the remaining $d - d_d$ dimensions through the shear relaxation process. Effectively, the sharing reduces the overall rate of the diffusion in the d_d directions, where the diffusion is allowed. The reduction factor is directly related to the ratio of the diffusion dimensions d_d to the total dimensions, which is $d = 3$. The reduction factors are, therefore, 3/3, 2/3, and 1/3 (or d_d/d), for sphere, cylinder, and disc, respectively.

Figure 9 shows that the relaxation time of a gel is shape-dependent, and the results of these experiments agree with the theoretical predictions (see Table 1).

Kinetics of Gel Phase Transitions

The kinetic theory (128, 131, 132) predicts that near the critical point (K = 0), the relaxation time approaches infinity. This has been observed

Before Swelling

Collective Diffusion ⇕ Relative motion between
Time > 0   Network and water

Shear Relaxation ⇕ No relative motion between
Time = 0   Network and water

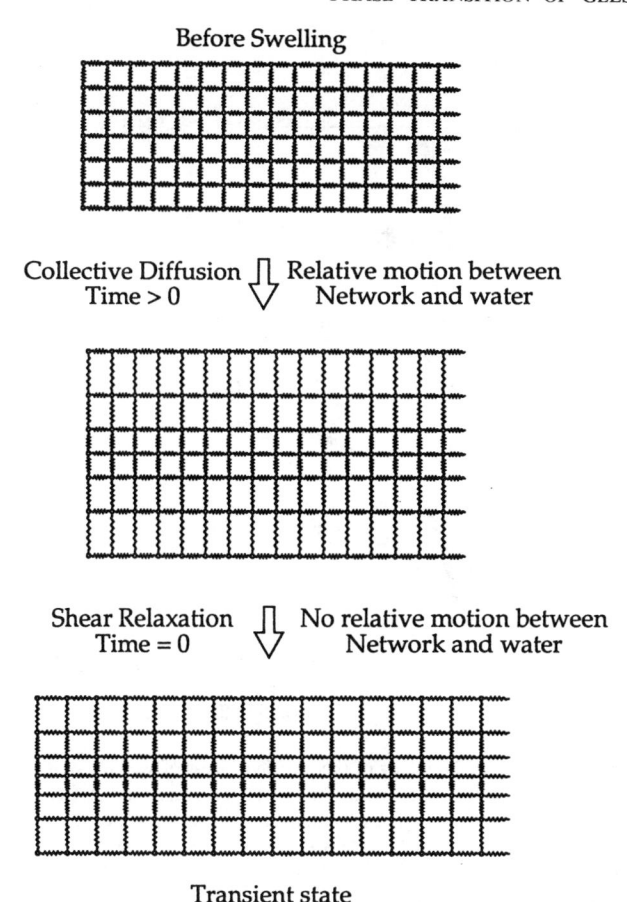

Transient state

Figure 8 An infinitesimally small swelling process of a cylindrical gel network can be considered as a two-step process. The diffusion process makes the gel thicker, but builds up shear stress and shear energy. Then the shear relaxation process takes place to minimize the shear energy. The second process is practically instantaneous since there is no relative motion between network and solvent (from Reference 129).

experimentally (23, 50). For a discontinuous phase transition, the kinetic process is highly nonlinear. Because of the large change of the network density, the elastic moduli, K and μ, and the friction coefficient f cannot be treated as constants any more. A recent study has shown that the relaxation time of the volume phase transition is a strong function of the initial and final state (95).

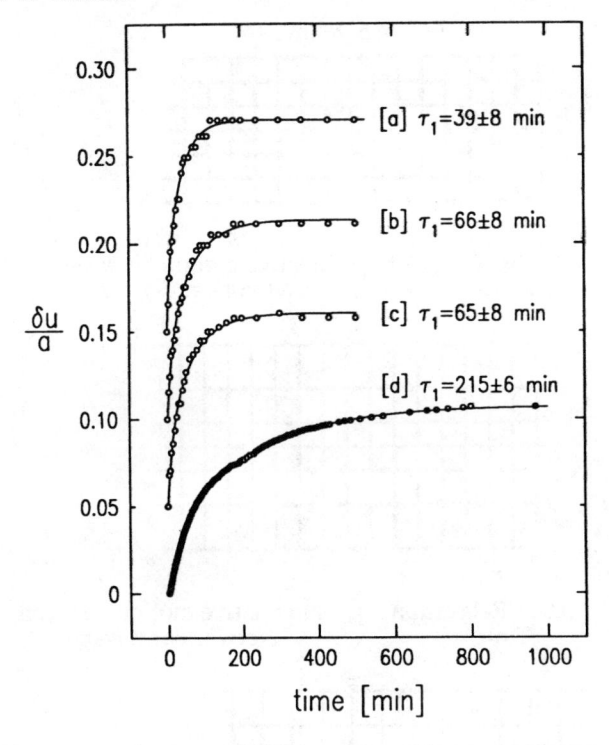

Figure 9 Experimental measurements of the kinetic process of swelling of acrylamide gels. (*a*) Diameter of a short cylindrical gel with equal length and diameter; (*b*) diameter of a long cylinder with length roughly thirty times the diameter; (*c*) the length of the long cylinder of (*b*); (*d*) the thickness of a disc with the side twenty times the value of the thickness. Each of the top three curves has been shifted from the one below it by 0.05 for a clear view (from Reference 129).

The size dependence of the relaxation time ($\tau \sim r^\varepsilon$) reported in Reference 127 is weaker, with a power $\varepsilon = 1.7 \pm 0.1$. A similar power was reported in Reference 132. The smaller power, however, seems to be the result of the structural heterogeneity introduced into small gels (sub-millimeter) prepared by inverse suspension polymerization. The surface and bulk portions of such gels are known to have different structures.

CRITICAL BEHAVIOR OF GELS

Associated with the volume phase transition of gel network systems, gels show critical behavior near the critical point at which the discontinuity of

Table 1 Collective diffusion coefficients and relaxation times of gels with different geometries[a]

	Theory		Experiment	
	D_a/D_o	$\tau_1/\tau_{1,\text{sphere}}$	$\tau_{1,\text{exp}}$ (min)	$\tau_1/\tau_{1,\text{sphere}}$
Sphere	1	1	39 ± 8	1
Cylinder	2/3	2.0, 1.9	$(66, 65)\pm 8$	1.7 ± 0.3
Disc	1/3	5.7, 5.0	215 ± 6	5.5 ± 0.9

[a] Kinetic behavior of a gel strongly depends on the gel geometry. D_a and τ_1 denote the overall collective diffusion coefficient and swelling-shrinking time of gels, respectively. $\tau_{1,\text{sphere}}$ denotes the swelling time for a spherical gel. The last two columns are experimental results. The diameter of the cylinder was 1.35 mm with aspect ratio equal to 30. The thickness of the disc was 1.33 mm with aspect ratio equal to 20. The paired numbers in the second column correspond to $\mu/K = 0$ (*left*) and $\mu/K = \infty$ (*right*). The two values of τ_1 for the cylinder in column three represent the relaxation times of swelling along radial and axial directions.

the volume change disappears. The study of the critical point is of great theoretical and practical interest. It was known decades ago that the critical exponents of many systems cannot be explained by the classical mean-field theories (133). The discrepancy was not understood until the renormalization group theory was introduced (134). According to the renormalization group theory, the critical exponents are related to only a few parameters of the system and are independent of the details. Totally different systems can belong to the same universality class and have identical exponents. It is therefore important to determine the critical exponents and the possible universality class to which the system may belong.

The state of a gel is determined by three variables; the reduced temperature τ, the osmotic pressure π, and the network density ϕ. In most gel experiments, the gels are immersed in a large volume of solvent, thus the external osmotic pressure is kept at zero. Depending on the chemical and physical composition of the network, the zero external pressure can be higher or lower than the critical osmotic pressure of the network. If it is higher, the isobar curve is continuous, and if lower, it is discontinuous. The critical point of a gel can be conveniently approached by fixing any one of the three variables at the critical value, and varying one of the other two. One can fix the gel volume and approach the critical point from the positive osmotic pressure region. Approaching from the negative osmotic pressure domain results in a gel shrinkage. Or one can choose a system that has a critical osmotic pressure $\pi_c = 0$ and approach the critical point along the isobar path by varying the temperature (26, 89). The $\pi_c = 0$ condition can be achieved by changing the chemical or physical structure of the network (see Figure 5) or solvent (29). For N-isopropylacrylamide (NIPA) gel we find that for BIS = 2.6, the critical osmotic pressure is approximately zero. The asterisk * in Figure 5 denotes the gel composition we used as the critical sample in the following experiments on critical behaviors.

The critical density fluctuations of poly(acrylamide) gel were first observed in the scattered light intensity by Tanaka et al (23) and Hochberg et al (135). The NMR study (136) of poly(acrylamide) gels in an acetone/water mixture showed that the proton-lattice relaxation time diverges and the effective diffusion coefficient approaches zero near the critical acetone concentration. Near the critical point, the shear modulus is proportional to the density of the network and does not show any anomaly (29, 97). The bulk modulus, in contrast, approaches zero at the critical point (29).

The singularity of response functions, such as specific heat and osmotic compressibility near the critical point, can be characterized by the exponents of their temperature (or equivalent parameter) dependence. The critical exponents of specific heat α, spatial correlation length ν, and scattered light intensity γ, are conventionally defined along the critical isochore. Since the isochore path is the same as the coexistence path, along the isochore we have $\Delta\rho \sim (\Delta T)_\rho^\beta$. In gel experiments, however, the most convenient path is the critical isobar. Along the critical isobar, we have $\Delta\rho \sim (\Delta T)_\pi^{1/\beta\delta}$. Accordingly, the exponents related with ΔT should be rescaled by the factor $1/\beta\delta$. Thus the exponent α_π measured along the isobar is related to the conventional exponent α through $\alpha_\pi = \alpha/\beta\delta$. This result can also be obtained from the scaling argument of the free energy (89).

Using the scaling laws (137), if one knows α_π and δ, then the other exponents can be obtained through the relations such as $\alpha = 2\alpha_\pi\delta/(\delta+1+\alpha_\pi\delta)$.

Critical Isobar

From Figure 5 we find that the non-ionic NIPA gel, with BIS concentration equal to 3.45 mg/cc, passes through the critical point (89). Whether an isobar is continuous or discontinuous is relatively insensitive to the concentration of NIPA monomers. In all the experiments on the gel critical behavior, the samples we used were taken from a large gel with BIS and NIPA concentrations 3.45 mg/cc and 140 mg/cc, respectively.

Figure 10a is the swelling ratio of a sample that passes through the critical point. Figure 10b is a replot of Figure 10a ($T < T_c$ only) with a log-log scale. The critical values T_c and ρ_c are chosen so that the line for $T > T_c$ is parallel to $T < T_c$. The curve in Figure 10a defines the critical

PHASE TRANSITION OF GELS 265

(a)

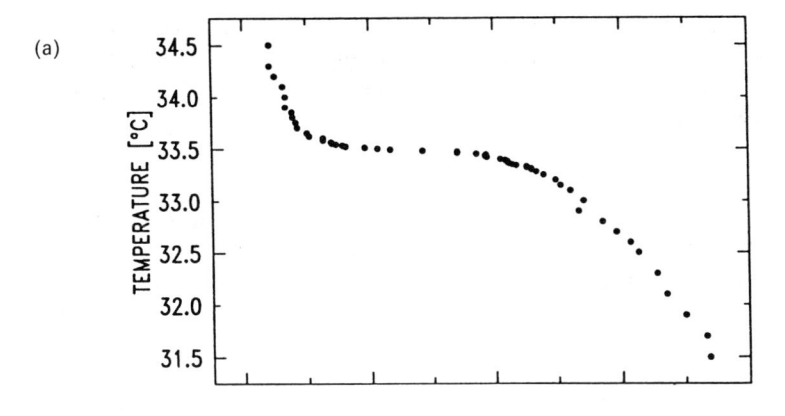

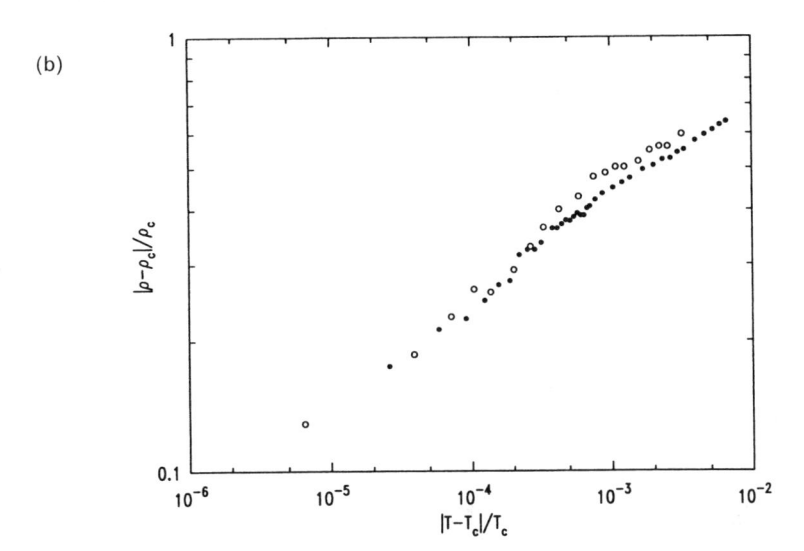

(b)

Figure 10 (*a*) Critical isobar of NIPA gel. (*b*) Density vs temperature relationship from Figure 10*a*. The slope is the exponent δ (from Reference 89).

exponent δ through $\Delta\rho \sim \Delta T^{1/\delta}$. This can be seen from the fact that the derivative of the curve is the thermal expansion coefficient, which diverges as $\Delta T^{-\gamma}$. This gives $\Delta\rho \sim \Delta T^{-(\gamma+1)}$. Using the scaling laws, one can readily show that $\gamma + 1 = 1/\delta$.

Mean-field theory predicts that $\delta = 3$ (137). From Figure 10*b* we find that the exponent δ is 4.2 ± 0.5. This is much larger than the mean field value of 3, but close to the value experimentally obtained for Ising systems (4.3 for ferromagnets). T_c and ρ_c determined for this sample are 33.47°C and 0.568 mg/cc.

Specific Heat

The critical behavior of the specific heat is one of the most intensively studied response functions both theoretically and experimentally. Specific heat studies often give more conclusive results partly because the exponents are quite different for different universality classes. Not only the specific heat exponent, but also the ratio of the amplitudes and the quantities related with the correction-to-scaling term, are informative (138).

In this study, the heat capacity $C = \Delta Q/\Delta T$ of the sample (holder + solvent + gel) were measured along the critical isobar path. The gel sample consisted of hundreds of small pieces of crushed gels of linear size ~ 0.7 mm. The specific heat of the network was obtained using the mass fraction of the network and the heat capacity of the blank sample (for details, see 89).

The measured specific heat is shown in Figure 11. The sample was the same as the one used to obtain Figure 10. We noticed that except for the

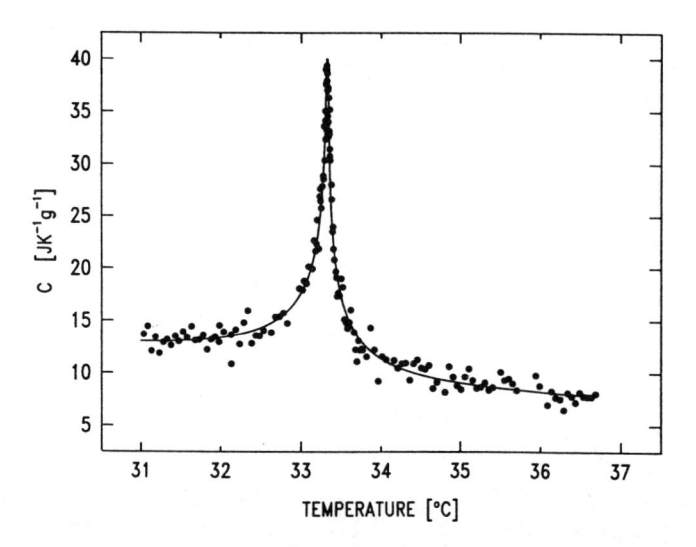

Figure 11 Specific heat of N-isopropylacrylamide gel near the critical point along the critical isobar (from Reference 128).

slow background, the curves are symmetric about the critical temperature (89). The fact that the data are symmetric about T_c is a direct result of the path taken to approach the critical point. Along the critical isobar, the system never enters the two-phase coexistence region, and hence the amplitudes above and below T_c should be the same. For an Ising system along the isochore, the system passes from one-phase to two-phase regions, and thus the amplitudes above and below the critical point differ by a factor of 2 (137).

The data can be well-fitted to a logarithmic function. This coincides with the prediction made by Moore & de Gennes (139, 140) on the coil-globule transition of a single polymer chain. The mean-field theory, in general, predicts a step-wise discontinuity in the specific heat rather than a singular peak. The T_c determined from this measurement is 33.32°C.

Collective Diffusion Constant

Figure 12 shows the collective diffusion coefficient as a function of $(T-T_c)$ of the sample used in obtaining Figure 10 and Figure 11. As the critical temperature is approached, D goes to zero (see also Reference 55).

Tanaka et al (21) showed that the scattered intensity I and the collective diffusion constant D are related to the shear modulus μ and bulk modulus K of the gel through

$$I \propto \frac{kT}{K+4\mu/3},\qquad\qquad 21.$$

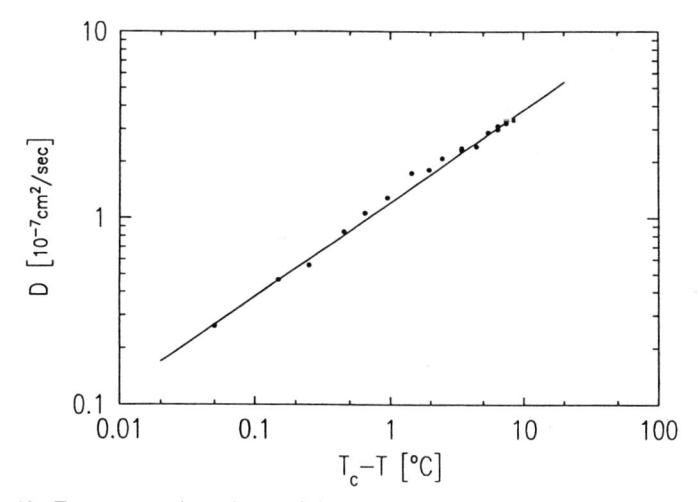

Figure 12 Temperature dependence of the collective diffusion coefficient of NIPA gel as determined by dynamic light scattering spectroscopy (191).

and

$$D = \frac{K+4\mu/3}{f}. \quad 22.$$

If the shear modulus μ is negligible (or if μ goes to zero by the same power as or faster than K near the critical point), then the scattered light intensity I and D diverge with power γ_π and v_π, respectively (24). From Figure 8 we find that $v_\pi = 0.45 \pm 0.07$. Again this value disagrees with the mean field result, which is $1/3$ ($1/2$ along the isochore) (137).

Discussion

In a liquid-gas system, the critical point is identified as the point at which the bulk modulus K vanishes (137). In the case of gels, there are two moduli K and μ. The critical point measured from the isotherm or isobar corresponds to the point where the bulk modulus $K = 0$. At this point, however, the shear modulus μ usually is not zero, although it has a step singularity at a discontinuous transition (29, 97).

The observed critical phenomena are still not well understood. The most controversial aspect of the understanding comes from the shear modulus. At the point where $K = 0$, both the shear modulus μ and the longitudinal modulus $K+4\mu/3$ are finite. As has been pointed out by Onuki, near the critical point the fluctuations should be suppressed and hence no short wavelength fluctuations should be observed (141, 142). The critical divergence of the fluctuations as observed in the dynamic light scattering experiments should originate from $K+4\mu/3 = 0$. However, the experimental results (29, 97) show that near the critical point, the value of μ is comparable with a value far from the critical point. From Reference 24, we find that the change of the intensity is about 200 times. As seen in Figure 13, D decreases by a factor of 50. We need $K \sim -4\mu/3$ to explain the large change of the scattered intensity I and the collective diffusion constant D. It is difficult to believe that near T_c, K can have such a large negative value. Therefore, what we could expect following Onuki's assumption is that I and D change by a much more moderate factor, rather than 200 times.

Another unsolved problem is the location of the singularities. We measured the thermal expansion coefficient $\alpha = (1/V)(dV/dT)_\pi$ and the swelling relaxation time τ of NIPA gel near the critical point (Figure 13). This sample has a continuous phase transition upon change of temperature. The thermal expansion coefficient α is obtained by taking the derivative of its swelling ratio curve. The relaxation time τ was measured by tracing the evolution of the length of the gel after a small temperature jump was given to the sample. The two curves in Figure 13 were obtained at the same time,

PHASE TRANSITION OF GELS 269

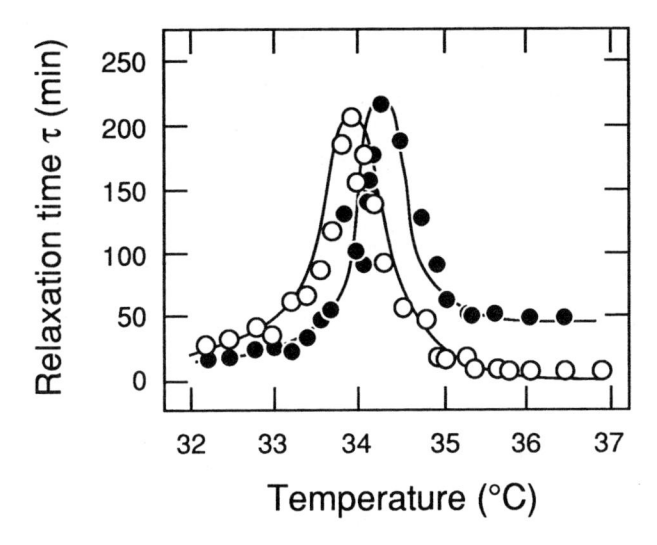

Figure 13 Thermal expansion coefficient α and the swelling relaxation time constant τ. They peak at different temperatures (from Reference 89).

therefore no aging effect, if any, needs to be considered between these two curves. Clearly, the two curves peak at different temperatures. The result was reproduced with a number of other samples.

Since $\alpha = (1/V)(dV/dT)_\pi \sim K^{-1}(d\pi/dT)_v$ is proportional to the inverse of the bulk modulus K, the peak in α corresponds to the critical point. Following the swelling kinetics theory, τ should also diverge at $K = 0$. Thus the two peaks in Figure 13 should be located at the same position. However, the actual peak τ occurs at higher temperature by 0.4°C (Figure 13). Clearly something is missing in our current understanding of the nature of the gel system. Note that in this experiment K is always positive, since the transition was continuous.

APPLICATIONS

Gels exist widely in biological systems and food products, often as thickening agents, e.g. starch and gelatin (143). The use of gels as a sieving matrix in electrophoresis and chromotography has been established for several decades (144). The hydrolic fracturing technique used in the petroleum and natural gas industry uses gels to open and crack rocks and soils (145). The gel beads mixed with soil can effectively control the moisture necessary for plants (146). The organic solvent (oil) based gels have been used to seal electric components and keep them from corrosion by water moisture. The ability of gel to exchange gas and moisture with the eye makes long-wearing soft contact lenses possible (147). The swelling property of the gel network makes the material ideal to be used in the absorption technologies. Well-known examples are baby diapers and sanitary napkins. In all these applications, however, only the swelling properties of gels are used.

Reversible volume transitions in response to various types of stimuli indicate that gels can be used as smart materials, and have attracted industrial scientists, engineers, and medical doctors to the gel research (F. Ilmain & T. Tanaka, in preparation). In the section below we discuss some of the potential applications of the gels related with gel volume phase transitions.

Mechanical Devices

The contraction and expansion of gel fibers provides a means of converting chemical energy into a mechanical energy, which can be used to develop artificial muscles and actuators. The early work on this subject was pioneered by Steinberg et al (159) and Sussman & Katchalsky (160), using contraction and relaxation of collagen fibers, which can be induced by changing solvent condition. Recent studies have shown that the efficiency of a gel engine increases with the load and the degree of crosslinking (161). Cross-linked poly(vinyl alcohol) fiber can induce a high strength reaction (162). De Rossi and his colleagues have developed an artificial sphincter using electric-sensitive gels (163). A group in the Toyota Research Laboratories developed an interesting fish made of gels that swims in water driven by an electric field (164). They also developed a three-finger hand that can pick up a quail egg. Suzuki devised a model gel arm that can raise a Coke bottle (165). Osada and his groups devised a seesaw balance that oscillates in response to the shrinking and swelling of gels driven by an electric field (166–170). A good review on the related topics is given by Osada (166).

Solvent Purification

When a gel swells from the shrinken state to the swollen state, it can absorb a tremendous amount of solvent. The final swelling ratio depends on the solvent (168). In the case of a mixed solvent, however, the degree of absorption depends on the chemical structure and molecular weight of the solvent molecules to be absorbed (169). It was found that the absorption efficiency of a gel decreases as the molecular weight of the solvent increases. Such a selective absorption provides a way of using polymer gels to extract solvents from aqueous solutions (169). The absorbed solvent can be disposed of or collected by squeezing the gel using the collapsing phase transition.

It is also possible to design the network with a strong selective absorption ability. Kokufuta et al, showed that hydrophobic gels can be used to extract detergent molecules (173). This concept can also be used in designing oil cleaning products using hydrophobic gels.

Pharmaceutical Applications

The diffusion of a large guest molecule trapped in a gel network depends on the state and the structure of the network (171–176). Naturally diffusion is prohibited when the gel is collapsed, but enhanced when the gel is swollen. Based on this principle, specific condition-sensitive, therefore target-sensitive, drug delivery systems have been designed (177).

The activity of enzymes trapped in a gel matrix can be controlled by varying the state of the gel. Hoffman and his colleagues have shown that temperature cycling (called a hydrolic pump) of thermal hydrogels can enhance the enzyme productivity dramatically (177–180). It was also shown that the swelling behavior of a gel network can also be modified by the existence of the trapped enzymes (180). Kokufuta et al designed a gel using poly(N-isopropylacrylamide) in which some enzymes are incorporated that undergoes a phase transition when the enzymatic reaction takes place (181).

There are basically two kinds of drug-release mechanisms related to gel volume change. One acts by squeezing, where the drug is released because of the fast initial shrinking of the network (177). In the swollen state, the drug molecules are dissolved in the solvent and remain inside the gel because of the small diffusion rate. When the gel shrinks suddenly, the drug molecules are squeezed out together with solvent, thus causing a drug release burst. An example of this mechanism using pH as the controlling parameter is shown in Figure 14a (179).

The other mechanism relates to a hydrophobic drug trapped in a hydrophobic network, in which the drug diffusion rate and crystallization degree are strongly dependent on the network density (183–186). Figure 14b shows a typical result obtained by using a temperature-sensitive gel (183). Siegel et al observed a similar result using pH as the controlling parameter (185). In the collapsed state, the diffusion rate is low and the crystallization of the drug molecules is high, which results in the off state of the release. The swollen state provides a freer environment for the drug molecules and they can diffuse out easily.

Using hydrogel membranes, Siegel & Firestone showed that it is possible to design human implantable self-regulating insulin delivery systems that are sensitive to glucose concentration (186). Grimshaw & Grodzinsky

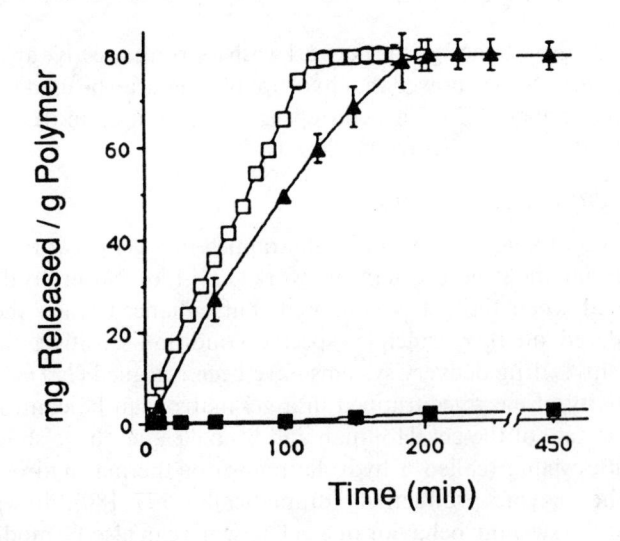

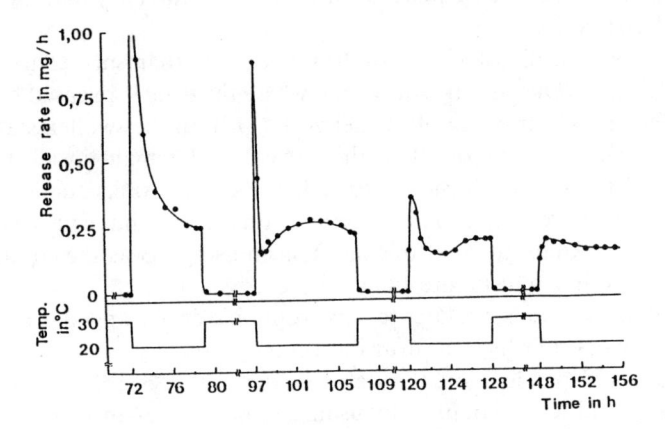

Figure 14 (*a*) Release kinetics of caffeine from cross-linked MMA/DMA copolymer hydrogels for various values of released medium pH. Initial caffeine loading was approximately 8%. Open square is for pH 3, solid triangle for pH 5, and solid square is for pH 7 (Figure 1 of Reference 186). (*b*) Pulsatile release rate of indomethacin in response to a step-wise temperature change between 20 and 30°C. The N-isopropylacrylamide gel in a phosphate-buffered saline was used (Figure 2 of Reference 122).

reported that it is possible to selectively control the transport of proteins across polyelectrolyte gel membranes (187).

Readers interested in the topic of drug delivery systems upon the dissolution of the gel matrix are directed to Reference 188.

CONCLUSION

Since the theoretical prediction twenty-five years ago and the experimental observation of the gel volume phase transition fifteen years ago, great progress has been made in our fundamental understanding of the phenomenon and its applications. However, there are still many areas to be studied and explored. We do not have a satisfactory equation of state that can be used to quantitatively analyze the phase behavior of gels. Incorporation of an extensive density change of an elastic body appears to lead to a fascinating new area (189; K. Sakimoto & M. Doi, submitted). The critical behavior of the gel is not well understood. For example, we have not determined the spinodal line nor followed the process of spinodal decomposition. From an application perspective, the toxicity, durability, response time, strength, biodegradability, and many other practical questions remain to be answered.

Recent findings of the phase transitions driven by each of the fundamental biological forces suggest that merely an improvement of presently available theories would not be enough to describe a gel system and, more generally, polymer systems. By combining these biological forces new multiple phases have been discovered in some gel systems. The findings have opened an entirely new prospect in the field between polymer sciences and biology. The principle on which the structure and function of proteins are created, based on the one-dimensional sequence of the amino acids, is one of the major riddles in modern science. The studies of the phase transitions in gels may provide a vital physical insight into this question.

ACKNOWLEDGMENT

This work was supported by National Science Foundation grant No. DMR 89-20401.

Literature Cited

1. Flory, P. J. 1953. *Principles of Polymer Chemistry*. Ithaca, NY: Cornell Univ. Press
2. Staudinger, H. 1920. *Ber. Dtsch. Chem. Ges.* 53: 1073–85
3. Meyer, K. H., von Susich, G., Valk, E. 1932. *Kolloid Z.* 59: 208–16
4. Guth, E., Mark, H. 1934. *Monats. Chem.* 65: 93
5. Kuhn, W. 1934. *Kolloid Z.* 68: 2–15

6. Kuhn, W. 1936. *Kolloid Z.* 76: 258–71
7. Wall, F. T. 1942. *J. Chem. Phys.* 10: 132–34
8. Flory, P. J., Rehner, J. Jr. 1943. *J. Chem. Phys.* 11: 512–20
9. Treloar, L. R. G. 1943. *Trans. Faraday Soc.* 39: 36, 241
10. Hermans, J. J. 1947. *Trans. Faraday Soc.* 43: 591
11. James, H. M. 1947. *J. Chem. Phys.* 15: 651–68
12. James, H. M., Guth, E. 1947. *J. Chem. Phys.* 15: 669–83
13. James, H. M., Guth, E. 1949. *J. Polym. Sci.* 4: 153–82
14. James, H. M., Guth, E. 1943. *J. Chem. Phys.* 11: 455–81
15. Staverman, A. J. 1982. *Adv. Polym. Sci.* 44: 73–101
16. Mark, J. E., Erman, B. 1988. *Rubberlike Elasticity, Molecular Primer*. New York: Wiley
17. Kuhn, W., Hargitay, B., Katchalsky, A., Hisenberg, H. 1950. *Nature* 165: 514–16
18. Flory, P. J. 1942. *J. Chem. Phys.* 10: 51–61
19. Huggins, M. L. 1941. *J. Chem. Phys.* 9: 440
20. Dusek, K., Patterson, D. 1968. *J. Polym. Sci.* 6: 1209–16
21. Tanaka, T., Hocker, L. O., Benedek, G. B. 1973. *J. Chem. Phys.* 59: 5151–59
22. de Gennes, P. G. 1979. *Scaling Concepts in Polymer Physics*. Ithaca, NY: Cornell Univ. Press
23. Tanaka, T., Ishiwata, S., Ishimoto, C. 1977. *Phys. Rev. Lett.* 38: 771–74
24. Tanaka, T. 1978. *Phys. Rev. Lett.* 40: 820–23
25. Amiya, T., Tanaka, T. 1987. *Macromolecules* 20: 1162–64
26. Hirotsu, S., Hirokawa, Y., Tanaka, T. 1987. *J. Chem. Phys.* 87: 1392–95
27. Tanaka, T., Fillmore, D., Sun, S.-T., Nishio, I., Swislow, G., et al. 1980. *Phys. Rev. Lett.* 45: 1636–39
28. Ilavsky, M. 1982. *Macromolecules* 15: 782–88
29. Hirotsu, S. 1987. *J. Phys. Soc. Jpn.* 56: 233–42
30. Tanaka, T. 1981. *Sci. Am.* 244: 124–38
31. Tanaka, T., Nishio, I., Sun, S.-T., Ueno-Nishio, S. 1982. *Science* 218: 467–69
32. Suzuki, A. 1990. *4th Gel Symp.* Tokyo
33. Mamada, A., Tanaka, T., Kungwatchakun, D., Irie, M. 1990. *Macromolecules* 23: 1517–19
34. Suzuki, A., Tanaka, T. 1990. *Nature* 346: 345–47
35. Clark, A. H., Ross-Murphy, S. B. 1985. *Br. Polym. J.* 17: 164–68
36. Stauffer, D. 1985. *Introduction to Percolation Theory*. London: Taylor & Francis
37. Stauffer, D. 1979. *Phys. Rep.* 54: 1–74
38. Stauffer, D., Coniglio, A., Adam, M. 1982. *Adv. Polym. Sci.* 44: 103–58
39. Candau, S., Bastide, J., Delsanti, M. 1982. *Adv. Polym. Sci.* 44: 27–71
40. des Cloiseaux, J. 1974. *Phys. Rev. A* 10: 1665–69
41. Daoud, M., Cotton, J. P., Farnoux, B., Jannink, G., Sarma, G., et al. 1975. *Macromolecules* 8: 804–18
42. des Cloiseaux, J. 1975. *J. Phys.* 36: 281–91
43. de Gennes, P. G. 1976. *Macromolecules* 9: 587–93
44. Noda, I., Kato, N., Kitano, T., Nagasawa, M. 1981. *Macromolecules* 14: 668–76
45. Wang, L., Bloomfield, V. A. 1990. *Macromolecules* 23: 194–99
46. Daoud, M., Bouchaud, E., Jannink, G. 1986. *Macromolecules* 19: 1955–60
47. Munch, J. P., Candau, S., Herz, J., Hild, G. 1977. *J. Phys.* 38: 971–76
48. Geissler, E., Hecht, A. M., Duplessix, R. 1982. *J. Polym. Sci. Polym. Phys. Ed.* 20: 225–33
49. Davidson, N. S., Richards, R. W., Maconnachie, A. 1986. *Macromolecules* 19: 434–41
50. Tanaka, T. 1985. In *Dynamic Light Scattering*, ed. R. Pecora, pp. 347–62. New York: Plenum
51. Candau, S. J., Young, C. Y., Tanaka, T., Bastide, J. 1979. *J. Chem. Phys.* 70: 4694–4700
52. Horkay, F., Zrinyi, M. 1982. *Macromolecules* 15: 1306–10
53. Richards, R. W., Davidson, N. S. 1986. *Macromolecules* 19: 1381–89
54. Geissler, E., Hecht, A. M., Horkay, F., Zrinyi, M. 1988. *Macromolecules* 21: 2594–99
55. Geissler, E., Horkay, F., Hecht, A. M., Zrinyi, M. 1989. *J. Chem. Phys.* 90: 1924–29
56. Benoit, H., Decker, D., Duplessix, R., Picot, C., Rempp, P., et al. 1976. *J. Polym. Sci. Polym. Phys. Ed.* 14: 2119–28
57. Candau, S. J., Ilmain, F., Schosseler, F., Bastide, J. 1989. *Mater. Res. Soc. Symp. Proc.* 177: 3–15
58. Dusek, K., Prins, W. 1969. *Adv. Polym. Sci.* 6: 1–102
59. Weiss, N., Van Vliet, T., Silberberg, A. 1979. *J. Polym. Sci. Phys.* 17: 2229–40
60. Bastide, J., Picot, C., Candau, S. 1981. *J. Macromol. Sci. Phys. B* 19: 13–34

61. Richards, E. G., Temple, C. J. 1971. *Nature* 230: 92–96
62. Hsu, T., Ma, D., Cohen, C. 1983. *Polymer* 24: 1273–78
63. Flory, P. J. 1977. *J. Chem. Phys.* 66: 5720–29
64. Pusey, P. N., van Megen, W. 1989. *Physica A* 157: 705–41
65. Bansil, R., Gupta, M. K. 1980. *Ferroelectrics* 30: 63–71
66. Mallam, S., Hetch, A., Geissler, E., Pruvost, P. 1989. *J. Chem. Phys.* 91: 6447–54
67. Weiss, N., Silberberg, A. 1975. *Polym. Prepr. Am. Chem. Soc., Div. Polym. Chem.* 16: 289–92
68. Weiss, N., Silberberg, A. 1977. *Br. Polym. J.* 9: 144–50
69. Tokita, M. 1990. Preprint
70. Mallam, S., Horkay, F., Hetch, A., Geissler, E. 1989. *Macromolecules* 22: 3356–61
71. Hecht, A. M., Geissler, E. 1978. *J. Phys.* 39: 631–38
72. de Gennes, P. G. 1976. *Macromolecules* 9: 594–98
73. Bastide, J., Buzier, M., Boué, F. 1987. In *Polymer Motion in Dense Systems*, ed. D. Richter, T. Springer, pp. 112–20. New York: Springer-Verlag
74. Bastide, J., Picot, C., Candau, S. J. 1979. *J. Polym. Sci. Polym. Phys. Ed.* 17: 1441–56
75. Nagy, M. 1985. *Coll. Polym. Sci.* 263: 245–65
76. Bastide, J., Leibler, L. 1988. *Macromolecules* 21: 2649–51
77. Bastide, J., Boué, F., Audebert, R., Leibler, L. 1988. In *Biological and Synthetic Polymer Networks*, ed. O. Kramer, p. 277. The Netherlands: Elsevier Sci.
78. Klonowski, W. 1985. *J. Appl. Phys.* 58: 2883–92
79. Tran-Cong, Q., Nagaki, T., Nakagawa, T., Yano, O., Soen, T. 1989. *Macromolecules* 22: 2720–23
80. Deleted in proof
81. Nishio, I., Swislow, G., Sun, S.-T., Tanaka, T. 1982. *Nature* 300: 243–44
82. Swislow, G., Sun, S.-T., Nishio, I., Tanaka, T. 1980. *Phys. Rev. Lett.* 44: 796–98
83. Zrinyi, M., Horkay, F. 1989. *Macromolecules* 22: 394–400
84. Zrinyi, M., Horkay, F. 1984. *Macromolecules* 17: 2805–11
85. Marchetti, M., Prager, S., Cussler, E. L. 1990. *Macromolecules* 23: 1760–65
86. Otake, K., Inomata, H., Konno, M., Saito, S. 1990. *J. Chem. Phys.* 91: 1345–50
87. Konak, C., Bansil, R. 1989. *Polymer* 30: 677–90
88. Hasa, J., Ilavsky, M., Dusek, K. 1975. *J. Polym. Sci. Polym. Phys.* 13: 253–62
89. Li, Y., Tanaka, T. 1989. *J. Chem. Phys.* 90: 5161–66. The critical density given in this paper is not correct. The correct value should be 0.568 g/cc.
90. Siegel, R. A., Firestone, B. A. 1988. *Macromolecules* 21: 3254–59
91. Hirokawa, Y., Tanaka, T., Katayama, S. 1984. In *Life Sciences Research Rep. 31*, pp. 177–88. New York: Springer-Verlag
92. Beltran, S., Hooper, H. H., Blanch, H. W., Prausnitz, J. M. 1990. *J. Chem. Phys.* 92: 2061–66
93. Ohmine, I., Tanaka, T. 1982. *J. Chem. Phys.* 77: 5725–29
94. Osada, Y., Umezawa, K., Yamauchi, A. 1988. *Makromol. Chem.* 189: 597–605
95. Irie, M. 1986. *Macromolecules* 19: 2890–92
96. Deleted in proof
97. Ilavsky, M. 1981. *Polymer* 22: 1687–91
98. Katayama, S., Ohata, A. 1985. *Macromolecules* 18: 2781–82
99. Hirokawa, Y., Tanaka, T. 1985. *Macromolecules* 18: 2782–84
100. Ricka, J., Tanaka, T. 1984. *Macromolecules* 17: 2916–21
101. Firestone, B. A., Siegel, R. A. 1988. *Polym. Commun.* 29: 204–8
102. Pincus, P. 1976. *Macromolecules* 9: 386–88
103. Witten, T. A., Pincus, P. 1987. *Europhys. Lett.* 3: 315–20
104. Oosawa, F. 1971. *Polyelectrolytes*. New York: Marcel Dekker
105. Mannings, G. S. 1969. *J. Chem. Phys.* 51: 924–33
106. Schurr, J. M., Schmitz, K. S. 1986. *Annu. Rev. Phys. Chem.* 37: 271–305
107. Fuoss, R. M., Katchalsky, A., Lifson, S. 1951. *Proc. Natl. Acad. Sci. USA* 37: 579–89
108. Deleted in proof
109. Lee, H. B., Jhon, M. S., Andrade, J. D. 1975. *J. Colloid Interface Sci.* 51: 225–31
110. Takizawa, A., Kinoshita, T., Nomura, O., Tsujita, Y. 1985. *Polym. J.* 17: 747–52
111. Al-Issa, M. A., Davis, T. P., Huglin, M. B., Rego, J. M., Rehab, M. M., et al. 1990. *Makromol. Chem.* 191: 321–30
112. Ilavsky, M., Hrouz, J., Havlicek, I. 1985. *Polymer* 26: 1514–18
113. Kokufuta, E., Zhang, Y.-Q., Tanaka, T. 1992. *Phase Trans.* In press

114. Amiya, T., Hirokawa, Y., Hirose, Y., Li, Y., Tanaka, T. 1987. *J. Chem. Phys.* 86: 2375–79
115. Hirotsu, S. 1990. *Macromolecules* 23: 903–5
116. Otake, K., Inomata, H., Konno, M., Saito, S. 1990. *Macromolecules* 23: 283–89
117. Strauss, U. P. 1989. *Am. Chem. Soc. Symp. Ser.* 223: 317–24
118. Peer, W. J. 1989. *Am. Chem. Soc. Symp. Ser.* 223: 381–97
119. Otake, K., Inomata, H., Konno, M., Saito, S. 1990. *4th Gel Symp.* Tokyo
120. Iliopoulos, I., Audebert, R., Quivoron, C. 1987. *Am. Chem. Soc. Symp Ser.* 350: 72–86
121. Ilmain, F., Tanaka, T., Kokufuta, E. 1991. *Nature* 349: 400–1
122. Okano, T., Bae, Y. H., Jacobs, H., Kim, S. W. 1990. *J. Contrl. Rel.* 11: 255–65
123. Tanaka, T., Fillmore, D. 1979. *J. Chem. Phys.* 70: 1214–18
124. Peters, A., Candau, S. J. 1986. *Macromolecules* 19: 1952–55; 21: 2278–82
125. Chiarelli, P., De Rossi, D. 1988. *Progr. Colloid. Polym. Sci.* 78: 4–8
126. Tanaka, T. 1986. *Physica A* 140: 261–68
127. Sato-Matsuo, E., Tanaka, T. 1988. *J. Chem. Phys.* 89: 1695–1703
128. Li, Y., Tanaka, T. 1990. *J. Chem. Phys.* 92: 1365–71
129. Li, Y., Tanaka, T. 1991. Effect of shear modulus of polymer gels. In *Polymer Gels*, ed. D. De Rossi, K. Kajiwara, Y. Osada, A. Yamauchi. New York: Plenum
130. Candau, S. J., Peters, A. 1987. *Polym. Mater. Sci. Eng.* 57: 270
131. Onuki, A. 1988. *Phys. Rev. A* 38: 2192–95
132. Tanaka, T., Sato, E., Hirokawa, Y., Hirotsu, S., Peetermans, J. 1985. *Phys. Rev. Lett.* 55: 2455–58
133. Heller, P. 1967. *Rept. Prog. Phys.* 30: 731
134. Ma, S.-K. 1976. *Modern Theory of Critical Phenomena*. Menlo Park, CA: Benjamin/Cummings
135. Hochberg, A., Tanaka, T., Nicoli, D. 1979. *Phys. Rev. Lett.* 43: 217–19
136. Tabak, F., Corti, M., Pavesi, L., Rigamonti, A. 1987. *J. Phys. C* 20: 5691–5701
137. Stanley, H. E. 1971. *Introduction to Phase Transition and Critical Phenomena*. New York/Oxford: Oxford Univ. Press
138. Bervillier, C. 1986. *Phys Rev. B* 34: 8141–43, and references cited therein
139. Moore, M. A. 1977. *J. Phys. A* 10: 305
140. de Gennes, P. G. 1978. *J. Phys. Lett.* 39L: 299–301
141. Onuki, A. 1990. *J. Phys. Soc. Jpn.* 59: 3423
142. Onuki, A. 1988. *J. Phys. Soc. Jpn.* 57: 699–702, 1868–71
143. Kramer, O. 1986. *Biological and Synthetic Polymer Networks*. New York: Elsevier Sci.
144. Boyde, T. R. C. 1976. *J. Chromatogr.* 124: 219–30
145. Veatch, R. W. Jr. 1983. *J. Pet. Technol.* April: 677–87
146. Werner, G. 1984. *Allg. Vliesstoff-Rep.* 4: 178–82
147. Kossmelh, G., Klaus, N., Schafer, H. 1984. *Angew. Makromol. Chem.* 123/124: 2412–59
148. Deleted in proof
149. Deleted in proof
150. Deleted in proof
151. Deleted in proof
152. Deleted in proof
153. Deleted in proof
154. Deleted in proof
155. Deleted in proof
156. Deleted in proof
157. Deleted in proof
158. Deleted in proof
159. Steinberg, I. Z., Oplatka, A., Katchalsky, A. 1966. *Nature* 210: 568–71
160. Sussman, M. V., Katchalsky, A. 1970. *Science* 167: 45–47
161. Horkay, F., Zrinyi, M. 1989. *Makromol. Chem., Macromol. Symp.* 30: 133–43
162. Fujiwura, H., Shibayama, M., Chen, J.-H., Nomura, S. 1989. *J. Appl. Polym. Sci.* 37: 1403–14
163. De Rossi, D., Chiarelli, P., Buzzigoli, G., Domenici, C., Lazzeri, L. 1986. *Trans. Am. Soc. Artif. Intern. Organs* 32: 157
164. Shiga, T., Hirose, Y., Okada, A., Kurauchi, T. 1989. *197th Am. Chem. Soc. Polym. Prep.*, p. 310
165. Suzuki, M. 1989. *Polym. Prep. Jpn.* 46: 603–6
166. Osada, Y., Hasebe, M. 1985. *Chem. Lett.* 9: 1285–88
167. Irie, M., Kungwatchakun, D. 1985. *Makromol. Chem. Rapid Commun.* 5: 829–32
168. Yoshino, K., Nakao, K., Onoda, M., Sugimoto, R. 1989. *Jpn. J. Appl. Polym.* 28: L682–85
169. Yoshino, K., Nakao, K., Morita, S., Onoda, M. 1989. *Jpn. J. Appl. Polym.* 28: L2027–30
170. Osada, Y. 1987. *Adv. Polym. Sci.* 82: 1–46
171. Saitoh, K., Ozawa, T., Suzuki, N. 1976. *J. Chromatogr.* 124: 231–37

172. Freitas, R. F. S., Cussler, E. L. 1987. *Sep. Sci. Technol.* 22: 911–19
173. Kokufuta, E., Zhang, Y.-Q., Tanaka, T. 1991. *Nature* 351: 302–4
174. Gehrke, S. H., Cussler, E. L. 1989. *Chem. Eng. Sci.* 44: 559–66
175. Kou, J. H., Fleisher, D., Amidon, G. L. 1990. *J. Control. Rel.* 12: 241–50
176. Hasirci, V. N. 1982. *J. Appl. Polym. Sci.* 27: 33–41
177. Hoffman, A. S. 1987. *J. Contrl. Rel.* 6: 297–305
178. Park, T. G., Hoffman, A. S. 1988. *Appl. Biochem. Biotech.* 19: 1–9
179. Park, T. G., Hoffman, A. S. 1990. *Biotech. Bioeng.* 35: 152–59
180. Dong, L. C., Hoffman, A. S. 1987. *Am. Chem. Soc. Symp. Ser.* 350: 236–44
181. Kokufuta, E., Tanaka, T. 1991. *Macromolecules* 24: 1605–7
182. Hoffman, A. S., Afrassiabi, A., Long, L. C. 1986. *J. Contrl. Rel.* 4: 213–22
183. Bae, Y. H., Okano, T., Hsu, R., Kim, S. W. 1987. *Makromol. Chem. Rapid Commun.* 8: 481–85
184. Bae, Y. H., Okano, T., Kim, S. W. 1989. *J. Contrl. Rel.* 9: 271–79
185. Siegel, R. A., Falamarzian, M., Firestone, B. A., Moxley, B. C. 1988. *J. Contrl. Rel.* 8: 179–82
186. Siegel, R. A., Firestone, B. A. 1990. *J. Contrl. Rel.* 11: 181–92
187. Grimshaw, P. E., Grodzinsky, A. J. 1989. *Chem. Eng. Sci.* 44: 827–40
188. Wang, P. Y. 1987. *Polym. Mater. Sci. Eng.* 57: 400–3
189. Sekimoto, K., Suematsu, N., Kawasaki, K. 1989. *Phys. Rev. A* 39: 4912–14
190. Deleted in proof
191. Li, Y. 1989. *Structure and critical behavior of polymer gels.* PhD thesis. Mass. Inst. Technol. 176 pp.

PLASMA ION-ASSISTED EVAPORATIVE DEPOSITION OF SURFACE LAYERS

S. Pongratz and A. Zöller

Leybold AG, Siemens-Strasse 100, D-8755 Alzenau, Germany

KEY WORDS: ion deposition, plasma source, scratch-resistant, antireflection coatings, organic substrates

INTRODUCTION

Vacuum thin film processes are becoming increasingly important in modern high technology because of their possible use in changing the surface properties over a wide range while maintaining the properties of the bulk material. The requirements for temperature stability and low absorption and scattering losses are becoming ever more stringent for thin films used in optical applications. Furthermore, synthetic substrate materials are more and more used for ophthalmic lenses as well as for high precision optical components. Especially important is the fact that a rather soft substrate material can be covered with a hard surface layer, thus providing a high scratch-resistance without the brittleness of common hard materials.

In principle, the process technology for vacuum deposition of thin films can be subdivided in physical vapor deposition (PVD) (evaporation, sputtering) and chemical vapor deposition (CVD) with subgroups thermal CVD, PECVD, and photo CVD.

Evaporation, the classical method of producing thin coating layers, using the technique of electron beam evaporation provides a comparatively high rate of deposition. The coating of shaped or two-dimensional optical substrates also uses evaporation for reasons of better film distribution. Evaporation processes under vacuum require an arrangement of an evaporation source and the substrate to be covered such that the evaporated particle flux can reach the surface to be coated. According to mean free path conditions the evaporated material travels from the source in a vacuum of less than 1×10^{-4} mbar without any gas phase collisions toward the substrate surface where it condenses. Because of the straight line-directed flux of free atoms or molecules, these particles arrive at the substrate surface at a distinctive angle of incidence. The most prominent advantage of evaporation processes is the very high deposition rate achievable for many materials.

In comparison, sputtering delivers high packing density and high quality films with very good thickness uniformity on large flat substrates. Sputtering processes make use of the momentum transfer of accelerated particles when impinging onto the target surface. Atoms or molecules in the target surface are ejected by this momentum transfer. They travel from the target surface into the gas phase where they undergo collisions with other particles before landing upon the substrate. Diode sputtering, the earliest practical implementation of sputtering was not widely accepted.

Alternatively, magnetron sputtering applies a magnetic field in order to confine the plasma in front of the target. As a result, higher particle densities are achieved in the plasma, with resultant higher ion current densities arriving at the target. This produces higher sputter rates and consequently higher deposition rates. Because of the plasma confinement, the electron current toward the substrate surface is reduced, which results in a lower heat load at the substrate. Many papers have been published on the basic properties of sputtered optical layers (1–4).

Beside the conventional vacuum coating processes like sputter and evaporation, Plasma Enhanced Chemical Vapor Deposition (PECVD) is especially attractive for specific applications. In this process the reaction gas is fed into the reaction area through a shower head in the upper electrode. The plasma itself is generated by high-energy electrons, which transfer their energy to the gas molecules by formation of radicals, by excitation, dissociation, and by ionization. The highly reactive species are deposited on the substrate surface at a rate depending primarily on the gas mixture and the substrate temperature. The morphology of the film being deposited depends mainly on the ion energy, as well as the general process parameters (power, pressure, geometry, etc.). The additional ion impact leads to the desorption of weakly bonded fragments as well as to a densification of the film.

In general, research and development on the various thin film growth processes is directed toward avoiding undesirable effects at the interface between the substrate and the growing layer (unwanted contamination, lack of adhesion, etc.) and improving the structural quality of the films.

Both can be improved by use of sufficient substrate or deposition temperatures. But in many of the potential fields of application of thin film deposition technology, elevated substrate temperatures are undesirable. Thus ion-assisted and ion beam techniques are of increasing interest.

ION PLATING

Ion plating has been used for many years (5) as a means to provide additional energy to the substrate surface. The substrate is maintained at a negative potential and therefore attracts ions between the source and the substrate. These ions are either generated by the electron beam, which heats the evaporative source in a noble gas atmosphere, or they are generated by an auxiliary gas discharge electrode in the vacuum chamber.

A major drawback of the principal ion plating processes (6) is that nonconductive substrates cannot be biased with a DC voltage. However, employment of a low voltage reactive ion plating process (7) allows the deposition of single layer and multilayer oxide coatings onto unheated electrically isolating substrates.

ION BEAM-ASSISTED EVAPORATIVE DEPOSITION

Ion-assisted deposition is a well-known technique for improving the properties of thin films by influencing the microstructure of the growing layer and its adherence to the substrate (5–8).

A growing thermally evaporated film is bombarded by an energetic ion beam (E_{ion}: 100–800 eV). A wide range of materials and multilayer systems has already been investigated (8–11). In addition to the limited beam size of the available ion sources, hot filament cathodes (12) and extraction grids often exhibit limited lifetimes at high emission currents, especially when oxygen is introduced into the ion source (13). On the other hand, previous investigations with conventional ion sources have shown that ionized oxygen is necessary for many coating materials in order to obtain fully oxidized films with low absorption losses and high packing density (8).

ION CLUSTER BEAM DEPOSITION

With the ion cluster beam deposition technique, the problem of low energy ions has been convincingly solved (14). Using this process, the adhesive strength as well as the density of the layers can be obtained with low deposition temperatures. In general this technique shows promising results in the production of metal layers and low melting point compounds. However, the small diameter of the coating area does not allow production of large scale coatings.

ARC EVAPORATION

Higher plasma densities can be obtained with the aid of a powerful arc discharge. There are different types of arcs depending on the physical phenomenon at the electrodes. Glow cathode, cold cathode, and hollow-cathode methods are used. Normally the arc discharge is utilized for the generation of the vapor stream as well as for the production of the plasma. Therefore its role involves both the excitation and ionization of the vapor and of the reaction gas.

Several variations of the activated reactive evaporation by means of a low voltage arc discharge are described in the literature (15–18). At present, arc evaporation is used mainly for producing hard films of nitrides and carbides.

In summary, ion-based evaporation techniques result in improved film properties like increased adhesion, hardness, and packing densities. However, for economic industrial production it has been necessary to develop a plasma source that is able to irradiate large substrate areas with a high plasma density. Even with relatively high plasma densities, the temperature loading of the substrate is rather low compared with the other conventionally used methods.

PLASMA ION-ASSISTED DEPOSITION

As recently described in the literature (14), the plasma ion-assisted deposition (IAD) process can be performed in a standard box coating system with conventional thermal evaporation sources like electron beam guns, a rotating, large area substrate holder (up to ≈ 8000 cm^2), and a specially designed plasma source (Figure 1 *left*). The plasma source, based on a hot cathode, a cylindrical anode tube, and a solenoid magnet, is located in the center of the chamber bottom. A DC voltage between anode and cathode creates a glow discharge plasma with a hot electron emitter, supplied with a noble gas such as argon. Due to the magnetic field lines of the solenoid magnet that surround the anode tube, the mobility of the plasma electrons is strongly increased in the axial direction and strongly decreased in the

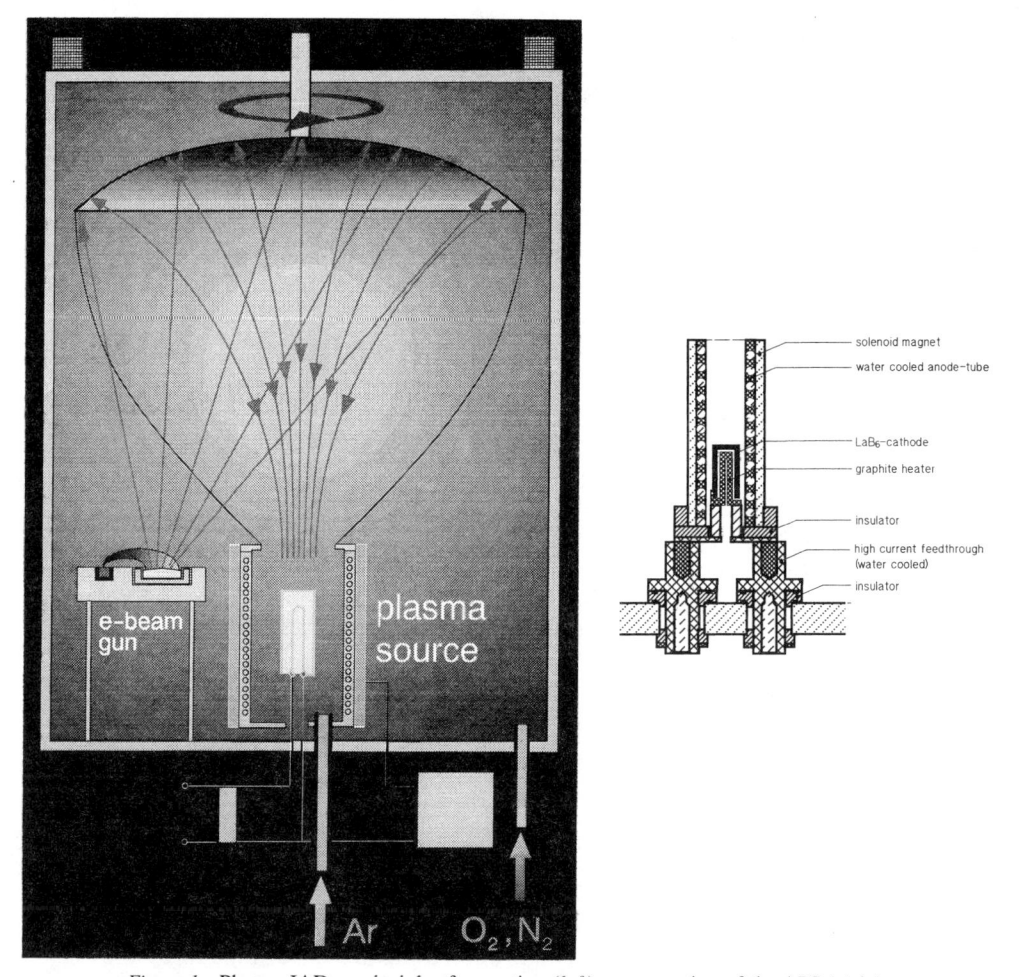

Figure 1 Plasma-IAD—principle of operation (*left*); cross-section of the APS (*right*).

radial direction. The electrons spiral along the magnetic field lines and the plasma extends in the direction of the substrate holder. A ring-shaped magnetic coil, located above the substrate holder, is used for producing a uniform layer of plasma on the substrate. The reactive gas is normally introduced directly into the coating chamber. In principle it is also possible to operate the source with the reactive gas itself, or a mixture of the noble and the reactive gases. The high plasma current density inside the chamber causes the reactive gas and the evaporant to become ionized and activated. Ionization of the reactive gas lowers the reactive gas concentration required to grow stoichiometric films.

The complete plasma source can be electrically isolated from the chamber ground. Because of the high mobility of the electrons, the plasma gets a positive self-bias potential relative to the chamber walls. Therefore the ion energy is mainly determined by the applied discharge voltage and an additional positive floating potential of the source relative to the chamber and also to the substrate holder. The electrons reflect on oscillatory paths between the source and the substrate holder. Therefore an effective ionization of the gas atoms and molecules is achieved. In this operational mode, the substrate holder does not need to be isolated. The magnitude of the floating potential depends on the applied discharge voltage, the partial pressures, and the strength of the magnetic fields. Thus the pressure and the magnetic fields can be used to control the ion energy at a given discharge current. Because the substrate holder is negatively charged relative to the plasma, the ions from the plasma are accelerated to the substrate and bombard the growing film while the electrons are reflected. Thus the momentum transfer effects, as described in the literature, densify the condensing film (8).

Because of the small acceleration distance between the plasma sheet and substrate holder, the ions are not affected by energy losses. In IAD processes with grid ion sources, where the source is located at the bottom of the coating chamber, the gas pressure is limited by scattering effects. In the case of plasma-IAD, the plasma is extended in the volume between the plasma source and the substrate holder with the ring-shaped magnetic coil and irradiates a large area with a high plasma density.

Since the plasma process is completely separated from the evaporation source, the choice of starting evaporation materials is not limited by the plasma generator. Thus it is possible in principle to evaporate metals, semiconductors, oxides, suboxides, fluorides, sulfides, etc.

ADVANCED PLASMA SOURCE

For further investigation of the plasma-IAD process we have developed an Advanced Plasma Source (APS) with some special features such as high plasma density, compatibility with oxygen, and low thermal radiation to the substrate area.

The APS is based on a large area LaB_6 cathode that is indirectly heated by a graphite filament heater to a temperature of $\sim 1500°C$ (Figure 1 *right*).

The emission current density of LaB_6 (15) is relatively high compared to other cathode materials like tantalum and tungsten. The heater power is fed by two water-cooled high current feedthroughs. The cathode is surrounded by a water-cooled anode tube. Because of the water cooling of the anode tube, the thermal load to the substrates caused by the plasma source is very low. The substrate temperatures are largely determined by the amount of heat radiated by the evaporation sources. Both cathode and anode tubes are isolated from the chamber ground. A solenoid magnet, which surrounds the anode tube, causes the extraction of the plasma into the coating chamber and to the substrate holder. The APS is operated with partial pressures between 1×10^{-4}–7×10^{-4} mbar of argon, or mixtures of argon and oxygen. The typical discharge current range is between 30 and 80 A, while discharge voltages of up to 100 V are used. Until now, for the majority of experiments, the plasma source has been supplied with pure argon while the oxygen flow was introduced directly into the coating chamber. Typical partial pressures are 2×10^{-4} mbar for argon and 4×10^{-4} mbar for oxygen in the case of reactive processes. Additional experiments have shown that it is also possible to feed the oxygen flow through the plasma source if the cathode temperature is higher than 1600°C. Oxidation of the surface at temperatures below 1500°C deactivates LaB_6 cathodes, but heating of the material to over 1600°C restores the electron emission.

The first measurements of the ion current density and the ion energy distribution were done with Faraday cups and Langmuir probes. Ion current densities of up to 0.5 mA/cm^2 were measured at a substrate holder of 5000 cm^2 area. The ion current density uniformity is approximately $\pm 10\%$ without using the magnetic coil above the substrate holder. Even with this non-uniform ion current density, a refractive index uniformity better than 0.35% for single layers of TiO_2 has been obtained. Figure 2 shows the transmission spectral curves of four substrates that were located on different radial positions of a 800 mm diameter substrate holder. The mean ion energy is equal to the anode voltage relative to the chamber ground (± 20–30 eV). Comparisons of ion current density measurements with and without oxygen have demonstrated that oxygen is ionized. At an argon partial pressure of $\sim 2 \times 10^{-4}$ mbar and a typical discharge current of 70 A, the ion current density was doubled because of an additional oxygen partial pressure of $\sim 4 \times 10^{-4}$ mbar. The oxygen was introduced directly into the coating chamber.

In a typical reactive process with partial pressures of $\sim 6 \times 10^{-4}$ mbar of argon and oxygen, and a discharge current of 70 A, the anode potential is ~ 70 V. In this case an ion current density of ~ 0.4 mA/cm^2 and a mean ion energy of 70 eV have been observed.

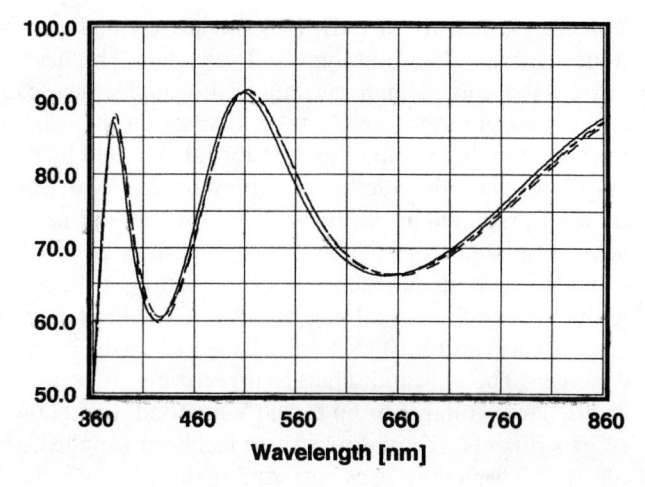

Figure 2 Transmission spectral curves of four substrates that are located on different radial positions of a 800 mm diameter substrate holder.

EXPERIMENTAL RESULTS

Experiments on single layers of TiO_2, Ta_2O_5, and SiO_2 demonstrate that the film properties can dramatically improve in comparison to conventional electron beam gun evaporation. The applicability of this method to the production of coatings on organic substrates has been demonstrated by the synthesis of a multilayer antireflection system on ophthalmic CR 39 lenses.

Titanium dioxide, one of the most important high-index materials for optical coatings in the visible and near infrared, is well suited to demonstrate the effect of the ion assistance during the film deposition. The structure of the films and therefore the packing density and the refractive index are strongly dependent on the partial pressure, the substrate temperature, the deposition rate and, in case of IAD, also on the ion energy and the ion current density (8). Figure 3 (*top*) shows an SEM micrograph of an electron beam-evaporated TiO_2 film deposited without ion assistance and without substrate heating. The typical columnar structure can clearly

Figure 3 SEM-micrographs of single layer TiO_2 deposited with conventional evaporation (n = 2.2) (*top*); deposited with plasma-IAD (n = 2.37) (*middle*); deposited with plasma-IAD (n = 2.6) (*bottom*).

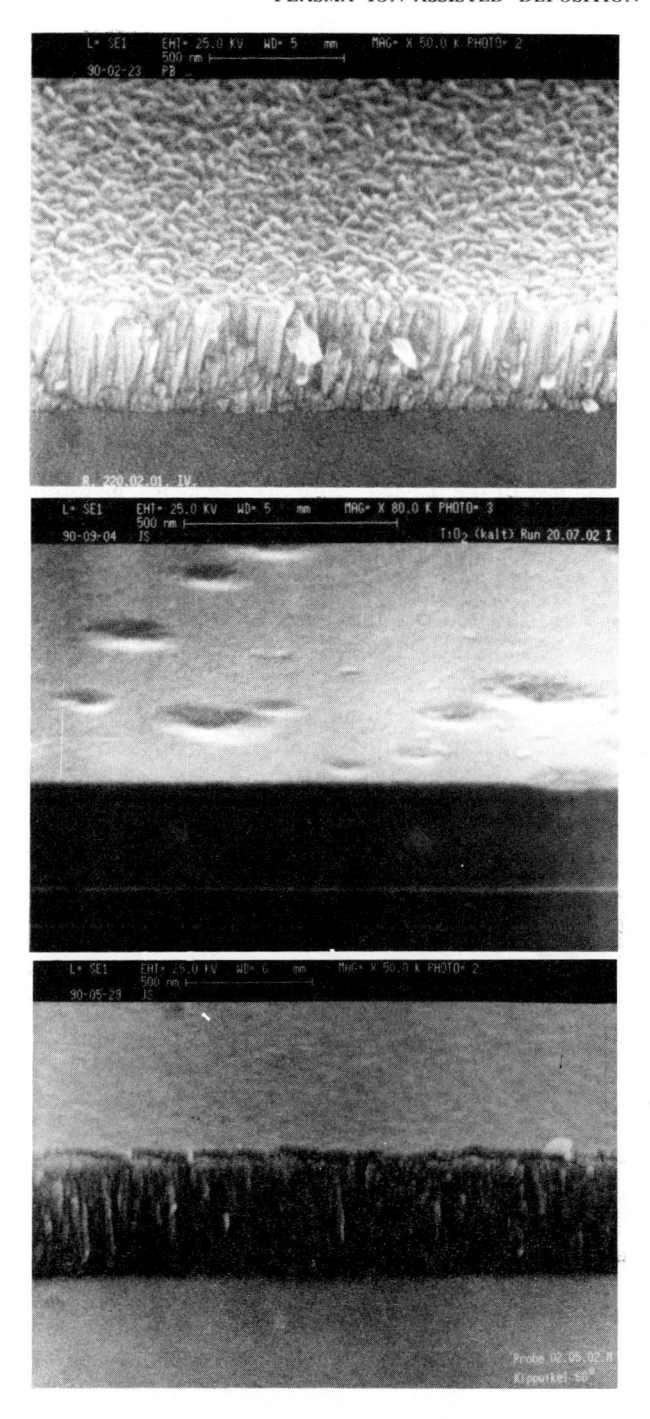

PLASMA ION-ASSISTED DEPOSITION 287

be seen. The refractive index for such films is normally ∼2.2 at 550 nm. Heating the substrates to 300°C improves the refractive index to 2.3.

With plasma-IAD two different structural modifications of TiO_2 can be obtained. Figure 3 (*middle*) shows the SEM micrograph of one film that was deposited with a rate of ∼0.2 nm/s and a plasma power of ∼5 kW without substrate heating. The columnar structure has disappeared completely. The interface between the substrate and the film could be seen only after a chemical etching step. The refractive index of this TiO_2 film was 2.37. The measured transmission and reflection curves (Figure 4, *top*) indicate a homogeneous, stoichiometric film. The properties of this modification are relatively insensitive to variations of the deposition rate and the plasma density. Therefore a good refractive index uniformity (<0.35%) over the substrate area (800 mm) was obtained, although the plasma density distribution was not optimized.

A large increase of the plasma density can increase the refractive index to 2.6. Figure 3 (*bottom*) shows the SEM picture of such a TiO_2 film ($n_{550} = 2.6$). It is easy to see that this film is structured. Because the refractive index (2.6) of this film is higher than that of the unstructured film (2.37), one can assume that the crystalline modification with the higher density (rutile) was obtained. As reported by Martin (8), rutile has been prepared by reactive evaporation at a very high oxygen partial pressure and elevated substrate temperatures, up to 1100°C. The grain size varies from 10 to 60 nm. During the deposition of this film, the partial pressure was increased in contrast to the modification described above in order to grow a stoichiometric film. Figure 4 (*bottom*) shows the measured reflection and transmission curves of the high index film. For these experiments, Ti_2O_3 was used as the starting material.

One method to verify the packing density and the structural stability is to compare the spectral characteristic of the sample directly after coating, after being kept under water, and after baking. Test results (14) indicate that the packing density of the TiO_2 films deposited with plasma-IAD is essentially higher than that of the conventionally evaporated films.

Besides titanium dioxide, tantalum pentoxide (Ta_2O_5) is another preferred coating material in the visible and near infrared wavelength range. Because of its low scattering surface and high laser damage threshold, Ta_2O_5 is one of the most widely used materials for laser coatings in the visible and near infrared wavelength range. In addition, the refractive index is well suited for broad band antireflection coatings.

For most applications, the conventional electron beam-evaporated Ta_2O_5 films must be baked in an air atmosphere after the deposition. This post-treatment improves the stoichiometry. With plasma-IAD, stoichiometric Ta_2O_5 films were grown without substrate heating and without

PLASMA ION-ASSISTED DEPOSITION 289

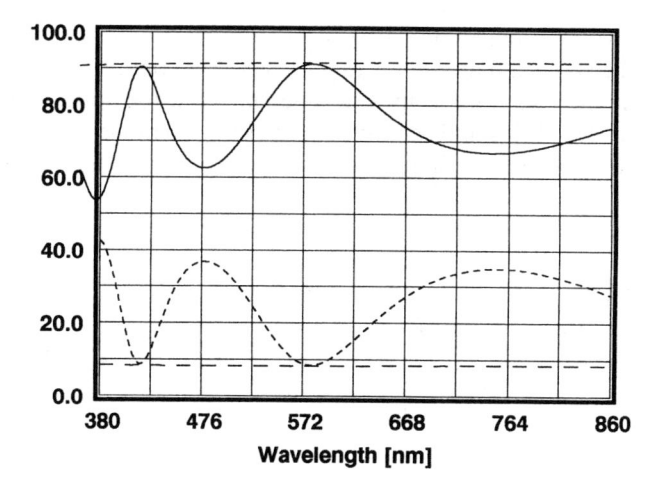

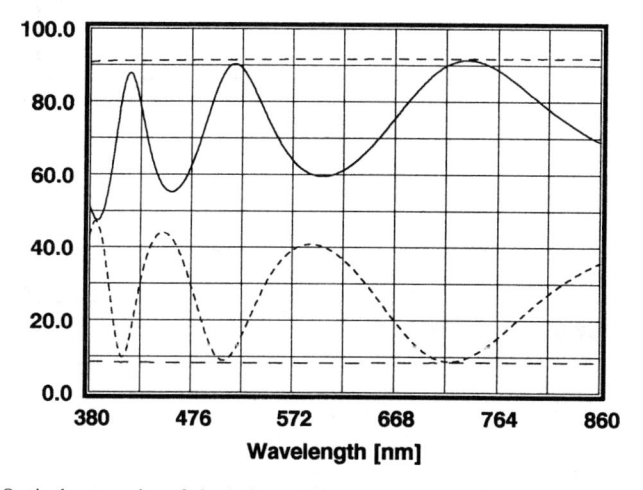

Figure 4 Optical properties of single layer TiO$_2$ deposited with plasma-IAD: n$_{550}$ = 2.37; d = 500 nm (*top*); n$_{550}$ = 2.6; d = 300 mm (*bottom*); —— transmittance (%); - - - reflectance (%).

post-treatment. In addition, the refractive index is greater than that found for conventionally evaporated Ta$_2$O$_5$ films. With conventional evaporation onto heated substrates, a refractive index of n$_{550}$ = 2.05 can be achieved after baking in the atmosphere. With plasma-IAD, a refractive index of n$_{550}$ = 2.12 was obtained without post-treatment. The measured transmission and reflection curve is shown in Figure 5 (*top*). Ta_2O_5 was used as the starting material for the plasma-IAD coatings.

Silica is the most important low index material for optical coatings in the ultraviolet, visible and near infrared wavelength range. In conventional evaporation processes, SiO_2 grows stoichiometrically if one uses SiO_2

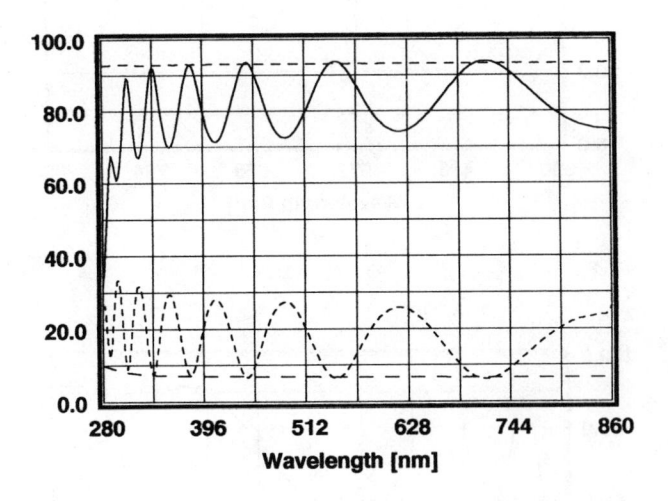

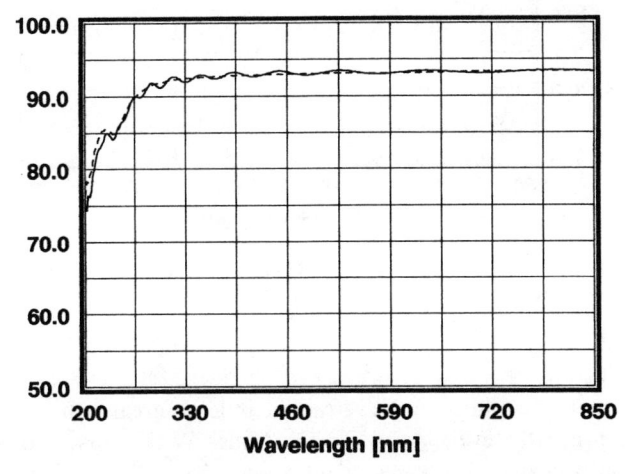

Figure 5 Optical properties of single layer Ta_2O_5 deposited with plasma-IAD (*top*) ($n_{550} = 2.12$; d = 500 nm). Single layer SiO_2 deposited with plasma-IAD onto a suprasil substrate (*bottom*). —— transmittance (%); --- reflectance (%).

PLASMA ION-ASSISTED DEPOSITION 291

as starting material, even without an additional oxygen inlet during the deposition. Only a small amount of oxygen in the 10^{-5} mbar range is required to avoid absorption losses in the ultraviolet wavelength range. With plasma-IAD the same UV-transparency can be achieved as with conventional evaporation. Figure 5 (*bottom*) shows the transmission curve of a 3500 nm-thick single-layer SiO_2, which was deposited with plasma-IAD onto a quartz substrate without additional oxygen. The growing film

Figure 6 SEM-micrograph of a single layer SiO_2 deposited with conventional electron beam gun evaporation onto a heated glass substrate (300°C) (*top*); deposited with plasma-IAD onto an unheated plastic substrate (*bottom*).

292   PONGRATZ & ZÖLLER

was bombarded with argon ions at a partial pressure of $\sim 2 \times 10^{-4}$ mbar. The high UV transparency indicates that there is no absorption caused by impurities from the plasma source. Figure 6 (*top*) shows the SEM picture of a conventional evaporated SiO_2 film deposited onto a glass substrate with additional substrate heating. In comparison, Figure 6 (*bottom*) shows an SEM micrograph of an SiO_2 film on a CR 39 substrate produced with plasma-IAD without substrate heating. The surface of the film that was

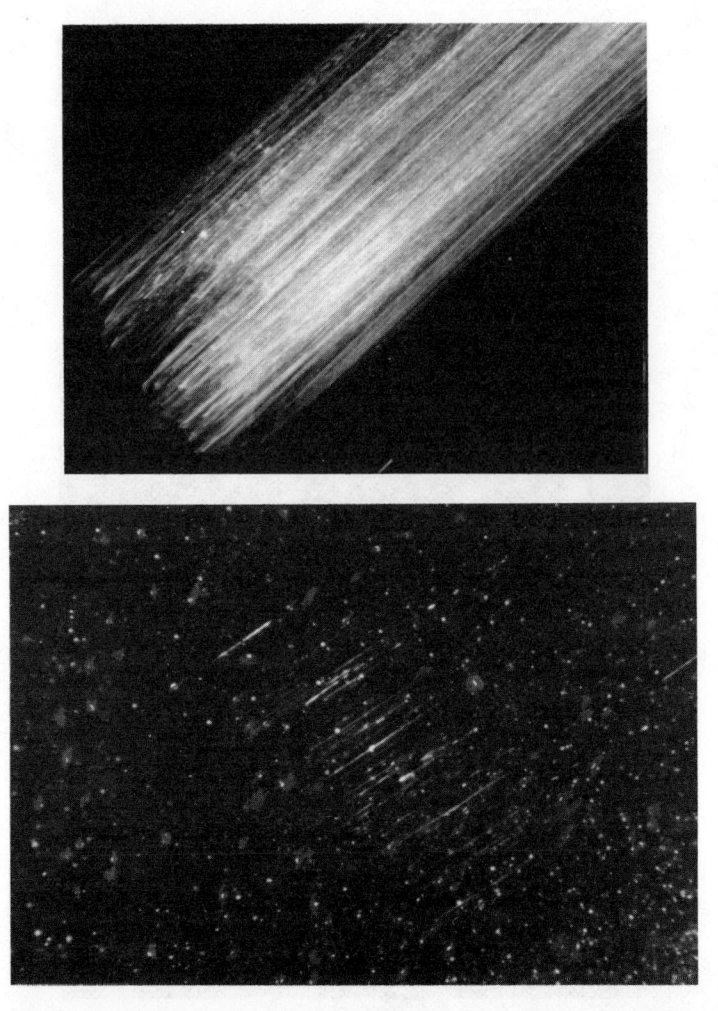

Figure 7 Tested areas of CR 39 lenses after abrasion test (500 g load, 20 strokes) looked at with a stereo microscope with dark field illumination: uncoated (*top*), coated with plasma-IAD (*bottom*).

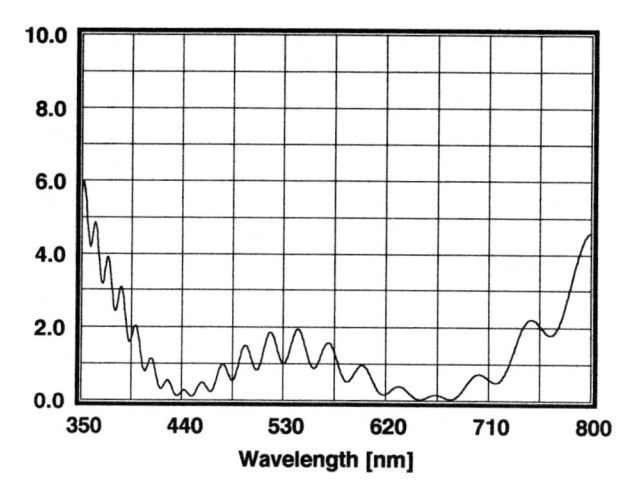

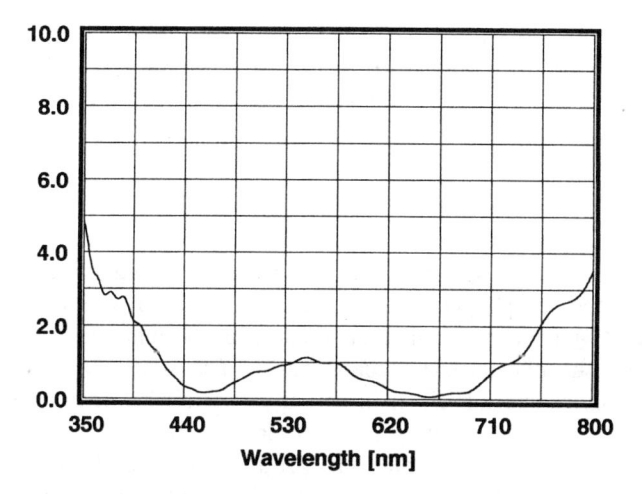

Figure 8 Reflection curves of one side of a CR 39 lens with a wear-resistant layer in combination with a broad band AR system: with oscillations caused by the wear-resistance layer (*top*); with attenuated oscillations (*bottom*).

deposited with plasma-IAD appears smoother than the conventional evaporated layer.

Synthetic substrate materials are gaining more and more importance in the ophthalmic industry. Wear-resistant coatings of these materials and some associated test methods have been described by various workers (16–19).

CR 39, an allyldiglycol carbonate, is the most commonly used material for ophthalmic plastic lenses. With plasma-IAD we developed a wear-resistant and broadband antireflective coating system on CR 39 lenses. The coating was tested with a salt water boiling test. The coated substrate was kept in a boiling salt solution (5% NaCl) for 2 min and then cooled-down in room temperature water. The coatings withstand 20 cycles of this test without any visible damage. Figure 7 shows an uncoated and a coated lens after an abrasion test. An abrader with a 500 g-load rubbed the surface of each sample for 20 strokes. In Figure 7 (*top*) the 5 mm-wide stress mark of the eraser in the center of the convex side of an uncoated CR 39 lens is shown. The photograph was made with the aid of a Wild Leitz stereo microscope type M 8 with dark field illumination (illumination time 8 min). Figure 7 (*bottom*) shows the comparable area of a coated lens. Only a few smooth scratches were visible after 20 min of illumination. It is obvious that the abrasion-resistance of the coated surface is improved substantially. Figure 8 (*top*) shows the reflection curve of this coating system. The superimposed oscillations are caused by the optical effect of a thick layer between the substrate and the antireflection coating system. With the aid of an additional layer, the oscillations were attenuated satisfactorily (Figure 8, *bottom*).

CONCLUSION

It has been shown that the plasma-IAD process has a great potential for various applications. The results on single-layer coatings of several oxides show that the film properties are fundamentally improved in comparison to conventional evaporation even without additional substrate heating. Therefore, the plasma-IAD technology is also useful for high quality coatings on organic substrates like wear-protective layers in combination with optically active multilayers. Furthermore, there is no limitation in the choice of evaporation materials produced by the plasma source. In conclusion, plasma-IAD is a promising technique that combines the economic advantages of conventional evaporation with the improved film properties of IAD.

Literature Cited

1. Coleman, W. J. 1974. *Appl. Opt.* 13: 946–50
2. Bulatov, N. N., Motovilow, O. A. 1975. *Sov. J. Opt. Technol.*, pp. 338–39
3. Kienel, G. 1981. *Thin Solid Films* 77: 213–24
4. Pawlewicz, W. T., Martin, P. M., Hays, D. D., Mann, I. B. 1982. *Proc. SPIE* 325: 105–16
5. Barany, I., Kienel, G. 1979. *Vak.-Technik.* 28: *Jahrgang/Heft* 6: 168–72

6. Mattox, D. M. 1964. *Electro-Chem. Technol.* 2: 295
7. Pulker, H. K. 1988. *Proc. SPIE* 1019: 138–47
8. Martin, P. J. 1986. *J. Mater. Sci.* 21: 260–84
9. Martin, P. J., MacLeod, H. A., Netterfield, R. P., Pacey, C. G., Sainty, W. G. 1983. *Appl. Opt.* 22: 178–84
10. Targove, J. D., Lehan, J. P., Linda, J. L., MacLeod, H. A., Leavitt, J. A., McIntyre, L. C. 1987. *Appl. Opt.* 26: 3733–37
11. Flory, F., Amra, C., Commandré, M., Pelletier, E., Albrand, G. 1989. *Proc. Int. Symp. Shanghai, China*, pp. 137–40
12. Kaufmann, H. R., Cuomo, J. J., Harper, J. M. E. 1982. *Vac. Sci. Technol.* 21: 725–36
13. Kaufmann, H. R. 1984. Fundamentals of ion source operation. *Commonwealth Sci. Corp. Virginia*, p. i
14. Matl, K., Klug, W., Zöller, A. 1991. *Mater. Sci. Eng. A* 140: 523–27
15. Goebel, D. M., Crow, J. T., Forrester, A. T. 1978. *Am. Inst. Phys. Rev. Sci. Instrum.* 49: 4
16. Becker, K. J. 1986. *Dtsch. Opt.* Issue 2
17. Masso, J. D. 1989. *SVC Proc. 32nd Ann. Meet. St. Louis*, pp. 237–40
18. Klug, W., Schneider, R., Zöller, A. 1990. *Proc. SPIE San Diego* 1323: 88–97
19. Obstfeld, H., Needham, C. M., Constantinides, M. 1991. *Opt. World* Issue 8

SUBJECT INDEX

A

Acid gases
  removal from natural gases, 47–48
Acrylamide gels
  swelling of
    experimental measurements of, 262
Affine network theory, 243, 246
Alkali halides
  return to equilibrium of kinetics of, 2–3
  superconductivity of, 2
Alumina
  microwave sintering of, 153, 163–64
Aluminosilicate glass-ceramics, 94–102
Amalgamation, 44
Arc evaporation, 282
Atomic composition/density
  X-ray attenuation coefficient and, 122

B

Back-scattered scanning electron microscopy, 145–46
Barium titanites
  microwave sintering of, 153
Barrier packaging
  free volume/free volume fraction and, 57
Beall, G. H., 91–118
Bismuth films
  superconductivity of, 2
Bisphenol-A polycarbonate
  mobility selectivity of, 56
Bonding
  diffusion, 23
  instantaneous liquid phase, 43
  liquid interface diffusion, 41
  Rohr, 41
  transient liquid insert metal diffusion, 42
  transient liquid phase, 23–45
    applications of, 41–45
    diffusion profiles for, 40
    grain size effect in, 40
    interface kinetics in, 38–40
    solidification front in, 40–41
    stages of, 24, 25, 27–38
    theory of, 26–38
Boron
  diffusion into silicon, 5

Boron carbide
  microwave sintering of, 165

C

Cahn-Hilliard equation, 211
Calcium aluminosilicate glass-ceramics, 101
Calcium-magnesium metasilicates, 92
Calcium metasilicates, 92
Calcium silicates, 93–94
Carbon dioxide
  permeability for 50/50 carbon dioxide-methane feed, 79
  recovery from natural gases, 47–48
Carnot cycle, 12
Cathodoluminescence method
  semiconductor imperfections and, 8
Ceramic joining
  microwave technology and, 166–67
Ceramic matrix composites
  X-ray tomographic microscopy and, 133–38
Ceramics
  microwave sintering of, 153–69
  See also Glass-ceramics
Ceramic sintering, 163–66
Chain-fluosilicates
  glass-ceramics based on, 103–8
Chemical vapor deposition, 279
  plasma enhanced, 280
Chemical vapor infiltration
  X-ray tomographic microscopy and, 134
Coherent solids
  Ostwald ripening in, 209–13
Coleman, M. R., 47–81
Computed tomography, 123–28
  instrumentation for, 128–32
Computer simulation
  Ostwald ripening and, 205
Conduction
  electromagnetic energy dissipation and, 156–57
Conductors
  superionic
    structural disorder in, 233–35
Continuity equation
  Ostwald ripening and, 201

Copper-silver system
  transient liquid phase bonding in, 25
Cordierite
  glass-ceramics based on, 99–100
Crystals
  disorder in
    reverse Monte Carlo modeling and, 232–36

D

Debye-Hückel theory, 256
Dental caries
  X-ray tomographic microscopy and, 145–47
Dentin
  X-ray tomographic microscopy and, 145–47
Dielectric constant
  material-microwave interactions and, 154–55
Dielectric heating
  material-microwave interactions and, 157–58
Diffusion bonding, 23
Diffusion solidification, 44
Diodes
  laser, 16
Diode sputtering, 280
Disilicates, 92
Dissolution
  transient liquid phase bonding and, 29–33
Drug delivery
  gels and, 271–73

E

Eagar, T. W., 23–45
Electromagnetic energy
  dissipation in crystalline dielectrics, 156–57
Electron-beam induced current collection
  semiconductor imperfections and, 8
Electronic polarization
  electromagnetic energy dissipation and, 156–57
Electron microscopy
  semiconductor imperfections and, 8
  Z-contrast, 171–94
    electronic/photonic materials and, 178–82

296

SUBJECT INDEX 297

imaging process in, 172–77
superconducting films/
superlattices and, 182–93
Electron-optics excitation
semiconductor imperfections and, 8
Enstatite
glass-ceramics based on, 94
composition and properties of, 95
Epitaxy
liquid-phase, 16
Ergodicity
gels and, 249
Eutectic systems
binary
transient liquid phase bonding in, 25–26
Extended X-ray absorption fine structure
structure in disordered materials and, 231–32

F

Fast ion conductors
structural disorder in, 233–35
Ferrites
microwave sintering of, 153
Flory-Huggins theory
gels and, 244, 251–54
Fluorcanasite
glass-ceramics based on, 106–8
Fluormica sheet silicates
glass-ceramics based on, 102–3
Fluosilicate glass-ceramics, 102–8

G

Gases
supercritical
hydrogen separation from, 47
Gas permeability
controlled, 47–81
Gas separation membranes, 47–81
Gels
applications of, 269–73
collective diffusion constants and, 267–68
critical behavior of, 262–69
critical isobars of, 264–66
defects in, 248–50
diffusion coefficients/relaxation times of, 263
Flory-Huggins theory and, 244, 251–54

hydrogen-bonding interactions and, 257–58
hydrophobic interactions and, 257
ionic interactions and, 254–57
kinetics of, 258–62
phase transitions of, 243–73
volume, 250-58
specific heat and, 266–67
Van der Waals interactions and, 257–58
Germanium
p-n junctions in, 11–12
Glass-ceramics, 91–118
acicular interlocking in, 114–15
aluminosilicate, 94–102
cellular membrane in, 109–11
coast-and-island structure in, 111–13
composition of, 92–108
dendritic crystallization in, 108–9
fluosilicate, 102–8
house-of-cards structure in, 113–14
lamellar twinning in, 116
microstructure of, 108–16
microwave sintering of, 153
relict structure in, 111
silicate, 92–94
ultra fine-grained, 109
Glasses
bond angle distribution in, 230
devitrification of, 91
glass-ceramic-forming temperature-viscosity curves for, 99
microwave technology and, 168
reverse Monte Carlo modeling and, 226–32
Grain size
transient liquid phase bonding and, 40
Granulation, 26

H

Hexafluoro-polycarbonate
permselectivity of, 61
Homogenization
transient liquid phase bonding and, 38
Howe, H. A., 217–41
Hydrocarbon vapors
separation from air, 48–49
Hydrogen
separation from supercritical gases, 47

Hydrogen-bonding interactions
gels and, 257–58
Hydrogen sulfide
removal from natural gases, 47
Hydrophobic interactions
gels and, 257

I

Instantaneous liquid phase bonding, 43
Ion beam-assisted evaporative deposition, 281
Ion cluster beam deposition, 281–82
Ionic interactions
gels and, 254–57
Ionic vibration
electromagnetic energy dissipation and, 156–57
Ion jump relaxation
electromagnetic energy dissipation and, 156–57
Ion plating, 281
Isothermal solidification
transient liquid phase bonding and, 34–38

K

Katz, J. D., 153–69
Kinetic equation
Ostwald ripening and, 199–200
Kinney, J. H., 121–50
Koros, W. J., 47–81

L

Lamellar twinning
in glass-ceramics, 116
Laser diodes, 16
Li, Y., 243–73
Linear photodiode arrays
computed tomography and, 128–29
Liquid interface diffusion bonding, 41
Liquid-phase epitaxy, 16
Liquids
reverse Monte Carlo modeling and, 223–26
Lithium silicates, 93
Luminescence
impurity detection and, 8
transistor dislocations and, 6

M

MacDonald, W. D., 23–45
Magic angle spinning nuclear magnetic resonance

SUBJECT INDEX

bond angle distribution in glasses and, 230
Magnesium metasilicates, 92
Magnesium silicates, 94
Magnetic disorder
 reverse Monte Carlo modeling and, 236–39
Magnetron sputtering, 280
Mass conservation equation
 Ostwald ripening and, 201
McGreevy, R. L., 217–41
Melting point depressant
 transient liquid phase bonding and, 23–25
Metal matrix composite failure
 X-ray tomographic microscopy and, 138–44
Metasilicates, 92
Metropolis Monte Carlo method, 218–19
Microwave cavities, 158
Microwave-material interactions, 154–58
Microwave radiation
 exposure standard for, 154
Microwave sintering, 153–69
 advantages of, 158–63
Microwave technology
 ceramic joining and, 166–67
 glasses and, 168
 powder synthesis and, 167
 sol-gels and, 167–68
Mineral tissues
 imaging of
  X-ray tomographic microscopy and, 144–47
Misfit dislocation multiplication, 11
Monochromator
 computed tomography and, 129
Monte Carlo methods, 218–19
Monte Carlo modeling
 reverse, 217–41
  advantages of, 222
  disorder in crystals and, 232–36
  glasses and, 226–32
  liquids and, 223–26
  magnetic disorder and, 236–39
  methods of, 218–23
Mullite glass-ceramics, 100-1

N

Natural gases
 carbon dioxide recovery from, 47–48
Nernst-Brunner theory, 30-31
Nichols, M. C., 121–50
Nickel-boron system
 transient liquid phase bonding in, 37–38
Nickel superalloys
 transient liquid phase bonding and, 42
Nitrogen
 separation from air, 47
 solubility selectivities in liquids, 54
Nonzero volume fraction systems
 Ostwald ripening and, 203–9
Nuclear magnetic resonance
 magic angle spinning
  bond angle distribution in glasses and, 230

O

Optical coatings
 tantalum pentoxide and, 288–90
 titanium dioxide and, 286–88
Optical fibers
 quartz
  microwave heating and, 168
Ostwald ripening, 197–214
 in coherent solids, 209–13
 continuity equation and, 201
 infinitely dilute limit in, 201–3
 kinetic equation and, 199–200
 mass conservation equation and, 201
 nonzero volume fraction systems and, 203–9
Oxides
 microwave sintering of, 163–64
Oxygen
 separation from air, 47
 solubility selectivities in liquids, 54

P

Passivated and emitter rear cell, 15
Penetrants
 diffusive jump through transient gap opening, 59
 zeolite sieving diameters of, 56
Pennycook, S. J., 171–94
Phantom network theory, 243
Photoconductivity
 in semiconductors, 18–20
Photodiode arrays
 computed tomography and, 128–29
Photon statistics
 microscopy X-ray imaging and, 123–24
Photovoltaic solar cells
 p-n junctions in, 12–15
Physical vapor deposition, 279
Pinhole scanning
 computed tomography and, 128
Plasma enhanced chemical vapor deposition, 280
Plasma-ion assisted deposition, 279–94
Plasticization, 48, 77–78
Poly(acrylamide) gels
 critical density fluctuations of, 264
Polycarbonates
 diffusivities/diffusivity selectivities of, 86
 fractional free volumes for, 62
 oxygen/nitrogen tradeoff for, 60
 permeabilities/permselectivities of, 82
 polymer structures of, 50-51
 solubilities/solubility selectivities of, 84
Polycarbones
 gas transport properties of, 69–71
Polyesters
 diffusivities/diffusivity selectivities of, 86
 permeabilities/permselectivities of, 82
 solubilities/solubility selectivities of, 84
Polyetherketones
 diffusivities/diffusivity selectivities of, 86
 permeabilities/permselectivities of, 82
 solubilities/solubility selectivities of, 84
Polyimides
 diffusivities/diffusivity selectivities of, 87
 permeabilities/permselectivities of, 83
 solubilities/solubility selectivities of, 85
Polymer membranes
 controlled permeability, 47–81
 gas transport through, 53
 solubility selectivities in, 54–56
 structure-permselectivity of, 57–81
Polymers
 diffusivities/diffusivity selectivities of, 86–87

SUBJECT INDEX 299

networks in solvents, 245–48
permeabilities/permselectivities of, 82–83
plasticization of, 78
solubilities/solubility selectivities of, 84–85
Polyphenyleneoxides
  diffusivities/diffusivity selectivities of, 87
  permeabilities/permselectivities of, 64–65, 83
  solubilities/solubility selectivities of, 85
Polypyrrolones
  diffusivities/diffusivity selectivities of, 86
  permeabilities/permselectivities of, 74–75, 82
  solubilities/solubility selectivities of, 84
Polystyrenes
  diffusivities/diffusivity selectivities of, 86
  permeabilities/permselectivities of, 82
  solubilities/solubility selectivities of, 84
Polysulfones
  diffusivities/diffusivity selectivities of, 86
  permeabilities/permselectivities of, 66–67, 82
  polymer structures of, 51–52
  solubilities/solubility selectivities of, 84
Poly(trimethyl-silyl-propyne) solubility selectivity of, 57
Pongratz, S., 279–94
Potassium fluorrichterite
  glass-ceramics based on, 103–4
Powder synthesis
  microwave technology and, 167

Q

Quartz solid solutions
  glass-ceramics based on, 95–97
  composition of, 96
Queisser, H. J., 1–21

R

Radiation
  microwave
    exposure standard for, 154
  synchrotron
    X-ray tomographic microscopy and, 121–50

Reflection high-energy electron diffraction, 184
Reverse Monte Carlo algorithm, 219–20
Reverse Monte Carlo modeling, 217–41
  advantages of, 222
  disorder in crystals and, 232–36
  glasses and, 226–32
  liquids and, 223–26
  magnetic disorder and, 236–39
  methods of, 218–23
Rohr bonding, 41
Rubber elasticity
  entropic nature of, 243

S

Scanning electron microscopy, 121
Schottky barriers, 15
Scintillator screen
  computed tomography and, 130
Semiconductors
  imperfections in
    scanning for, 8–11
  light-emitting
    liquid-phase epitaxy and, 16
  persistent photoconductivity in, 18–20
  solid-liquid inter-diffusion process and, 43–44
  transient liquid phase bonding and, 43–44
  Z-contrast electron microscopy and, 178–82
Silicate glass-ceramics, 92–94
Silicon
  boron diffusion into, 5
  p-n junctions in, 11–15
Silicon carbide
  microwave sintering of, 165
Silicon carbide matrix-Nicalon fiber composites
  X-ray tomographic microscopy and, 134
Silicon crystals
  photoluminescence spectrum of, 7
  small-angle grain boundaries in, 4
Silicon nitride
  microwave sintering of, 165–66
Small angle neutron scattering, 246
Solar cells
  p-n junctions in, 12–15

Sol-gels
  microwave technology and, 167–68
Solid-liquid inter-diffusion process, 43–44
Solid-liquid mixtures
  microstructures of, 198
Solids
  coherent
    Ostwald ripening in, 209–13
Solvents
  polymer networks in, 245–48
  purification of
    gels and, 270-71
Specific heat
  gels and, 266–67
Spodumene solid solutions
  glass-ceramics based on, 97–99
  composition of, 98
  thermal expansion of, 97
Sputtering, 280
Superalloys
  transient liquid phase bonding and, 42
Superconducting films
  Z-contrast electron microscopy and, 182–93
Superconductivity
  isotope effect of, 2
Supercritical gases
  hydrogen separation from, 47
Superionic conductors
  structural disorder in, 233–34
Superlattices
  Z-contrast electron microscopy and, 182–93
Surface films/layers
  deposition of, 279–94
    advanced plasma source and, 284–85
    arc evaporation and, 282
    ion beam-assisted evaporative, 281
    ion cluster beam, 281–82
    ion plating and, 281
    plasma-ion assisted, 282–84
Synchrotron radiation
  X-ray tomographic microscopy and, 121–50

T

Tanaka, T., 243–73
Tantalum pentoxide
  optical coatings and, 288–90
Tetramethyl-substituted polycarbonate
  permselectivity of, 61
Thick film diffusion equation

SUBJECT INDEX

transient liquid phase bonding and, 27
Thin film diffusion equation
  transient liquid phase bonding and, 26–27
Thin films
  vacuum deposition of, 279
Titanites
  microwave sintering of, 164
Titanium
  transient liquid phase bonding and, 41–42
Titanium diboride
  microwave sintering of, 166
Titanium dioxide
  microwave sintering of, 163
  optical coatings and, 286–88
Titanium-nickel systems
  transient liquid phase bonding in, 40
Tomographic microscopy, 121–50
  X-ray
    applications of, 132–47
    ceramic matrix composites and, 133–38
    hard mineral tissue imaging and, 144–47
    metal matrix composite failure and, 138–44
Tomography
  computed, 123–28
  instrumentation for, 128–32
Transient liquid insert metal diffusion bonding, 42
Transient liquid phase bonding, 23–45
applications of, 41–45
diffusion profiles for, 40
grain size effect in, 40
interface kinetics in, 38–40
solidification front in, 40-41
stages of, 24, 25, 27–38
theory of, 26–38
Transient liquid phase sintering, 44
Transistors
  dislocations in, 4–7
  p-n junctions in, 4
Transmission electron microscopy, 121
  semiconductor imperfections and, 8–11

U

Udimet 700
  diffusion welding of, 42
Uranium oxide
  microwave sintering of, 153

V

Van der Waals interactions
  gels and, 257–58
Vidicon array
  computed tomography and, 129
Voorhees, P. W., 197–214

W

Walker, D. R. B., 47–81
Widening
transient liquid phase bonding and, 34

X

X-ray attenuation coefficient
  atomic composition and density and, 122
X-ray diffraction
  network glass structures and, 228–31
X-ray tomographic microscopy, 121–50
  applications of, 132–47
  ceramic matrix composites and, 133–38
  hard mineral tissue imaging and, 144–47
  metal matrix composite failure and, 138–44

Z

Z-contrast electron microscopy, 171–94
  electronic/photonic materials and, 178–82
  imaging process in, 172–77
  superconducting films/superlattices and, 182–93
Zircaloy 2
  joining to 304 stainless steel, 42–43
Zirconia
  magnesium silicates and, 94
Zöller, A., 279–94

CUMULATIVE INDEXES

CONTRIBUTING AUTHORS, VOLUMES 18–22

A

Agrawal, D., 19:59–81
Ahn, B. T., 21:335–72
Alexopoulos, P., 20:391–420
Angus, J.C., 21:221–48
Assink, R. A., 21:491–513
Aucouturier, M., 18:219–56
Averbuch-Pouchot, M. T., 21:65–92

B

Beall, G. H., 22:91–119
Becher, P., 20:179–95
Beyers, R., 21:335–72
Bosman, G., 18:381–421
Brennan, S., 20:365–90
Brown, H. R., 21:463–89
Brown, S. B., 21:409–35
Brus, L. E., 19:471–95
Bube, R. H., 20:19–50

C

Chattopadhyay, K., 21:437–62
Chevallier, J., 18:219–56
Chu, D. S. N. G., 20:339–63
Coleman, M. R., 22:47–89
Collongues, R., 20:51–82

D

Dimiduk, D. M., 19:231–63
Dobbyn, R. C., 19:183–207
Dodson, B., 19:419–37
Drickamer, H. G., 20:1–17
Dudney, N. J., 19:102–20
Durif, A., 21:65–92
Dutta, D., 21:159–84

E

Eagar, T. W., 22:23–46
Ettmayer, P., 19:145–64

F

Fischer, T. E., 18:303–23
Flanagan, T. B., 21:269–304
Fleischer, R. L., 19:231–63
Friedel, J., 18:1–24
Fuoss, P., 20:365–90

G

Gösele, U. M., 18:257–82
Gourier, D., 20:51–82
Green, R. E. Jr., 20:197–217
Griffith, J. E., 20:219–44
Gunshor, R. L., 18:325–50

H

Häger, J., 19:256–93
Hench, L. L., 18:381–421; 20:269–98
Herlach, D., 21:23–44
Hirano, K., 18:351–80
Hirvonen, J. K., 19:401–17
Holt, J. B., 21:305–34
Hono, K., 18:351–80
Howe, M. A., 22:217–42

I

Ikeya, P. M., 21:45–63
Iyer, R. N., 20:299–338

K

Kahn-Harari, A., 20:51–82
Katz, J., 22:153–70
Kausch, H. H., 19:341–77
Kay, B. D., 21:491–513
Kelto, C., 19:527–50
Kingery, W. D., 19:1–20
Kinney, J. H., 22:121–52
Klein, M. B., 18:165–88
Klinowski, J., 18:189–218
Koch, C. C., 19:121–43
Kohl, P., 19:121–43
Koizumi, M., 18:47–73
Kolodziejski, L. A., 18:325–50
Koros, W. J., 22:47–89
Kuriyama, M., 19:183–207

L

Lantman, C., 19:295–317
Larché, F. C., 20:83–99
Lejus, M., 20:51–82
Li, Y., 22:243–77
Lundberg, R., 19:295–317

M

MacDonald, W. D., 22:23–46
MacKnight, W., 19:295–317
Macrander, A. T., 18:283–302
Martin, J. W., 18:101–19
McGreevy, R. L., 22:217–42
McKinsty, H., 19:59–81
Michel, C., 19:319–39
Miller, A. K., 19:439–69
Miyamoto, Y., 18:47–73
Mullen, R. A., 18:165–88

N

Nakahara, S., 21:93–129
Nemanich, R. J., 21:535–58
Nichols, M. C., 22:121–52
Nieh, T. G., 20:117–40
Niyama, E., 20:101–15
Nurmikko, A. V., 18:325–50

O

O'Sullivan, T., 20:391–420
Oates, W. A., 21:269–304
Ogarevic, V., 20:141–77
Ohnishi, H., 19:83–101
Okinaka, Y., 21:93–129
Orbach, R. L., 19:497–525
Ostermayer, F. W. Jr., 19:379–99
Otsuka, K., 18:25–45

P

Panish, M. B., 19:209–29
Panzer, S., 18:121–40
Parker, J. M., 19:21–41
Pawlikowski, G. T., 21:159–84
Pennycook, S., 22:171–95
Pickering, H. W., 20:299–338
Pongratz, S., 22:279–95
Pyzik, A. J., 19:527–50

Q

Queisser, H., 22:1–22

301

R

Ranganathan, S., 21:437–62
Raveau, B., 19:319–39
Rigney, D. A., 18:141–63
Roy, R., 19:59–81
Russell, T.P., 21:249–68
Rytz, D., 18:165–88

S

Schiller, S., 18:121–40
Shen, Y., 21:515–34
Shimizu, K., 18:25–45
Siegel, R. W., 21:559–78
Spiess, H. W., 21:131–58
Steigerwald, M., 19:471–95
Steiner, B., 19:183–207
Stephens, R. I., 20:141–77
Suganuma, K., 18:47–73
Sunkara, M., 21:221–48
Swalen, J. D., 21:373–408

T

Tadaki, T., 18:25–45
Tanaka, T., 22:243–77
Temkin, H., 19:209–29
Théry, J., 20:51–82
Thompson, C. B., 20:245–68
Timm, E. E., 19:527–50
Tirrell, M., 19:341–77
Tsao, J., 19:419–37

U

Usui, A., 21:185–219

V

Valley, G. C., 18:165–88
Van Vliet, C. M., 18:381–421
Vasconcelos, W., 20:268–98
Vivien, D., 20:51–82
Vogel, V., 21:515–34
von der Linde, D., 18:75–100
Voorhees, P. W., 22:197–215

W

Wadsworth, J., 20:117–40
Walker, R. B., 22:47–89
Walther, H., 19:265–93
Wang, Y., 21:221–48
Watanabe, H., 21:185–219
Wechsler, B., 18:165–88
Weil, R., 19:165–82
Weiss, R. A., 21:159–84
Winer, K., 21:1–21

Y

Yoon, D. N., 19:43–58

Z

Zöller, A., 22:279–95

CHAPTER TITLES, VOLUMES 18-22

PREFATORY CHAPTERS

Dislocations and Disclinations Past and Present: Some Personal Views and Reminiscences	J. Friedel	18:1-24
Ceramic Materials Science in Society	W. D. Kingery	19:1-20
Forty Years of Pressure Tuning Spectroscopy	H. G. Drickamer	20:1-17
Perfecting the Solid State	H. Queisser	22:1-22

EXPERIMENTAL AND THEORETICAL METHODS

Laser Spectroscopy of Ultrafast Solid-State Phenomena	D. von der Linde	18:75-100
Recent Advances in Solid-State NMR of Zeolites	J. Klinowski	18:189-218
Precision X-Ray Techniques for Semiconductors	A. T. Macrander	18:283-302
Dynamical Diffraction Imaging (Topography) with X-Ray Synchrotron Radiation	M. Kuriyama, B. W. Steiner, and R. C. Steiner	19:183-207
Use of Laser Techniques to Study the Dynamics of Molecular-Surface Interaction	J. Hager, H. Walther	19:265-93
Towards Unified Computer Models for Predicting Fracture of Solids	A. K. Miller	19:439-69
Fractal Phenomena in Disordered Systems	R. L. Orbach	19:497-525
Coherent Phase Transformations	F. C. Larché	20:83-99
Computer-Aided Design for Metal Casting	E. Niyama	20:101-15
Non-Destructive Evaluation of Materials	R. E. Green, Jr.,	20:197-217
Scanning Tunneling Microscopy	J. E. Griffith, G. P. Kochanski	20:219-44
Grain Growth in Thin Films	C. V. Thompson	20:245-68
Transmission Electron Microscopy of InP-Based Compound Semiconductor Materials and Devices	S. N. G. Chu	20:339-63
Surface Sensitive X-Ray Scattering	P. Fuoss, S. Brennan	20:365-90
Electron Spin Resonance (ESR) Microscopy in Materials Science	M. Ikeya	21:45-63
NMR Methods for Solid Polymers	H. W. Spiess	21:131-58
Metastable Growth of Diamond-like Phases	J. C. Angus, Y. Wang, M. Sunkara	21:221-48
The Characterization of Polymer Interfaces	T. P. Russell	21:249-68
The Adhesion Between Polymers	H. R. Brown	21:463-89
Study of Sol-Gel Chemical Reaction Kinetics by NMR	R. A. Assink, B. D. Kay	21:491-513
Air/Liquid Interfaces and Adsorbed Molecular Monolayers Studied with Nonlinear Optical Techniques	V. Vogel, Y. R. Shen	21:515-34
X-Ray Tomographic Microscopy (XMT) Using Synchrotron Radiation	J. H. Kinney, M. C. Nichols	22:121-52
Z-Contrast Transmission Electron Microscopy: Direct Atomic Imaging of Materials	S. Pennycook	22:171-95
RMC: Modelling Disordered Structures	R. L. McGreevy, M. A. Howe	22:217-42

PREPARATION, PROCESSING, AND STRUCTURAL CHANGE

Joining of Ceramics and Metals	M. Koizumi, K. Suganuma, Y. Miyamoto, M. Koizumi	18:47-73

CHAPTER TITLES

Thermal Surface Modification By Electron Beam High-Speed Scanning	S. Schiller, S. Panzer	
Gas-Source Molecular Beam Epitaxy	M. B. Panish, H. Temkin	19:209–29
Photoelectrochemical Methods for III-V Compound Semiconductor Device Processing	P. Kohl, F. W. Ostermayer, Jr.	19:379–99
Ion Beam Processing for Surface Modification	J. K. Hirvonen	19:401–17
Rapid Omnidirectional Compaction (ROC) of Powders	C. Kelto, E. E. Timm, A. J. Pyzik	19:527–50
Superplastic Ceramics	T. G. Nieh, J. Wadsworth	20:117–40
Gel-Silica Science	L. Hench, W. Vasconcelos	20:269–98
Containerless Undercooling and Solidification of Pure Metals	D. M. Herlach	21:23–44
Microstructure and Mechanical Properties of Electroless Copper Deposits	S. Nakahara, Y. Okinaka	21:93–129
Atomic Layer Epitaxy of III-V Electronic Materials	A. Usui, H. Watanabe	21:185–219
Self-Heating Synthesis of Materials	J. B. Holt, S. D. Dunmead	21:305–34
Molecular Films	J. D. Swalen	21:373–408
Chemical Processes Applied to Reactive Extrusion of Polymers	S. B. Brown	21:409–35
Growth of and Characterization of Diamond Thin Films	R. J. Nemanich	21:535–58
Transient Liquid Phase Bonding	W. D. MacDonald, T. W. Eagar	22:23–46
Microwave Sintering of Ceramics	J. Katz	22:153–70
Phase Transitions of Gels	Y. Li, T. Tanaka	22:243–77
Plasma Ion-Assisted Evaporative Deposition of Surface Layers	S. Pongratz, A. Zöller	22:279–95

PROPERTIES AND PHENOMENA

Sliding Wear of Metals	D. A. Rigney	18:141–63
Fast Diffusion in Semiconductors	U. M. Gosele	18:257–82
Tribochemistry	T. E. Fischer	18:303–23
Chemically Induced Interface Migration in Solids	D. N. Yoon	19:43–58
Materials Synthesis by Mechanical Alloying	C. C. Koch	19:121–43
The Structure of Electrodeposits and the Properties that Depend on Them	R. Weil	19:165–82
Polymer Interdiffusion	H. H. Kausch, M. Tirrell	19:341–77
Fatigue of Magnesium Alloys: A Comprehensive Review	V. Ogarevic, R. I. Stephens	20:141–77
Mechanisms and Kinetics of Electrochemical Hydrogen Entry and Degradation of Metallic Systems	R. N. Iyer, H. W. Pickering	20:229–338
Mechanical Properties of Thin Films	P. Alexopoulos, T. O'Sullivan	20:391–420
Nanophase Materials	R. W. Siegel	21:559–78
Ostwald Ripening of Two-Phase Mixtures	P. W. Voorhees	22:197–215

SPECIAL MATERIALS

Shape Memory Alloys	T. Tadaki, K. Otsuka, K. Shimizu	18:25–45
Aluminum-Lithium Alloys	J. W. Martin	18:101–19
Photorefractive Materials	G. C. Valley, M. B. Klein, R. A. Mullen, D. Rytz, B. Wechsler	18:165–88
Semimagnetic and Magnetic Semiconductor Superlattices	R. L. Gunshor, L. A. Kolodziejski, A. V. Nurmikko, N. Otsuka	18:325–50
New Perspectives of Silicon Carbide: An Overview with Special Emphasis on Noise and Space-Charge-Limited Flow	C. M. Van Vliet, G. Bosman, L. L. Hench	18:381–421
Fluoride Glasses	J. M. Parker	19:21–41

Very Low Thermal Expansion Coefficient Materials	R. Roy, D. Agrawal, H. McKinsty	19:59–81
Electroluminescent Display Materials	H. Ohnishi	19:83–101
Composite Electrolytes	N. J. Dudney	19:103–20
Hard Metals and Cermets	P. Ettmayer	19:145–64
Intermetallic Compounds for Strong High-Temperature Materials	R. L. Fleischer, D. M. Dimiduk, H. A. Lipsitt	19:231–63
Structural Properties of Ionomers	C. W. Lantman, W. J. MacKnight, R. D. Lundberg	19:295–317
Crystal Chemistry and Properties of Mixed Valence Copper Oxides	B. Raveau, C. Michel	19:319–39
Synthesis and Electronic Structure of Quantum Semiconductor Nanoclusters	M. L. Steigerwald, L. E. Brus	19:471–95
Materials for Photovoltaic Applications	R. H. Bube	20:19–50
Magnetoplumbite-Related Oxides	R. Collongues, D. Gourier, A. Kahn-Harari, M. Lejus, J. Théry, D. Vivien	20:51–82
Recent Advances in Whisker-Reinforced Ceramics	P. Becher	20:179–95
Defects in Hydrogenated Amorphous Silicon	K. Winer	21:1–21
Crystal Chemistry of Oligophosphates	M. T. Averbuch-Pouchot, A. Durif	21:65–92
Molecular Composites and Self-Reinforced Liquid Crystalline Polymer Blends	G. T. Pawlikowski, D. Dutta, R. A. Weiss	21:159–84
The Palladium-Hydrogen System	T. B. Flanagan, W. A. Oates	21:269–304
Thermodynamic Considerations in Superconducting Oxides	R. Beyers, B. T. Ahn	21:335–72

STRUCTURE

Hydrogen in Crystalline Semiconductors	J. Chevallier, M. Aucouturier	18:219–56
Field Ion Studies of Structure of GP Zones	K. Hirano, K. Hono	18:351–80
Quasicrystals	S. Ranganathan, K. Chattopadhyay	21:437–62
Controlled Permeability Polymer Membranes	W. J. Koros, M. R. Coleman, R. B. Walker	22:47–89
Design and Properties of Glass-Ceramics	G. H. Beall	22:91–119

ANNUAL REVIEWS INC.

a nonprofit scientific publisher
4139 El Camino Way
P. O. Box 10139
Palo Alto, CA 94303-0897 • USA

ORDER FORM

ORDER TOLL FREE
1-800-523-8635
(except California)

FAX: 415-855-9815

Annual Reviews Inc. publications may be ordered directly from our office; through booksellers and subscription agents, worldwide; and through participating professional societies.
Prices are subject to change without notice.   ARI Federal I.D. #94-1156476

- **Individuals:** Prepayment required on new accounts by check or money order (in U.S. dollars, check drawn on U.S. bank) or charge to MasterCard, VISA, or American Express.
- **Institutional Buyers:** Please include purchase order.
- **Students: $10.00 discount** from retail price, per volume. Prepayment required. Proof of student status must be provided. (Photocopy of Student I.D. is acceptable.) Student must be a degree candidate at an accredited institution. Order direct from Annual Reviews. Orders received through bookstores and institutions requesting student rates will be returned.
- **Professional Society Members:** Societies who have a contractual arrangement with Annual Reviews offer our books at reduced rates to members. Contact your society for information.
- **California orders** must add applicable sales tax.
- **CANADIAN ORDERS:** We must now collect 7% General Sales Tax on orders shipped to Canada. Canadian orders will not be accepted unless this tax has been added. Tax Registration # R 121 449-029. Note: Effective 1-1-92 Canadian prices increase from USA level to "other countries" level. See below.
- **Telephone orders,** paid by credit card, welcomed. Call Toll Free **1-800-523-8635** (except in California). California customers use 1-415-493-4400 (not toll free). M-F, 8:00 am - 4:00 pm, Pacific Time. Students ordering by telephone must supply (by FAX or mail) proof of student status if proof from current academic year is not on file at Annual Reviews. Purchase orders from universities require written confirmation before shipment.
- **FAX: 415-855-9815   Telex: 910-290-0275**
- **Postage paid by Annual Reviews** (4th class bookrate). UPS domestic ground service (except to AK and HI) available at $2.00 extra per book. UPS air service or Airmail also available at cost. UPS requires street address. P.O. Box, APO, FPO, not acceptable.
- **Regular Orders:** Please list below the volumes you wish to order by volume number.
- **Standing Orders:** New volume in the series will be sent to you automatically each year upon publication. Cancellation may be made at any time. Please indicate volume number to begin standing order.
- **Prepublication Orders:** Volumes not yet published will be shipped in month and year indicated.
- **We do not ship on approval.**

ANNUAL REVIEWS SERIES *Volumes not listed are no longer in print*	Prices, postpaid, per volume		Regular Order Please send Volume(s):	Standing Order Begin with Volume:
	Until 12-31-91 USA & Canada / elsewhere	After 1-1-92 USA / other countries (incl. Canada)		

Annual Review of ANTHROPOLOGY
Vols. 1-16	(1972-1987)..................	$33.00/$38.00		
Vols. 17-18	(1988-1989)..................	$37.00/$42.00 ⎱ $41.00/$46.00		
Vols. 19-20	(1990-1991)..................	$41.00/$46.00 ⎰		
Vol. 21	(avail. Oct. 1992).............	$44.00/$49.00	$44.00/$49.00	Vol(s)._____ Vol._____

Annual Review of ASTRONOMY AND ASTROPHYSICS
Vols. 1, 5-14, 16-20	(1963, 1967-1976) (1978-1982)...............	$33.00/$38.00		
Vols. 21-27	(1983-1989)..................	$49.00/$54.00 ⎱ $53.00/$58.00		
Vols. 28-29	(1990-1991)..................	$53.00/$58.00 ⎰		
Vol. 30	(avail. Sept. 1992).............	$57.00/$62.00	$57.00/$62.00	Vol(s)._____ Vol._____

Annual Review of BIOCHEMISTRY
Vols. 30-34, 36-56	(1961-1965, 1967-1987)	$35.00/$40.00		
Vols. 57-58	(1988-1989)..................	$37.00/$42.00 ⎱ $41.00/$47.00		
Vols. 59-60	(1990-1991)..................	$41.00/$47.00 ⎰		
Vol. 61	(avail. July 1992)	$46.00/$52.00	$46.00/$52.00	Vol(s)._____ Vol._____

ANNUAL REVIEWS SERIES

Volumes not listed are no longer in print

Prices, postpaid, per volume

	Until 12-31-91 USA & Canada / elsewhere	After 1-1-92 USA / other countries (incl. Canada)	Regular Order Please send Volume(s):	Standing Order Begin with Volume:

Annual Review of BIOPHYSICS AND BIOMOLECULAR STRUCTURE
Vols. 1-11	(1972-1982)	$33.00/$38.00			
Vols. 12-18	(1983-1989)	$51.00/$56.00	$55.00/$60.00		
Vols. 19-20	(1990-1991)	$55.00/$60.00			
Vol. 21	(avail. June 1992)	$59.00/$64.00	$59.00/$64.00	Vol(s).____	Vol.____

Annual Review of CELL BIOLOGY
Vols. 1-3	(1985-1987)	$33.00/$38.00			
Vols. 4-5	(1988-1989)	$37.00/$42.00	$41.00/$46.00		
Vols. 6-7	(1990-1991)	$41.00/$46.00			
Vol. 8	(avail. Nov. 1992)	$46.00/$51.00	$46.00/$51.00	Vol(s).____	Vol.____

Annual Review of COMPUTER SCIENCE
| Vols. 1-2 | (1986-1987) | $41.00/$46.00 | $41.00/$46.00 | | |
| Vols. 3-4 | (1988, 1989-1990) | $47.00/$52.00 | $47.00/$52.00 | Vol(s).____ | Vol.____ |

Series suspended until further notice. Volumes 1-4 are still available at the special promotional price of $100.00 USA /$115.00 other countries, when all 4 volumes are purchased at one time. Orders at the special price must be prepaid.

Annual Review of EARTH AND PLANETARY SCIENCES
Vols. 1-10	(1973-1982)	$33.00/$38.00			
Vols. 11-17	(1983-1989)	$51.00/$56.00	$55.00/$60.00		
Vols. 18-19	(1990-1991)	$55.00/$60.00			
Vol. 20	(avail. May 1992)	$59.00/$64.00	$59.00/$64.00	Vol(s).____	Vol.____

Annual Review of ECOLOGY AND SYSTEMATICS
Vols. 2-18	(1971-1987)	$33.00/$38.00			
Vols. 19-20	(1988-1989)	$36.00/$41.00	$40.00/$45.00		
Vols. 21-22	(1990-1991)	$40.00/$45.00			
Vol. 23	(avail. Nov. 1992)	$44.00/$49.00	$44.00/$49.00	Vol(s).____	Vol.____

Annual Review of ENERGY AND THE ENVIRONMENT
Vols. 1-7	(1976-1982)	$33.00/$38.00			
Vols. 8-14	(1983-1989)	$60.00/$65.00	$64.00/$69.00		
Vols. 15-16	(1990-1991)	$64.00/$69.00			
Vol. 17	(avail. Oct. 1992)	$68.00/$73.00	$68.00/$73.00	Vol(s).____	Vol.____

Annual Review of ENTOMOLOGY
Vols. 10-16, 18	(1965-1971, 1973)				
20-32	(1975-1987)	$33.00/$38.00			
Vols. 33-34	(1988-1989)	$36.00/$41.00	$40.00/$45.00		
Vols. 35-36	(1990-1991)	$40.00/$45.00			
Vol. 37	(avail. Jan. 1992)	$44.00/$49.00	$44.00/$49.00	Vol(s).____	Vol.____

Annual Review of FLUID MECHANICS
Vols. 2-4, 7	(1970-1972, 1975)				
9-19	(1977-1987)	$34.00/$39.00			
Vols. 20-21	(1988-1989)	$36.00/$41.00	$40.00/$45.00		
Vols. 22-23	(1990-1991)	$40.00/$45.00			
Vol. 24	(avail. Jan. 1992)	$44.00/$49.00	$44.00/$49.00	Vol(s).____	Vol.____

Annual Review of GENETICS
Vols. 1-12, 14-21	(1967-1978, 1980-1987)	$33.00/$38.00			
Vols. 22-23	(1988-1989)	$36.00/$41.00	$40.00/$45.00		
Vols. 24-25	(1990-1991)	$40.00/$45.00			
Vol. 26	(avail. Dec. 1992)	$44.00/$49.00	$44.00/$49.00	Vol(s).____	Vol.____

Annual Review of IMMUNOLOGY
Vols. 1-5	(1983-1987)	$33.00/$38.00			
Vols. 6-7	(1988-1989)	$36.00/$41.00	$41.00/$46.00		
Vol. 8	(1990)	$40.00/$45.00			
Vol. 9	(1991)	$41.00/$46.00	$41.00/$46.00		
Vol. 10	(avail. April 1992)	$45.00/$50.00	$45.00/$50.00	Vol(s).____	Vol.____

ANNUAL REVIEWS SERIES

Volumes not listed are no longer in print

Prices, postpaid, per volume

		Until 12-31-91 USA & Canada / elsewhere	After 1-1-92 USA / other countries (incl. Canada)	Regular Order Please send Volume(s):	Standing Order Begin with Volume:

Annual Review of MATERIALS SCIENCE
Vols. 1, 3-12	(1971, 1973-1982)	$33.00/$38.00 ⎫	$68.00/$73.00		
Vols. 13-19	(1983-1989)	$68.00/$73.00 ⎭			
Vols. 20-21	(1990-1991)	$72.00/$77.00	$72.00/$77.00		
Vol. 22	(avail. Aug. 1992)	$72.00/$77.00	$72.00/$77.00	Vol(s)._____	Vol._____

Annual Review of MEDICINE
Vols. 9, 11-15	(1958, 1960-1964)				
17-38	(1966-1987)	$33.00/$38.00 ⎫			
Vols. 39-40	(1988-1989)	$36.00/$41.00 ⎬ $40.00/$45.00			
Vols. 41-42	(1990-1991)	$40.00/$45.00 ⎭			
Vol. 43	(avail. April 1992)	$44.00/$49.00	$44.00/$49.00	Vol(s)._____	Vol._____

Annual Review of MICROBIOLOGY
Vols. 20-41	(1966-1987)	$33.00/$38.00 ⎫			
Vols. 42-43	(1988-1989)	$36.00/$41.00 ⎬ $41.00/$46.00			
Vol. 44	(1990)	$40.00/$45.00 ⎭			
Vol. 45	(1991)	$41.00/$46.00	$41.00/$46.00		
Vol. 46	(avail. Oct. 1992)	$45.00/$50.00	$45.00/$50.00	Vol(s)._____	Vol._____

Annual Review of NEUROSCIENCE
Vols. 1-10	(1978-1987)	$33.00/$38.00 ⎫			
Vols. 11-12	(1988-1989)	$36.00/$41.00 ⎬ $40.00/$45.00			
Vols. 13-14	(1990-1991)	$40.00/$45.00 ⎭			
Vol. 15	(avail. March 1992)	$44.00/$49.00	$44.00/$49.00	Vol(s)._____	Vol._____

Annual Review of NUCLEAR AND PARTICLE SCIENCE
Vols. 12-37	(1962-1987)	$36.00/$41.00 ⎫			
Vols. 38-39	(1988-1989)	$51.00/$56.00 ⎬ $55.00/$60.00			
Vols. 40-41	(1990-1991)	$55.00/$60.00 ⎭			
Vol. 42	(avail. Dec. 1992)	$59.00/$64.00	$59.00/$64.00	Vol(s)._____	Vol._____

Annual Review of NUTRITION
Vols. 1-7	(1981-1987)	$33.00/$38.00 ⎫			
Vols. 8-9	(1988-1989)	$39.00/$44.00 ⎬ $43.00/$48.00			
Vols. 10-11	(1990-1991)	$43.00/$48.00 ⎭			
Vol. 12	(avail. July 1992)	$45.00/$50.00	$45.00/$50.00	Vol(s)._____	Vol._____

Annual Review of PHARMACOLOGY AND TOXICOLOGY
Vols. 2-3, 5-27	(1962-1963, 1965-1987)	$33.00/$38.00 ⎫			
Vols. 28-29	(1988-1989)	$36.00/$41.00 ⎬ $40.00/$45.00			
Vols. 30-31	(1990-1991)	$40.00/$45.00 ⎭			
Vol. 32	(avail. April 1992)	$44.00/$49.00	$44.00/$49.00	Vol(s)._____	Vol._____

Annual Review of PHYSICAL CHEMISTRY
Vols. 11-21, 23-38	(1960-1970, 1972-1987)	$34.00/$39.00 ⎫			
Vols. 39-40	(1988-1989)	$40.00/$45.00 ⎬ $44.00/$49.00			
Vols. 41-42	(1990-1991)	$44.00/$49.00 ⎭			
Vol. 43	(avail. Nov. 1992)	$48.00/$53.00	$48.00/$53.00	Vol(s)._____	Vol._____

Annual Review of PHYSIOLOGY
Vols. 19-49	(1957-1987)	$34.00/$39.00 ⎫			
Vols. 50-51	(1988-1989)	$38.00/$43.00 ⎬ $42.00/$47.00			
Vols. 52-53	(1990-1991)	$42.00/$47.00 ⎭			
Vol. 54	(avail. March 1992)	$46.00/$51.00	$46.00/$51.00	Vol(s)._____	Vol._____

Annual Review of PHYTOPATHOLOGY
Vols. 3-25	(1965-1987)	$33.00/$38.00 ⎫			
Vols. 26-27	(1988-1989)	$38.00/$43.00 ⎬ $42.00/$47.00			
Vols. 28-29	(1990-1991)	$42.00/$47.00 ⎭			
Vol. 30	(avail. Sept. 1992)	$46.00/$51.00	$46.00/$51.00	Vol(s)._____	Vol._____

Annual Review of PLANT PHYSIOLOGY AND PLANT MOLECULAR BIOLOGY
Vols. 14-23, 26-29	(1963-1972, 1975-1978)				
31-38	(1980-1987)	$33.00/$38.00 ⎫			
Vols. 39-40	(1988-1989)	$36.00/$41.00 ⎬ $40.00/$45.00			
Vols. 41-42	(1990-1991)	$40.00/$45.00 ⎭			
Vol. 43	(avail. June 1992)	$44.00/$49.00	$44.00/$49.00	Vol(s)._____	Vol._____

ANNUAL REVIEWS SERIES
Volumes not listed are no longer in print

Prices, postpaid, per volume

	Until 12-31-91 USA & Canada / elsewhere	After 1-1-92 USA / other countries (incl. Canada)	Regular Order Please send Volume(s):	Standing Order Begin with Volume:

Annual Review of PSYCHOLOGY
- Vols. 4, 5, 8, 10, 13-24 (1953, 1954, 1957, 1959, 1962-1973)
- 26-30, 32-38 (1975-1979, 1981-1987) ... $33.00/$38.00
- Vols. 39-40 (1988-1989) .... $36.00/$41.00 } $40.00/$45.00
- Vols. 41-42 (1990-1991) .... $40.00/$45.00
- Vol. 43 (avail. Feb. 1992) .... $43.00/$48.00 $43.00/$48.00   Vol(s). _____   Vol. _____

Annual Review of PUBLIC HEALTH
- Vols. 1-8 (1980-1987) .... $33.00/$38.00
- Vols. 9-10 (1988-1989) .... $41.00/$46.00 } $45.00/$50.00
- Vols. 11-12 (1990-1991) .... $45.00/$50.00
- Vol. 13 (avail. May 1992) .... $49.00/$54.00 $49.00/$54.00   Vol(s). _____   Vol. _____

Annual Review of SOCIOLOGY
- Vols. 1-13 (1975-1987) .... $33.00/$38.00
- Vols. 14-15 (1988-1989) .... $41.00/$46.00 } $45.00/$50.00
- Vols. 16-17 (1990-1991) .... $45.00/$50.00
- Vol. 18 (avail. Aug. 1992) .... $49.00/$54.00 $49.00/$54.00   Vol(s). _____   Vol. _____

SPECIAL PUBLICATIONS

Prices, postpaid, per volume

	Until 12-31-91 USA & Canada / elsewhere	After 1-1-92 USA / other countries (incl. Canada)	Regular Order Please send:

The Excitement and Fascination of Science
- Volume 1 (1965) Softcover .... $25.00/$29.00 $25.00/$29.00   _____ Copy(ies).
- Volume 2 (1978) Softcover .... $15.00/$19.00 $25.00/$29.00   _____ Copy(ies).
- Volume 3 (1990) Hardcover .... $90.00/$95.00 $90.00/$95.00   _____ Copy(ies).

(Volume 3 is published in two parts with complete indexes for Volumes 1, 2, and both parts of Volume 3. Sold as a two-part set only.)

Intelligence and Affectivity:
Their Relationship During Child Development, by Jean Piaget
(1981) Hardcover .... $8.00/$9.00 $8.00/$9.00   _____ Copy(ies).

Send To: ANNUAL REVIEWS INC., a nonprofit scientific publisher
4139 El Camino Way • P. O. Box 10139
Palo Alto, CA 94303-0897 USA

Please enter my order for the publications indicated above. Prices are subject to change without notice.

Date of order _____   ☐ Proof of student status enclosed

Institutional purchase order No. _____   ☐ California order, must add applicable sales tax.

Individuals: Prepayment required in U.S. funds or charge to bank card below.

☐ Canadian order must add 7% GST.

☐ Amount of remittance enclosed $ _____

☐ Optional UPS shipping (domestic ground service except to AK or HI), add $2.00 per volume. UPS requires a street address. No P. O. Box, APO, or FPO.

Or Charge my ☐ VISA

☐ MasterCard ☐ American Express

Account Number _____   Exp. Date _____ / _____

Signature _____

Name _____
       Please print

Address _____
        Please print

_____ Zip Code _____

_____ Send free copy of current *Prospectus* ☐

Area(s) of interest   ARI Federal I.D. #94-1156476